Straight Talk about Cosmetic Surgery

Yale University Press Health & Wellness

A Yale University Press Health & Wellness book is an authoritative, accessible source of information on a health-related topic. It may provide guidance to help you lead a healthy life, examine your treatment options for a specific condition or disease, situate a healthcare issue in the context of your life as a whole, or address questions or concerns that linger after visits to your healthcare provider.

Thomas E. Brown, Ph.D., *Attention Deficit Disorder: The Unfocused Mind in Children and Adults*

Ruth H. Grobstein, M.D., Ph.D., *The Breast Cancer Book: What You Need to Know to Make Informed Decisions*

James W. Hicks, M.D., *Fifty Signs of Mental Illness: A Guide to Understanding Mental Health*

Steven L. Maskin, M.D., *Reversing Dry Eye Syndrome: Practical Ways to Improve Your Comfort, Vision, and Appearance*

Mary Jane Minkin, M.D., and Carol V. Wright, Ph.D., *A Woman's Guide to Menopause and Perimenopause*

Mary Jane Minkin, M.D., and Carol V. Wright, Ph.D., *A Woman's Guide to Sexual Health*

Arthur W. Perry, M.D., F.A.C.S., *Straight Talk about Cosmetic Surgery*

Catherine M. Poole, with DuPont Guerry IV, M.D., *Melanoma: Prevention, Detection, and Treatment,* 2nd ed.

E. Fuller Torrey, M.D., *Surviving Prostate Cancer: What You Need to Know to Make Informed Decisions*

Barry L. Zaret, M.D., and Genell J. Subak-Sharpe, M.S., *Heart Care for Life: Developing the Program That Works Best for You*

Straight Talk about Cosmetic Surgery

ARTHUR W. PERRY, M.D., F.A.C.S.

With a foreword by

MICHAEL F. ROIZEN, M.D.,

author of *RealAge* and *YOU: The Owner's Manual*

Yale University Press / New Haven and London

The information and suggestions contained in this book are not intended to replace the services of your physician or caregiver. Because each person and each medical situation is unique, you should consult your own physician to get answers to your personal questions, to evaluate any symptoms you may have, or to receive suggestions for appropriate medications.

The author has attempted to make this book as accurate and up to date as possible, but it may nevertheless contain errors, omissions, or material that is out of date at the time you read it. Neither the author nor the publisher has any legal responsibility or liability for errors, omissions, out-of-date material, or the reader's application of the medical information or advice contained in this book.

Published by the foundation established in memory of William Chauncey Williams of the Class of 1822, Yale Medical School, and of William Cook Williams of the Class of 1850, Yale Medical School.

Designed by Mary Valencia. Set in Stone by Tseng Information Systems, Inc. Printed in the United States of America by Vail-Ballou Press.

Library of Congress Cataloging-in-Publication-Data
Perry, Arthur W.
Straight talk about cosmetic surgery / Arthur W. Perry with a foreword by Michael F. Roizen.
p. cm. — (Yale University Press health & wellness)
Includes bibliographical references and index.
ISBN 978-0-300-11999-2 (cloth : alk. paper) — ISBN 978-0-300-12104-9 (pbk. : alk. paper)
1. Surgery, Plastic. 2. Consumer education. I. Title.
R118.P47 2007
617.9′52—dc22

2007001333

A catalogue record for this book is available from the British Library.

The paper in this book meets the guidelines for permanence and durability of the Committee on Production Guidelines for Book Longevity of the Council on Library Resources.

10 9 8 7 6 5 4 3 2 1

I dedicate this book to my wife, Bedonna, who has stood by me through thick and thin, and to my children, Ben, Meredith, and Julia, who have tolerated my workaholism, sometimes to the exclusion of their chemistry homework. It is also dedicated to my parents, Michael and Harriet Perry. Dad, at age 84, is still a practicing dentist—what a role model! Mom was president of the Town of Highlands (N.Y.) Board of Education while I was growing up. Her job was invaluable in keeping me from being suspended for my hijinks. She taught me ethics, responsibility, and politics.

Contents

Color plates follow page 174 and are shown also as black-and-white images throughout the text.

Foreword

Pssst . . . come this way . . . come to Thailand for a quick tummy tuck . . . or how about an eyebrow procedure or a nose job down this alley? How about in this fancy hotel? Quick, you, get a little Botox at lunchtime, or maybe in an afternoon break, and take away those wrinkles.

Yes, cosmetic treatments and cosmetic surgery are as common as pitches for second mortgages. But this increase in the popularity of cosmetic surgery has brought forth deception and practitioners of less-than-sound training.

How can you as a nonprofessional in this field learn what is real and who is a reputable practitioner? How can you get the face you want? Why wouldn't you trust the doctor in an upscale shopping center, or with a beautiful office where every member of the staff wears white? You do want to look younger, don't you?

Separating truth from hype, and reality from serious side effects of surgery, is what *Straight Talk about Cosmetic Surgery* is all about. Most bottles of herbal pills such as Coenzyme Q10 contain NO Coenzyme Q10. Is that just inferior manufacturing? Since twenty-eight of the thirty preparations tested have no active ingredients, it is actually betrayal of the public. In the case of Coenzyme Q10, that isn't so troublesome; you are only getting a sugar pill instead of the real thing. No bad side effect, just a partial wallet transfer.

But with cosmetic surgery you *can* have bad effects. Arthur Perry tells you how to avoid the fakes, the falsehoods, and the hucksterism that have plagued a profession in which anyone can buy a board certification that looks real, and enough equipment to look truly professional. So much cosmetic surgery is consumer driven, performed in private offices beyond the scrutiny of the watchdogs of organized medicine, the Joint Commission on Accreditation of Healthcare Organizations, and the certification of formal boards accredited by the American Board of Medical Specialties, that it is easy to be betrayed. It is time to shed some light on how to obtain excellence in cosmetic surgery.

Dr. Perry answers this need. *Straight Talk about Cosmetic Surgery* isn't a "how-to" guide to cosmetic surgery. It is a toolbox. After you read it, you'll be ready for the barrage of promotional material and hype from today's slick advertisers. It will give you the knowledge and power to choose doctors and procedures that are likely to make you look the way you want to look. If you want to look young, this book will do for you what *YOU: The Owner's Manual* does for those who want to stay healthy: it will empower you with easy-to-understand knowledge and action steps.

Some of you may wish to read every chapter, soaking up the interesting details and historical tidbits about cosmetic surgery. Others will skip straight to the procedures they want to consider. Either way, this book can be your ready consultant.

When I became chairman of the University of Chicago's anesthesia department in 1985, the 28-year-old Arthur Perry had just started as Dr. Thomas Krizek's first plastic surgery resident. As we worked together to improve surgery on burn patients, I saw his thirst for knowledge. That trait is evident in this book. He shares his knowledge with you easily and tells it like it is.

Do you need help to understand creams, lotions, and procedures? Well, I certainly do! The incredible number of creams, treatments, and operations purported to help your appearance can be bewildering. The surgeries are practiced not only by people trained in plastic surgery but by anyone who has a medical degree and hangs out a shingle. I have seen orthopedic surgeons sued so much for their incompetence that they finally decided they were going to perform Botox and plastic surgery procedures. Think their training was adequate?

You will be an empowered consumer and patient after reading this book, and you will learn how to find someone who really has been trained and is skilled at doing what he or she purports to do. Whether you plan to look like Sean Penn or Madonna, or just look as you did five or ten years ago, *Straight Talk about Cosmetic Surgery* will make you much more likely to succeed.

Michael F. Roizen, M.D.
Chair, Cleveland Clinic Division of Anesthesiology, Critical Care Medicine and Comprehensive Pain Management

Preface

In this era of *Extreme Makeover, The Swan, I Want a Famous Face, Miami Slice, Plastic Surgery 90210,* the fictional *Nip/Tuck,* and an epidemic of other television shows about plastic surgery, there is still no easy way to obtain unbiased information about cosmetic surgery. Each show has a spin. Every website has a bias. Newspaper and magazine articles and television stories are actually advertisements. Physicians and hospitals send out press releases, hoping to hook a story and land some extra business. Twenty-four-hour television and radio stations need to fill time. Plastic surgery has become entertainment. Producers push the limits and dangle the carrot of fame in front of gullible and sometimes greedy plastic surgeons. With the lure of fame and fortune, plastic surgeons "play the game" and often become part of the problem.

I wrote *Are You Considering Cosmetic Surgery?* in 1997, before the field exploded. Now in 2007, the book is ready for an update. But my question-and-answer format is no longer enough for this new era. So much has happened. So much has changed. *Straight Talk about Cosmetic Surgery* is my answer to the need for a straight-shooting look at cosmetic surgery.

Medicine is an ever-changing science. This book can give you as a prospective patient the information you need to ask your doctor appropriate questions and begin to understand this complex field. It is not meant to provide specific medical advice, and should not be used as a sole source of information. The choice to undergo a cosmetic procedure or surgery or skin care program should be made only after consulting your own physician.

Cosmetic Surgery Can Improve One's Appearance and Self-Esteem

Surgery is a deadly serious medical endeavor. Pain is endured, complications occur, and people die. It is not a form of entertainment. We're now in the midst of an unprecedented popularity of cosmetic surgery. Between 1997 and 2005, the number of surgical procedures in the United States rose from 973,000 to 2.1 million, a 119 percent increase, according to the American Society for Aesthetic Plastic Surgery. For nonsurgical procedures, the numbers are even more staggering. In 1997, 1.2 million procedures were performed. In 2005, the number rose to 9.3 million! Cosmetic surgery is popular for good reasons. There are more plastic surgeons than ever in the United States and around the world. We have economic prosperity. Baby boomers are aging but are refusing to look old. And, of course, cosmetic surgery has become more refined. There are more procedures to choose from. There are less invasive procedures. And, finally, the relative costs have declined.

Plastic Surgery Is the Surgery of Change

It is the surgery that alters your appearance. Derived from the Greek word *plastikos,* to mold or form, society has become familiar with plastic surgery through *People* magazine and television shows. *Cosmetic surgery* is the branch of plastic surgery that deals with normal tissue, as opposed to injured or diseased tissue. In this book, the terms "plastic surgery" and "cosmetic surgery" are used interchangeably.

Plastic Surgery Is a Patient Driven, Financially Lucrative Field—A "Perfect Storm" for Abuse

Hucksters capitalize on the public's thirst for knowledge, products, and services that will keep everyone young and healthy. In cosmetic surgery, an innovation or a new surgical procedure is often performed on just a few patients before the enterprising surgeon contacts a television station. After the appearance, his office is bombarded with phone calls. New patients, more surgery, and more dollars flow in. Viewers call other plastic surgeons' offices inquiring whether those doctors perform the new procedure. The surgeons scamper to learn and offer the procedure even before studies determine whether the procedure really works.

Contrast this plastic surgeon's scenario with that of the cardiologist. For heart problems, carefully controlled studies are performed, overseen by institutional review boards (IRBs). These studies are "double-blinded"; neither the patient nor the researcher knows who received the experimental treatment. The treatment groups are mixed around, further validating the study. The results are published in peer-reviewed medical journals, academic magazines that publish only studies with merit. Finally, an expert group or a committee at the National Institutes of Health evaluates the various studies and decides whether doctors should adopt the new treatment. This whole process is a far cry from the recommendations made on the "6:00 News."

This Book Will Help Sort Through the Bias, the Spin, and the PR

My patients know me as a straight shooter. I tell it like it is. Plastic surgery is a service I offer, not sell. If, at the end of a consultation, the information given to the patient causes her to decline surgery, then I've done my job. I believe that plastic surgery procedures should be treated like any other area of medicine; that is, only tried and proven procedures should be performed. Surgery, treatments, and skin care regimens should be backed by science. Anything less constitutes some degree of fraud. Whether purposeful or by neglect, this fraud can hurt patients at most, and cost them money at least.

Straight Talk about Cosmetic Surgery is a mixture of fact and opinion. As such, there will be valid disagreements between me and other physicians, and between me and manufacturers of medicines, products, and equipment. Information I provide may indeed prove to be untrue as additional studies are performed in this world of evolving medicine.

When I completed my residency in plastic surgery at the University of Chicago in 1987, there were only a limited number of cosmetic surgery procedures I could offer. At that time, when someone had fine wrinkles of the face and sagging jowls, the surgical solution was a face-lift and a deep chemical peel. Twenty years later, we have a variety of skin care drugs that can stall or even reverse signs of aging, we can peel in at least three different depths, we can suction fat and cause skin redraping, we can fill wrinkles with an increasing number of chemicals, and we can shrink the skin with lasers. And then there's surgery. We can lift the face minimally or maximally, or with the use of instruments called endoscopes. We can use less invasive methods, pull the face up with threads, or support it in a multitude of ways.

The explosion of methods in plastic surgery has been accompanied by a huge increase in the number of plastic surgeons in the United States. In 1987, for instance, there were just over 2,000 board-certified plastic surgeons. By 2007, there were about 5,700. And this figure is compounded with the practice of non-plastic surgeons performing cosmetic surgery. Today otolaryngologists, ophthalmologists, general surgeons, dermatologists, and even dentists want to get into the act.

Each Chapter Provides the Information Learned in a Consultation

In addition to the standard procedures, you will find chapters on new and upcoming procedures. I discuss the all-too-frequent scams in plastic surgery—the unproven machines, creams, and procedures. One chapter discusses the truths and myths of skin creams. Others address weight-loss surgery and cosmetic dentistry, areas that border on cosmetic surgery. I've talked with the world's experts in these fields and incorporated their thoughts here.

How difficult it must be for the average consumer to understand what is truth and what is hype in cosmetic surgery. It is difficult even for plastic surgeons to stay on top of new developments and comprehend which have a scientific basis and which represent a corporation's "profit center."

This book gives you the tools you need to make informed decisions about cosmetic surgery and cosmetic medicine treatments. When you finish reading, you'll be able to choose between a proven technology and one that has no merit. You'll be a more powerful consumer, armed with knowledge. You'll understand whom to choose as your doctor and what questions to ask.

I wish to disclose that I was previously a paid consultant to the Neostrata Corporation and am an ongoing consultant to the Ethicon and Consumer Product divisions of Johnson & Johnson, Inc. I have also developed the Dr. Perry's Skin Solution family of products for New Vitality Corporation. I have no financial relationships with companies mentioned in this book other than L'Oréal, Neostrata, Procyte, Coapt, Medicis, and Caleel + Hayden, advertisers on my WOR radio program.

Some will read this book through like a novel. Others will select specific sections. The chapters are independent of one another and can be read in any order. Throughout, I define medical terms and then show the technical term in parentheses. Such words, and others, are further defined in the glossary that appears at the back of the book. While I am not trying to teach the reader to be a doctor, in order to understand this field you will need a little vocabulary.

Statistics about the popularity of procedures are available for the United States, but are spotty for the rest of the world. Interestingly, the popularity of procedures varies by country. For instance, while the nasal surgery known as rhinoplasty is the most common procedure in Israel, Jordan, Lebanon, Sweden, and Turkey, it doesn't even make the top three in Argentina, Australia, and France. However, the field of plastic surgery has become global. Thousands of plastic surgeons from around the world attend and present papers at the two annual meetings of the American Society of Plastic Surgeons and the American Society for Aesthetic Plastic Surgery, and they publish in the major American journals. The U.S. experience therefore largely mirrors that in the rest of the world.

Fees mentioned in this book are typical in the United States and are current as of 2005. You can expect an increase of about 5 percent each year. New procedures, or procedures that involve new technology or products, often have disproportionately higher fees. And of course, fees vary widely in different regions of the country.

In this book I apply masculine terms to plastic surgeons and feminine terms to patients, because the majority of plastic surgeons are men and the majority of cosmetic surgery patients are women. This convention is used only to facilitate the flow of the book; there are plenty of female plastic surgeons and a growing percentage of male patients. When viewing photographs, unless otherwise indicated, the preoperative photo will always appear on the left and the postoperative photo will appear on the right.

Acknowledgments

Many thanks to nurse Pam Mayers, R.N., C.N.O.R., who has worked with me for seventeen years as office manager, circulating nurse, patient care assistant, and in many other capacities. Her dedication allowed me to concentrate on surgery and patients, in addition to books and radio shows. My thanks to Michael Roizen, M.D., for his foreword to this book. I first met Dr. Roizen twenty-two years ago in the operating room in the middle of the night at the University of Chicago; he has been an inspiration both as a physician and as a medical writer. I thank my father, Michael M. Perry, D.D.S., Class of 1947, but up to the minute on dentistry, for his invaluable contribution to the chapter on cosmetic dentistry and for his life and career guidance.

My gratitude to my plastic surgery mentor, the surgical legend Thomas J. Krizek, M.D., F.A.C.S., cannot be measured. Dr. Krizek took me in as a medical student doing skin graft research on rats at the College of Physicians and Surgeons of Columbia University. He nurtured and guided me, ultimately accepting me as his first plastic surgery resident at the University of Chicago. Without him, I might still be selling hot dogs at Army football games.

I owe much to plastic surgeons and role models Thomas Baker, M.D., Gerald Colman, M.D., Robert Goldwyn, M.D., Howard Gordon, M.D., Larry Gottlieb, M.D., Steven Lynch, M.D., Robert Parsons, M.D., and Larry Zachary, M.D., for their early influences on my career. At Albany Medical College it was Chief Resident Dr. Lynch, reminding me of Hawkeye Pierce of M*A*S*H, who crystallized my early interest in plastic surgery by letting me, an eager freshman, cut my first skin graft. And I am grateful to Drs. Robert L. Friedlander and Alan Miller for their support of my journalistic endeavors, tolerating the creation of the *Albany Medical Nexus*.

Dr. Al Cram, Dr. Christine Petti, and Dr. Phyllis Chang were the most supportive co-plastic surgery residents at the University of Chicago I could imagine. We spent countless hours together in the operating room sewing up tendons and reassembling breasts.

I thank the thousands of patients whom I have had the pleasure of treating. I am particularly grateful to those who were kind enough to allow their photographs to be used in this book.

And thanks to my agent, Rita Rosenkranz, who shepherded this project and made it happen, to Jean Thomson Black at Yale University Press for her support and nurturing, and to Jessie Hunnicutt, also at Yale. Thanks go out to medical illustrator Craig Luce for his drawings, and to Drs. Darrick Antell and Joseph Capella for the use of their photographs. And much appreciation goes to Noah Fleishman, my coach and engineer at WOR radio, and to Jonathan Greenhut of New Vitality. And I can't leave out Benjamin, my aspiring filmmaker son and part-time radio show producer, for his cartoon.

I

Cosmetic Surgery

1

An Overview

Cosmetic surgery is a huge field. Skin care, noninvasive procedures, and surgery are all components. Surgery can change one's appearance into something that nature never intended: bigger breasts, bigger lips, a smaller nose, less fat around the thighs. And surgery can restore one's appearance to what it was before the wrinkles, sags, and gray hair. Face-lifts, eyelid and brow-lifts, even breast lifts and tummy tucks fall into this category.

Cosmetic Surgery Has Been Intertwined Throughout History with Reconstructive Surgery

One hundred fifty years ago it was difficult to separate the two fields, and it's still tough today. Society arbitrarily distinguishes between cosmetic and reconstructive, but who really knows when a nose is no longer merely large, but deformed? Who decides that a belly overstretched by fat is a cosmetic problem, but one overstretched by pregnancy is a reconstructive issue? When does the sagging eyelid skin cross the line from a "sign of aging" to a medical problem? And why is it cosmetic when both breasts fail to develop, but reconstructive when only one develops?

The modern era of cosmetic surgery began in the mid-nineteenth century with the first attempts at reshaping the nose (*rhinoplasty*). Later in that century, protruding ears were first corrected and breasts made smaller. Early cosmetic surgery was highly controversial; some considered it unimportant and unethical. This attitude continues to this day, with subtle looks of disdain from residency program directors directed at trainees interested in a career in cosmetic surgery.

Reconstructive surgery is considered by some to be a higher calling and a more noble exercise.[1]

By the start of the twentieth century, cosmetic surgery of the eyelids, injection of filler into wrinkles, and the first face-lifts were performed. The two world wars unfortunately provided surgeons with a tremendous amount of reconstructive surgery. Surgeons learned many cosmetic surgical techniques while repairing battlefield injuries.

It was not until the 1960s that plastic surgery really caught fire as a method for celebrities and the affluent to look younger and more beautiful. By the 1960s, teenagers were having their noses reshaped and their parents were having their faces lifted. Surgical techniques improved in the 1980s, as plastic surgeons increased in numbers and were able to improve their results. Horrific "nose jobs" of the 1960s prompted more conservative surgery. Liposuction, heralded by some as important as space travel and computer science, marked the greatest cosmetic surgical advance of the 1980s. Enormously popular, this type of plastic surgery was both ridiculed and revered. The unprecedented current popularity of cosmetic surgery began in the 1990s, with a multitude of new techniques and a media-charged group of plastic surgeons ready to change the appearance of America.

Skin Care Is the First Step in Improving Appearance

Cosmetics coat the surface of the skin and hide imperfections. The most popular cosmetic drugs, the "cosmeceuticals" glycolic acid and vitamin C, penetrate the skin and improve appearance. The true pharmaceuticals, tretinoin (Retin-A) and hydroquinones (fading creams), alter your genetic destiny and can reverse signs of aging. These drugs work slowly, generally taking months or years to show any effect.

Facials and microdermabrasion clean the skin and make you feel good and, in fact, may decrease acne and later scars and wrinkles. Clean skin is the foundation for healthy skin.

When the Problem Exceeds Creams, It's Time for Procedures

Increasing depths of chemical peeling—from the most superficial "lunch-hour" peels, to midlevel "weekend" peels, to the deepest laser peels—lessen wrinkles and sun damage and thicken the skin. A variety of new chemicals can be injected into the remaining wrinkles.

And When Office Procedures Are Not Enough, We Take to the Knife

When the skin has been significantly overstretched, and when wrinkles become sags, it is time for surgery. Procedures such as brow-lifts, eyelid lifts, face-lifts, breast lifts, and tummy tucks all remove extra skin and tighten remaining skin.

When a person is unhappy with her nose, lips, ears, size of her breasts, or amount of fat on her hips, cosmetic surgery can change the appearance of normal features.

The Cosmetic Surgery Industry Is Born

Only the rich could afford cosmetic surgery a hundred years ago. Cosmetic surgery today is at the height of its popularity; in 2005 about 2.1 million surgical procedures and 9.3 million nonsurgical procedures were performed in the United States. Cosmetic surgery is not just an American phenomenon. It is wildly popular in South America, Europe, Africa, Turkey, and Japan.

People have cosmetic surgery to improve their appearance. It makes sense, then, that they will be happier after cosmetic surgery. I helped Marlene Rankin, Ph.D., and other plastic surgeons with a 1999 study that found that depression decreased after cosmetic surgery. According to the lead author, Dr. Gregory Borah, chief of plastic surgery at the Robert Wood Johnson Medical School in New Jersey, "the most surprising finding of this study was that patients were actually *happier than expected.*"[2]

Cosmetic surgery is a unique field within medicine. It is the only field that doesn't deal with disease. The patients are all healthy. The surgery is all planned; only *really* large noses need to be corrected as an emergency. Cosmetic surgery is one of the few fields of medicine that is not covered by insurance companies.

Top Ten Countries Performing Cosmetic Surgery

Rank	Country	Rank	Country
1	United States	6	Germany
2	Mexico	7	Brazil
3	Argentina	8	South Africa
4	Spain	9	Turkey
5	France	10	Japan

Source: International Society of Aesthetic Plastic Surgery

Beauty Is Not Just Skin Deep

In fact, how you look alters others' perceptions of you. Like it or not, more attractive people have a lot of advantages in life. According to David Sarwer, Ph.D., a psychologist at the University of Pennsylvania, "studies have shown that both men and women who are physically attractive have more friends, better educational opportunities, better job offers, and even better medical care."[3] Attractive people are more successful in life because other people prefer to deal with them.

Society continually redefines what is considered beautiful. Until the last century, overweight women were regarded as beautiful. While the rest of the world was starving, the well-fed were usually the rich. Full-figured women were the actresses and models in the first half of the twentieth century, but by the 1960s the very thin model Twiggy redefined beauty. Today, a thin woman with full breasts and muscular definition is emerging as the "look."

Dr. Sarwer teaches that facial attractiveness is based on youthfulness, symmetry, and *averageness*. We all want to be normal. Except during the teen years, when we are searching for who we are, we want to blend in with the crowd. The most beautiful women in our society have petite faces with small mouths and full lips, a small jawline, and pronounced eyes and cheekbones.

Interestingly, our society accepts imperfections more readily in men than in women. Wrinkles on an actress's face end her career. Wrinkles on an actor's face make him more interesting. Think Paul Newman. And while Harrison Ford has a noticeable scar on his chin, I can't think of any female actress with a similar imperfection.

Even Dogs Are Getting into the Act

Yes, *dogs*. In Los Angeles, canines are having cosmetic surgery. They are having breast reductions, rhinoplasties, face-lifts, eyelid lifts, and even labial reduction surgeries, according to Dr. Alan J. Schulman, veterinarian to the dogs of the stars. "Many veterinary surgeons refer to the cosmetic or plastic procedures as 'reconstructive dermatologic surgeries,' as they are rarely performed for vanity or cosmetic reasons alone," he told me. While each of the procedures has a benefit to the health of the dog, I'm sure some of the California crowd send Fido in for vanity.

2

The Cosmetic Surgery Consultation

It all begins here. You're unsure whether you really want to have surgery. You're worried about being too vain. You're worried about what your spouse, friends, and coworkers will say. You watch those terrible television shows. You go on the Internet and read about the procedure. Your interest is piqued. Then you start looking for a surgeon.

How a Doctor Becomes a Plastic Surgeon

It is important to choose the *right* plastic surgeon. Certainly, the surgeon's medical school and residency training give you information about the doctor's basic intelligence. Chances are good that if the doctor trained at Harvard, he's not stupid. But that's only the beginning.

The training of a plastic surgeon is arduous. Only the best graduate from medical school and only the top surgical residents are accepted into plastic surgery residency programs. Most are at the top of their class and were likely inducted into Alpha Omega Alpha, the medical honor society.

To become a plastic surgeon, a doctor must obtain an M.D. (Doctor of Medicine) or a D.O. (Doctor of Osteopathy) degree. Following this, there are several paths the doctor may follow. Many doctors now spend five to six years in an integrated plastic surgery residency, although some complete three years of general surgery and two years in plastic surgery. Alternatively, otolaryngologists (ear, nose, and throat specialists), orthopedic surgeons (bone doctors), neurosurgeons (brain surgeons), urologists, and oral surgeons with medical as well as dental degrees may enter the two-year plastic surgery residency.

The marathon plastic surgery residency is a grueling learning process. The doctor learns how to take care of patients before, during, and after surgery during the internship, the first year. He learns how to suture and how to stop bleeding, and perfects his skills through animal surgery and computer simulations. In the operating room, interns, under the watchful eye of "attendings" (the fully trained surgeons), perform portions of procedures.

With advancing years comes more responsibility. The education includes rotations in ophthalmology, orthopedics, otolaryngology, and even the medical specialty, dermatology.

By the fourth year of residency, the doctor is immersed in plastic surgery. To gain experience in cosmetic surgery, the fifth- or sixth-year chief resident runs his own clinic, performing discounted-fee surgery. While the new doctor gains experience, this program allows lower income patients to afford cosmetic surgery.

After completion of his residency, the new plastic surgeon may start his own practice, or he may join an established plastic surgeon or the faculty of a medical school or multispecialty group. Alternatively, he may continue his marathon education with another year or two of training, called a fellowship. It can be in cosmetic surgery, hand and microsurgery, burn surgery, craniofacial surgery, or head and neck cancer.

Board Certification Is a Critical Credential

The next hurdle is to pass the "boards." The candidate takes a written examination during the first year in practice and an oral exam the next year. The written exam tests book knowledge, and the oral exam determines how the surgeon would handle clinical scenarios.

Considering how long it takes to become a plastic surgeon—four years of college, four years of medical school, and at least five years of residency—it is hard to believe that many aspiring surgeons (13 to 23 percent) flunk the boards each year. But these are not "rubber stamp" exams. These doctors go back to their communities, in some degree of depression, to ponder their fate and plan for the exam next year. Each year the American Board of Plastic Surgery certifies about 200 new plastic surgeons. Since 1937, 7,003 plastic surgeons have become board certified.

Board certification is the minimum criterion you should use in selecting a plastic surgeon. And don't be confused between boards and societies. Societies are merely professional associations—glorified clubs—that vary in their selection criteria. Most just want the membership fees and will admit almost anyone.

The American Medical Association, as well as state and county medical societies, only require a medical degree to qualify for membership. The American Society of Plastic Surgeons and the American Society for Aesthetic Plastic Surgery have stringent membership requirements, including a plastic surgery residency and certification by the American Board of Plastic Surgery.

Hospital Privileges Are an Important Consideration

To operate in a hospital, a surgeon needs to be granted "privileges." There are a wide variety of hospitals ranging from medical-school teaching hospitals to small community hospitals.

Some surgeons gain privileges simply by joining a politically connected group of surgeons. If a surgeon brings millions of dollars of revenue into a hospital, it is unlikely that the hospital will prevent him from bringing an associate onto the hospital staff. There are hospitals with "open staffs" that will take any properly trained doctor and then there are "closed staff" hospitals that will only take on a new physician if he joins a practice of a physician already on the staff. These hospitals will keep a physician from gaining privileges if he is deemed to be a competitor. Not fair, but financially difficult for the new physician to challenge.

A properly trained, board-certified plastic surgeon will usually be able to attain surgical privileges in at least one hospital in the area. If a surgeon has no hospital privileges and operates only in his office, I would be concerned. Certainly, there are surgeons who are extremely competent and are either politically boxed out of an area or choose to operate only in their office or a surgicenter. But the vast majority of legitimate plastic surgeons have hospital privileges. If yours doesn't, ask why.

But don't just ask whether a doctor has hospital privileges. *Make sure that the doctor has privileges at a hospital in exactly the procedure that you want performed.* For instance, a dermatologist may have privileges to take care of medical problems of the skin in a hospital. But he may *not* have privileges to perform liposuction or a face-lift in a hospital. That is crucial information because the hospital provides an important level of scrutiny. If they won't let the doctor perform the procedure in their hospital, why would you let him perform the procedure on *you* in his office?

Professional Societies Have Some Weight in Your Decision

Beyond credentials and training, plastic surgeons who are upstanding members of their communities are eligible for membership in the American Society of Plastic Surgeons (ASPS), the American Society for Aesthetic Plastic Surgery (ASAPS), and the American College of Surgeons (ACS). If your plastic surgeon is a member of these organizations, there is a certain assurance of quality. These organizations only admit board-certified surgeons, and only surgeons who follow their rules of conduct and ethics. The American College of Surgeons awards the letters F.A.C.S. to surgeons: these letters, written after the letters M.D., mean "Fellow, American College of Surgeons" and again give some assurance that the surgeon meets certain criteria.

Remember, these credentials are only starting points. Not every member of these groups is a perfect surgeon, ethically or technically. However, if your surgeon is *not* a member of these groups, you should be concerned.

"Wannabe" Plastic Surgeons

In this era of declining insurance reimbursement, physicians are scampering to find ways to supplement their income. Some otolaryngologists, dermatologists, and ophthalmologists are performing plastic surgery. Even primary care physicians such as family doctors and obstetricians are getting into the act. Usually these non–plastic surgeons call themselves cosmetic surgeons. Shockingly, nurses now want to inject fillers and Botox, and laser wrinkles independent of surgeons! They are taking *one- or two-day courses* hoping to learn to perform chemical peels, Botox injections, laser hair removal, collagen and Restylane injections, and other procedures.

In New Jersey one obstetrician advertised a *three-day course* for obstetricians in breast augmentation, liposuction, and abdominoplasties! With minimal or no formal aesthetic training in a residency program or understanding of the skin and beauty, these doctors are taking advantage of their medical licenses and opening cosmetic practices. The ad for the three-day course pointed out that "the investment in time and resources to learn aesthetic procedures is minimal." Just what the public wants to hear! And, according to the ad, "Upon completion of these seminars, the physician will have sufficient knowledge and hands-on training to safely and effectively perform cosmetic procedures in the office setting." I spent over two years in plastic surgery residency following four years of general surgery learning how to do these procedures. I guess I wasted a lot of time. I should have just taken the three-day course!

Inadequately trained surgeons have been around for a long time. In the 1950s surgeons returning from World War II started new practices after taking two-week courses in rhinoplasty.[1]

Certainly, non–plastic surgeons may be capable of performing some cosmetic surgery within the anatomic boundaries of their specialties. Some ophthalmologists legitimately perform cosmetic eyelid surgery, but how do they justify performing breast augmentations, as one near my office does? Ear, nose, and throat doctors may be capable of performing rhinoplasty, and even other facial cosmetic surgery, but how is it that they are trained to do liposuction of the thighs, as another in New Jersey does? And dermatologists are not surgeons: their area is not recognized by the American College of Surgeons as a surgical specialty. They may do minor procedures, but their field is *medicine of the skin.* Even with a one-year "dermatologic surgery" fellowship, their training is vastly different from that of a real surgeon. Yet again, all over the world dermatologists are performing liposuction, face-lifts, and breast augmentations.

Family doctors? They treat colds, sprained ankles, diabetes, and hypertension—and now they inject Botox. Where does this come from? Think money. Plain and simple.

It gets worse. Dentists are now claiming they have the right to do cosmetic surgery. California has *already* granted them this privilege. Laser and drug companies are egging them on, offering them lasers for wrinkle, acne, and hair removal and cellulite reduction systems with ultrasound. Where will it end? Ask your legislator.

Beware of Lots of Letters after the Doctor's Name

During my ten years as a governor's appointee to the New Jersey State Board of Medical Examiners, I noticed that many fraudulent doctors have quite a few letters after their names. In fact, multiple letters may be the telltale sign of an inadequately trained doctor trying to impress the public. Certainly your doctor should have the letters M.D. (Medical Doctor) or D.O. (Doctor of Osteopathy) after his name. He might even have one or more master's degrees (M.S., M.P.H., M.B.A. or M.A.), but these degrees are irrelevant to plastic surgery. He might also have a Ph.D., impressive but again irrelevant medically. Patients are well advised to select a plastic surgeon with the letters F.A.C.S. (Fellow, American College of Surgeons) after his name. These letters are awarded only to surgeons who are certified by one of the *surgical* boards of the American Board of Medical Specialties. Any other letters must be carefully scrutinized. I have seen people use letters from medical societies and virtually self-made societies. Some of the letters are awarded after completion of weekend courses! Most of the time, the multilettered doctors are not *real* plastic surgeons. I have even seen doctors who joined the International College of Surgeons just so they could add F.I.C.S. (Fellow, International College of Surgeons) to their names.

Consider the Plastic Surgeon's Ethics

When you choose a plastic surgeon, you are choosing a person who literally will have your life in his hands. The decisions made prior to, during, and after your surgery can affect your health. They can also cost you your life.

Besides the educational background, surgical training, board certification, and hospital privileges, you can check with your State Board of Medical Examiners, or national licensing boards in other countries to see if any negative action has been taken against the doctor. You can check to see the number of malpractice suits the doctor has *lost.* Not how many he has been named in—doctors who perform difficult surgery on complex cases may be sued more often than those who perform simple cases. You can also ask other doctors what they know about the surgeon.

Then you must determine the ethical flavor of the surgeon. Do the surgeon's advertisements make ridiculous claims? Do they advertise machines, as opposed to credentials? Claims of superiority are unethical and often illegal. Claims that a doctor has the "most experience" in a procedure are questionable. How does that doctor know he has the most experience? If a doctor advertises that he is the creator of a new procedure, ask if this procedure has been published in a peer-reviewed medical journal. According to the ASPS, photographs of models should be used only if they have undergone the specifically advertised procedure *and* if the plastic surgeon has performed that procedure on that model.

Beware of *The Swan*

The American Board of Plastic Surgery (ABPS), the ASPS, the ASAPS, and the International Society of Aesthetic Plastic Surgery all have codes of ethics. The board's mission is to assure the public that certified plastic surgeons meet its standards. They say it is unethical for plastic surgeons to offer surgery as a prize in a charity raffle, fund-raising event, or contest. Plastic surgery procedures should not be used as entertainment. That means that plastic surgeons who appear on shows such as *The Swan* or *On-line Surgery* could lose their board certification.

Once You Have Researched the Background of Your Surgeon, It's Time to Schedule the Consultation

If your first telephone encounter with your surgeon's office begins with an extended period on hold, this will color your experience. The personality of the office reflects that of the surgeon. If employees are impersonal and brusque, probably the doctor also will be. But don't cancel your appointment yet; give him the benefit of the doubt. I've read articles in women's magazines that tell patients to look at the décor of the office, and even at what books the doctor has on the shelves, to help determine if this is the right doctor for them. It is more important that the office be spotless and neat. The surgeon's taste is one ingredient, but not a critical one.

Once you meet with the doctor, be prepared to divulge all medical information. Don't hold back. Remember that this consultation is extremely confidential. I have had patients who came in for facial rejuvenation consultations and stated that they had never had surgery. When I found their prior face-lift scars, it seriously undermined our doctor-patient relationship. Remember, trust goes both ways. While a patient must respect and trust the surgeon, the surgeon must be able to rely on the truth of the patient's answers.

During the consultation, the doctor will review your medical history and ask what you would like changed. It is helpful to be direct. Ethical plastic surgeons do not want to suggest procedures. They want to know exactly what bothers you, so that they can direct the consultation and specifically address your concerns.

After a Physical Examination, Digital Photographs Will Be Taken

Almost all modern plastic surgeons use computer imaging. Some older surgeons still draw on Polaroid photographs. The surgeon will review your photos and ask what troubles you. He will then alter the photos digitally and discuss possible changes. The surgical or nonsurgical solutions to your problems and their risks and benefits will be discussed. He should show you photos of other patients. Typical results, less-than-average results, and a few "wows" should be shown.

Be Wary of Photographs

Of course it would be nice to believe everything you see, particularly if the photos come from a properly trained physician. Unfortunately, not all photos can be trusted. When you are shown before and after photographs, make sure that the lighting is the same. Look for the position of shadows. If the light is bounced off the ceiling, surface irregularities and wrinkles will be highlighted. If the flash is bounced straight on, these minor irregularities will disappear. Every celebrity knows this rule.

Look at the width of the face when assessing facial photos to assure that no distortion has occurred due to differing lenses or camera positions. Finally, check to be sure that the ears are at the same height relative to the nose. By tilt-

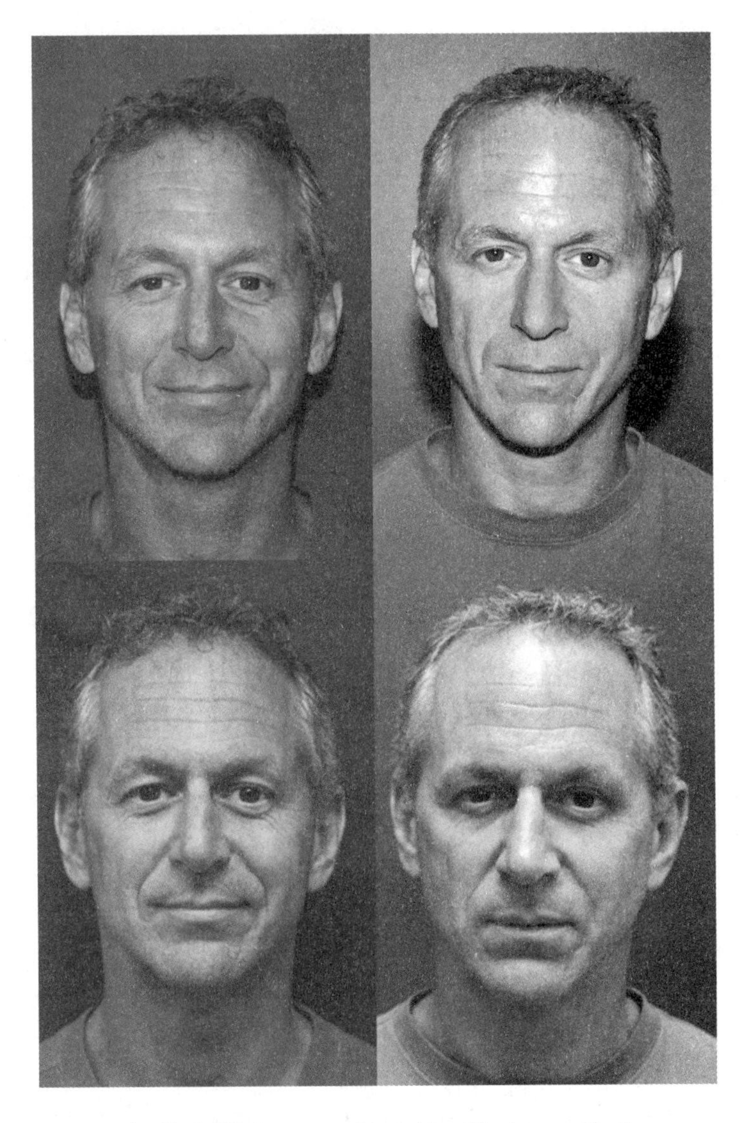

Beware: the flash fills in sags and wrinkles. The top and bottom photos were taken on the same days! The top photos were taken with full flash; on the bottom, the flash was bounced off the ceiling. On the left, the patient's face before upper eyelid lift, browpexy, and fat grafting to the nasolabial fold. The photos on the right were taken a few months after the procedures. See color plates.

ing the head up, the jowls can magically disappear in photos. Unfortunately, with the advent of digital photography and image altering programs, sophisticated digital manipulation can lead to deceptive practices.

The Purpose of the Consultation

You should consult with only the surgeon. If a nurse, a physician's assistant, a secretary, or a marketer leads any part of the discussion, make your way to the next doctor on your list.

An ethical plastic surgeon will discuss the pros and cons of a procedure. He has a legal and ethical responsibility to explain all the risks and benefits of the procedure, and the possible alternatives. You can then give what is called an informed consent. (If there were no risks to surgery, the decision would certainly be more straightforward.) If in the course of the consultation you decide not to have the procedure, the surgeon has done his job. Remember, a plastic surgeon offers a service; he does not sell a procedure.

At the conclusion of the consultation, the office staff will let you know what your surgery will cost. You'll be advised of fees for the surgeon, the facility, the anesthesiologist, and any related expenses.

It's Decision Time

By now you should have enough information about the risks, the benefits, and the costs to allow you to decide whether to have surgery. If the answer is yes, you need to decide if the surgeon with whom you have consulted is the right surgeon for you. You may immediately know the answer. If not, the forms included on the next pages will help you organize your thoughts and make your selection. Copy them and use one for each plastic surgeon you visit.

Robert Goldwyn, M.D., professor of plastic surgery at Harvard Medical School and a chief of mine during residency, has said that plastic surgeons would have less stress if they operated on people they liked. Personality conflict between the surgeon and patient prior to surgery is unpleasant, but becomes worse if there is a complication after surgery. It goes both ways. If you choose a surgeon you like, the entire experience will be more pleasant.

It's Time to Schedule the Procedure

Before surgery, you may need to see an internist for medical clearance. Cosmetic surgery is not *medically* necessary, so you may not have surgery if you have medical problems such as diabetes or heart disease. The surgeon is responsible for protecting you from yourself, and he may not operate if you have issues that in-

Cosmetic Surgery Consultation Comparison

Date of Consultation ____________ **Name of Surgeon** ____________

Reason I'm seeing the doctor ____________

Procedures suggested by surgeon ____________

Photos of patients viewed Yes/No Able to speak to other patients Yes/No

Other patient's name/phone ____________

Fees

Surgeon ____________

Facility ____________

Anesthesiologist ____________

Implants ____________

Other miscellaneous ____________

Total ____________

Surgeon Information

Medical school ____________

Plastic surgery residency program ____________

Fellowships or other residencies ____________

Board certified Yes/No

Which board ____________

Recommended by (doctor, friend, etc.) ____________

Facility Information

Hospital (which one) ____________

Surgicenter (which one) ____________

Office operating room ____________

If office operating room, is facility accredited by:
- ☐ American Association of Ambulatory Surgical Facilities (AAAASF)
- ☐ Accreditation Association for Ambulatory Healthcare (AAAHC)
- ☐ Medicare
- ☐ Other ____________

Anesthesia By

- ☐ Anesthesiologist ____________
- ☐ Nurse anesthetist ____________

My thoughts about the surgeon (personality, rapport, training, skills)

____

____

Cosmetic Surgery Preparation Sheet

- □ Date of preop testing ______
- □ Date of preop plastic surgeon visit ______
- □ Date of medical clearance ______
- □ Date of ophthalmologist/psychiatrist/other clearance ______
- □ Date of surgery ______
- □ Date of postop appointments ______

- □ Prescriptions obtained
- □ Prescriptions filled
- □ Preoperative clearance by internist
- □ Preoperative clearance by ophthalmologist
- □ Other preop clearance
- □ Lab testing completed

crease your chance of complications. If you have depression or anxiety, for example, a complication could trigger severe depression or even suicide.

A University of Pennsylvania study found that 19 percent of cosmetic surgery patients had a psychiatric disorder, but only 4 percent of patients who had reconstructive procedures had mental problems. Patients who want cosmetic surgery undoubtedly have at least some body-image issues.[2]

The Surgical Consultation Is a Two-Way Street

While you are learning from the surgeon, he is also interviewing you. Plastic surgeons try to weed out patients who have a high likelihood of being displeased after surgery. According to Dr. Goldwyn, the following types of patients might be expected to be unhappy after surgery:[3]

- The patient who writes a long letter to arrange the initial consultation
- The rude or pushy patient
- The slovenly patient
- The patient who tries to exhibit control over the surgeon
- The patient who comes on a little strong or disparages other plastic surgeons
- The patient who lies about her medical history
- The patient who is unsure of what she needs
- The patient whose problem is too small for the surgeon to see easily
- The uncooperative patient

- The perfectionist
- The doctor shopper
- The person who wants multiple operations
- The patient who wishes to please someone else
- The paranoid or depressed patient
- The psychiatric patient whose psychiatrist opposes the surgery
- The man who wants an operation not commonly performed in men
- The celebrity patient

A patient who had a bad result from prior surgery is more likely to be unhappy with subsequent surgery.

Plastic surgeons are trained to identify these types of patients and may refuse to operate on them. I worry about patients who seem unhappy during their consultation and do not laugh or smile. Patients who are on mood-altering medications may be asked to see a psychiatrist prior to surgery and have a letter of clearance sent to the surgeon.

Body Dysmorphic Disorder (BDD)

Two percent of people have this psychiatric abnormality. A small abnormality in appearance of the skin, hair, or nose, or an imagined problem, becomes an obsession. This can make the person miserable. They often have many cosmetic surgical procedures but are never satisfied. Patients with BDD do not benefit from cosmetic surgery, since perfection will never be achieved. Even if a specific procedure solves a specific problem, these patients will find another issue. They need to see psychiatrists, not plastic surgeons.

Your Second Consultation

Usually you will see your plastic surgeon twice before surgery. During the second visit, the surgeon reviews the upcoming procedure, examines you, and takes formal photographs. Consent forms are signed. You are given prescriptions for pain medicine and possibly sleeping pills for after surgery. Lab tests and medical clearance are arranged.

Consultation Fees

Most legitimate plastic surgeons charge a fee for their consultations. The ones who don't usually spend very little time with you during the consultation or are very new in practice. You get what you pay for. If you are given a *free* consultation, realize that it is a marketing session that will be spent for the most part watching a DVD or speaking with staff members. The average plastic surgeon who charges for a consultation barely meets his expenses with the fee. In business terms, it is called a "loss leader." Average overhead in a plastic surgeon's office ranges from $100 to $300 per hour. My fee for a consultation is $200, and I barely break even. I know that patients are reluctant to spend more because they often see several plastic surgeons before making a decision. From the plastic surgeon's point of view, however, if a patient is not willing to pay $200 for a consultation, she may not be a serious candidate for cosmetic surgery and may well be wasting the surgeon's time. Back in my first year of practice, when I did free consultations, I felt that the consultation was an afternoon's entertainment for many patients.

Surgical Fees

Surgical fees vary widely among surgeons. Some plastic surgeons are simply better than others: they can charge more for the surgery. Their results demonstrate their capabilities. Price shopping is not the best way to choose a plastic surgeon. Fees also vary between geographic areas, although the differences are less than they used to be.

Sometimes patients think that plastic surgeons' fees are excessive. In a world where actors, baseball players, and corporate presidents make millions of dollars, think about the importance of your appearance and your health when considering the appropriateness of fees. Thomas J. Krizek, M.D., F.A.C.S., former chief of plastic surgery at several medical schools, was fond of telling the following story:

> A man brought his sickly auto in to his auto mechanic. After hearing the symptoms, the mechanic told him it would cost $400 to fix the problem. Happy that the bill did not seem excessive, the man watched as his car was put up on the rack. The mechanic took a rubber mallet and banged four times underneath the car. He put the car back on the ground and announced that it was fixed. Outraged, the customer said, "$400 for 4 bangs on my car—that's $100 a bang. I could have done that! Your fee is outrageous." The mechanic replied, "Oh, I didn't charge you $100 a bang. It was $1 a bang. The $396 was for knowing *where* to bang."

The moral of this story is that the surgical fee not only pays for the actual procedure and aftercare, but for the ten years of training, the judgment developed over decades of training and practice, and the wealth of experience obtained over a career.

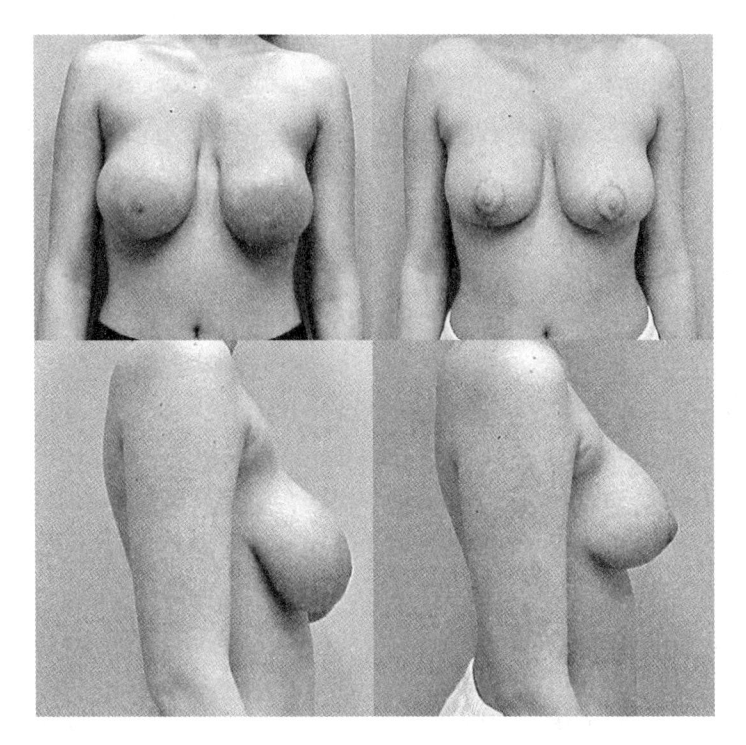

This 22-year-old woman had surgery in a "factory." Her thirty-minute implant procedure necessitated a five-hour salvage procedure. At two months postop on the right, smaller implants had been placed and her breasts lifted.

Warning: Enticements to Resist

Plastic Surgery Factories

New graduates or surgeons who skipped the ethics lectures in medical school sometimes work at plastic surgery factories, which depend on volume to pay for their ads. I've seen patients who had horrific fifteen-minute rhinoplasties and thirty-minute liposuctions that left them hiding their damaged bodies. I recall a 22-year-old girl who went to a free breast-augmentation consultation with a plastic surgeon who placed huge ads in most of the region's yellow pages. Her time spent with the "surgeon" was less than five minutes. The surgeon chose the size of the implants without asking what she wanted, then placed them with the simplest and quickest possible method. He used dissolving sutures, despite the fact that they result in worse scars. (That way he didn't have to see her after surgery.) Her surgery took less than half an hour! *Surgery isn't a race.* It takes me two hours to perform a breast augmentation. The photos here show her abysmal results. She sat in my office crying, with breasts that looked 60 years old. Her implants were too large and were placed in front of the muscle, a mistake if the patient has significant drooping.

Plastic Surgery Advertising

After a decade as chairman of the advertising committee for the New Jersey State Board of Medical Examiners, I've seen it all. False advertising is pervasive in plastic surgery. Physicians of all types advertise themselves as "cosmetic surgeons," although their training is usually not in plastic surgery. Ads claim "board certification" although the name of the board is often not mentioned—usually signifying a non-ABMS (American Board of Medical Specialties) board, otherwise known as a self-designated board.

False claims abound. My favorite ad was labeled "laser," showing spectacular before and after results. In reality, the patient was not a laser patient: she had had a chemical peel by another doctor who published her photos in his textbook. The photos were "borrowed" without permission from the textbook and labeled with a different procedure. The reader was to assume the spectacular result was that of the doctor who placed the ad. The New Jersey board issued a public reprimand to the doctor and fined him $18,000. I can assure you, this is not an isolated case. Let the buyer beware.

When a physician advertises a product or a machine such as a laser, I cringe. When he uses photos of models, I wince. When a celebrity endorses the doctor but there is no disclaimer that the celebrity received compensation or a free procedure in return for the ad, this is a violation of the law. When the physician shows photos, I worry: who can't show a few good photos? But are they really that doctor's? And when the ad states that the procedure is "scarless," I know that *a short scar is not scarless,* but do you?

Plastic Surgery Vacations

I am always amazed at the stories of people who go to other countries for plastic surgery. Shortsighted people look for cheap cosmetic surgery in places where you wouldn't drink the water. An increasingly popular cosmetic surgery option is the all-inclusive vacation that encompasses an interesting destination, a hotel, meals, and—incidentally—a face-lift or other procedure. How bizarre. Countries that are impoverished are now cosmetic surgical destinations. Mexico, Peru, Costa Rica, the Dominican Republic, Africa, Poland, India, and others are hosting cosmetic surgery and actively recruiting patients.

This concept flies in the face of any rational thought process. Some programs

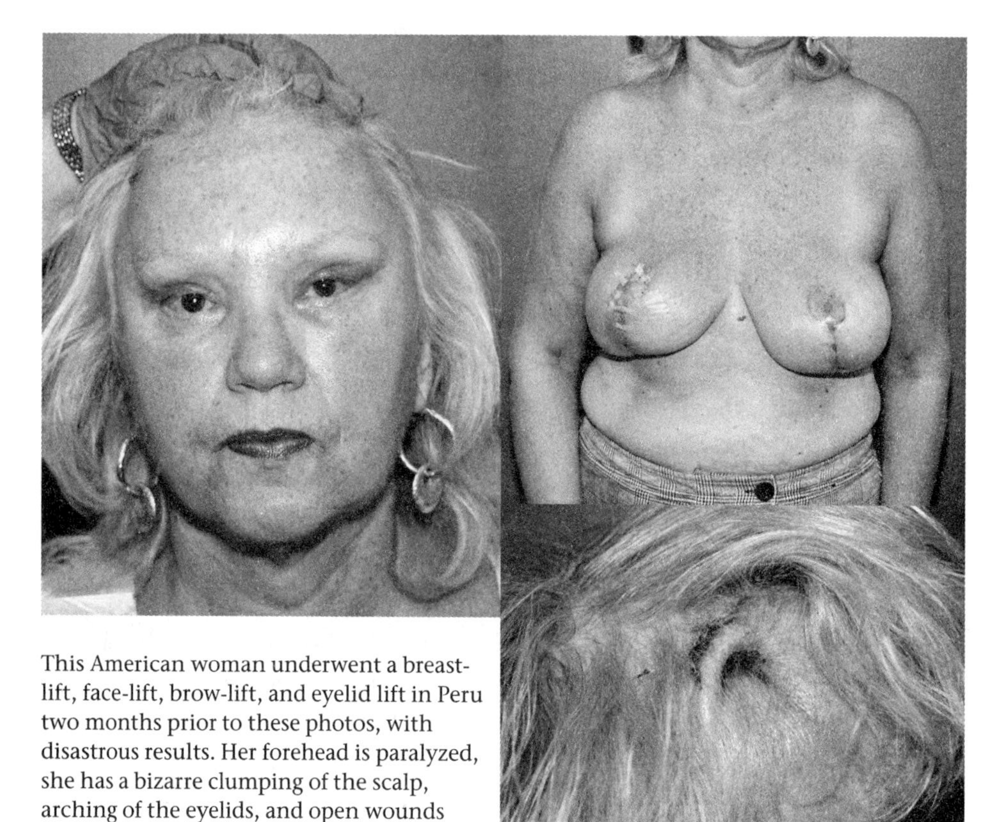

This American woman underwent a breast-lift, face-lift, brow-lift, and eyelid lift in Peru two months prior to these photos, with disastrous results. Her forehead is paralyzed, she has a bizarre clumping of the scalp, arching of the eyelids, and open wounds on her breasts and scalp. See color plates.

do not even disclose the name of the plastic surgeon. A travel agent screens most people and decides whether they are appropriate for surgery. I can just see the back room of the travel agency: "Let's see now, is this person an appropriate candidate for a face-lift? Check to see if her credit card goes through . . . yes? Okay, she's a candidate." Could it be far from this?

I have a patient who was actively recruited to have cosmetic surgery in Peru. The surgeon actually traveled to the United States and spoke to a small group of women. This particular individual underwent a brow-lift, an eyelid lift, a face-lift, and a breast reduction in one long procedure and was sent home from the "clinic" the same day. She came to see me a month after her surgery, asking me to help her in any way possible. Her brow-lift had been performed so bizarrely that she looked as if she had antlers. She had open wounds on her scalp and had lost much of her hair. Her forehead muscles were paralyzed. Her eyes were pulled so far up and to the side that she did not look human. Her face-lift showed no improvement and she had an asymmetric smile from a second nerve injury. The sensation in one earlobe had been destroyed. Despite a prior bout with breast cancer, a lumpectomy, and radiation, a breast reduction had been performed. She was told that the doctor "found something" in her breast but that it was nothing to worry about. This poor woman was maimed beyond what could be easily repaired. She had permanent injuries with no recourse whatsoever. She admitted that she had made a huge mistake, and her boyfriend began calling her Katrina, after the disastrous hurricane of 2005.

The lure of a surgical vacation is the price. Faceliftmexico.com advertises a face-lift, eyelid lift, and optional brow-lift for $6,000. In the United States, this surgery would cost more than $10,000. In Poland, doctors in border towns are recruiting German patients via procedures that are one-third their cost in Germany. I'm all for saving money on a car or a television. But medical procedures are not the place to cut corners. When you're dealing with your health and your appearance, there are no second chances.

There is a big difference in the techniques and results of different surgeons. The Mexican surgeon might indeed be terrific. But how can you check on the qualifications of a surgeon in another country? In fact, surgery in a far distant region is a dicey proposition. Instruments may not be properly sterilized, the surgeon's training and credentials cannot be verified, and *aftercare* may be an *afterthought.* Unapproved drugs and implants that undergo rigorous scrutiny in your home country might be used in another country. And how about malpractice insurance? Try suing someone in Thailand after you've returned back home to Madrid.

Patients who have had complications following surgery may be stranded. Who will care for the infection, the internal bleeding, the blistered, dead skin? Flights back to the country of surgery are not advisable when medical complications have occurred. And finding a surgeon in your home country who would be willing to care for someone else's disaster, and assume all the liability, may be impossible. The U.S. Centers for Disease Control and the New York City Department of Health counted nineteen serious infections in people traveling to the Dominican Republic for cosmetic surgery during a ten-month period in 2003–2004. A favorite topic of discussion among plastic surgeons at meetings is the rising number of people who limp into their offices requesting care after complications from plastic surgery vacations.

Superlatives

The best, the top, the most experienced, the only—these are just some of the superlatives that doctors use when describing themselves. Sometimes these statements are made by advertising or public relations agencies hired by the surgeon.

Sometimes the surgeon himself makes the claims. In either case, beware. When a doctor makes statements such as these, I worry about his ethics. Ethics are a character trait. If a doctor lies in advertisements, he'll lie to you.

There really is *no* quantifiable way to measure one surgeon's skills and compare them to those of another surgeon. Surgery is not like a sporting event, with measurable head-to-head competition. There is no easy way to assess who is the best or most experienced surgeon in a particular procedure.

Medical Spas

Medical spas, also called medspas, are the newest craze. They combine facials, massages, and seaweed wraps with medical procedures such as microdermabrasion, chemical peels, and even Botox and laser hair and laser spider-vein removal. Some even whiten teeth.

There are about fifteen hundred medical spas in the United States, bringing in about $1 billion annually, according to the International Medical Spa Association. These numbers increased threefold from 2004 to 2006. Medical-type treatments now make up over half of spa revenue.

Medspas are supposed to be physician supervised. Even though the doctor is meant to be *on site* full-time, the level of supervision varies widely. Some physicians use their names and medical licenses to allow other personnel—a nurse, an aesthetician, even an unlicensed worker—to perform procedures. The profit motive at medspas is tremendous, not only on the part of the spas, but also for the companies that manufacture their equipment.

There have been at least two deaths at medspas. Yes, deaths. In December of 2004, a 22-year-old woman seized after a lidocaine numbing cream was applied to her legs prior to a laser hair-removal appointment at a North Carolina spa. She died the next week. The lidocaine reached toxic levels and poisoned her. The medical director, an ear, nose, and throat doctor, had his license suspended for six months because he did not take a history and perform a physical prior to prescribing the anesthetic. Apparently, nonmedical personnel at the spa were selling 10-percent lidocaine gel without a prescription. This tragedy is strikingly similar to one that occurred in Arizona in January 2002. A woman suffered a seizure after applying anesthetic cream to her legs and lived in a coma for two years before dying in November 2004.

The American Society for Dermatologic Surgery found eight patients in 2001 who had complications from nonphysicians performing cosmetic dermatologic surgery procedures. And that was *before* the boom in medspas. Imagine how many people have been injured now.[4]

When one chooses a physician, the educational background of the doctor is paramount. Procedures such as laser hair removal, microdermabrasion, and chemical peels have well-known side effects that are respected by physicians. Careful examination of the backgrounds of nonphysicians performing these procedures in medspas can reveal some shocking facts. Nurses receive minimal or no formal training in skin care in nursing school. Skin care specialists, specific types of aestheticians (cosmetologists), receive six hundred total hours of training in New Jersey, including training in facials, makeup application, and waxing. Their training is often during high school. Education in skin physiology, biochemistry, and the use of lasers and drugs on the skin is minimal.

Why would anyone undergo medical treatment by minimally educated, unsupervised, nonmedical personnel? Because the fees for a nonphysician are a fraction of those for a doctor. But it is better not to have a procedure than to have one that is botched. The cost of treating complications such as scarring, pigment changes, or more serious medical problems, could dwarf the cost of the original procedure.

According to David Goldberg, M.D., dermatologist and director of Mount Sinai Medical School's dermatologic laser research, nurses and physician's assistants should be able to perform these procedures under the direct supervision of physicians. This means that the doctors are "on site and immediately available," he says. Problems occur when there is no physician in sight. Dr. Goldberg's patients indicated in a survey that they would rather be treated by a physician at a higher fee, than by a nonphysician at a lower fee. Another problem occurs, he says, when physicians such as gynecologists, family practitioners, or even dentists take advantage of their licenses and perform procedures they are not properly trained to do.

The Most Dangerous People Are Those Who Don't Know What They Don't Know

A California company will now send a physician assistant to your business office to inject Botox or wrinkle fillers! Seemingly minor procedures can result in disastrous complications, and even the lightest peels can result in scarring. Allergic reactions can be immediate and deadly, with no time to call the "supervising physician" at his office across town. Shame on state medical boards that allow unsupervised medical procedures. And shame on physicians whose greed gets in the way of good medical judgment. My advice? Insist on a physician—you deserve no less.

Your Anesthesia

It is senseless to spend countless hours researching plastic surgeons without considering the other person who will have your life in his hands. The person giving you sedation or general anesthesia is just as important as the surgeon. Some plastic surgeons use board-certified anesthesiologists and others use nurse anesthetists. Without doubt, there are lousy anesthesiologists and excellent nurse anesthetists. Personally, I am a great believer in education and training. An anesthesiologist is an M.D. Following graduation from college and medical school, he has taken at least four years of specialty training in anesthesiology. Board certification means he has passed both the written and oral examinations. A nurse anesthetist is a registered nurse (R.N.) who must have graduated from a two-year master's degree program.

Most nurse anesthetists work in hospitals, under the supervision of anesthesiologists. If there is a problem, an anesthesiologist will be in the room within seconds. If your surgery is being performed in an office operating room and your nurse anesthetist has a problem, she will look to the plastic surgeon for assistance. With due respect to plastic surgeons, they do not train in anesthesia and medical emergencies and should not be supervising the anesthetist. Acknowledging this, New Jersey has outlawed nurse anesthetists from independently administering general anesthesia in office operating rooms. Other states and countries will follow.

I feel much more comfortable with a person as well trained as possible administering the anesthesia. Therefore, clearly, I want a board-certified anesthesiologist in my operating room with me.

The Facility Is Very Important, Too

Most people do not expend a lot of effort considering where their surgery is performed. Surgery is performed around the world in facilities without oversight. The options for surgery include a hospital, a surgicenter, and a doctor's office. If you are healthy, any of these locations is probably acceptable. If you choose to be operated on in the doctor's office, be sure that the facility is accredited by an independent national organization. Examples include the American Association for the Accreditation of Ambulatory Surgical Facilities (AAAASF), the Accreditation Association for Ambulatory Health Care (AAAHC), the Joint Commission on Accreditation of Healthcare Organizations (JCAHO), and the federal government through the Medicare program. The American Society of Plastic Surgeons requires its members to operate only in such accredited facilities.

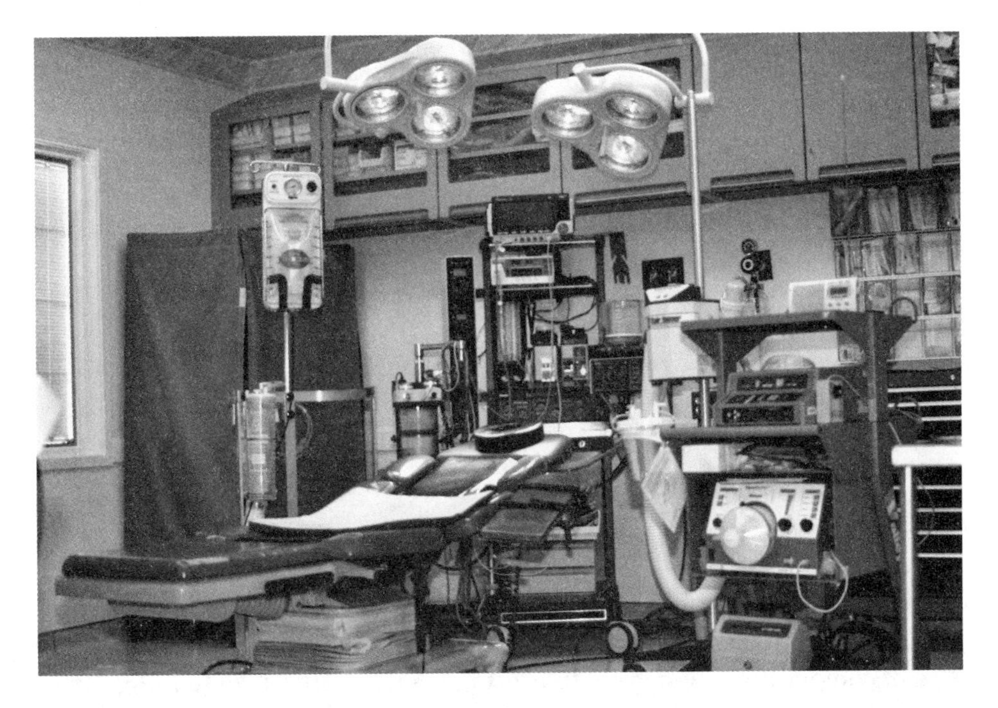

A proper operating room has a modern anesthesia machine and monitors, a defibrillator, appropriate lighting, a patient warming device, and a variety of other equipment.

If the facility you are considering is not accredited, steer clear. If the surgeon doesn't operate in another facility or hospital that *is* accredited, choose another doctor on your list.

Generally speaking, if you are healthy and your surgery will take less than four hours under general anesthesia or six hours under sedation, and if the facility is appropriately accredited and properly staffed, then an office operating room or a free-standing surgicenter may be the right decision for you. These facilities vary, but have the potential to be quieter and even cleaner than hospitals. You may receive friendlier, more personal care in an office. If you have medical issues, or feel uncomfortable with an office or surgicenter setting, then insist that the surgery be performed in a hospital. Finally, if you are tall or overweight, strongly consider a hospital. Most offices do not have the extra personnel necessary to lift heavy patients. The complication rate in accredited outpatient plastic surgery facilities is no different from that in a hospital.[5]

3

The Risks

As cosmetic surgery becomes more and more popular, more and more reports of complications and even disasters will surface. Here is a sampling of the record: A rhinoplasty death in a 17-year-old girl in New Jersey in 2004. Another rhinoplasty death in New York in March 2005. A flurry of liposuction deaths in Florida in 2002. Literally hundreds of deaths across the country from liposuction over a five-year period in the 1990s. An abdominoplasty death in New Jersey in 2003.

To keep surgery safe, we must not abandon common sense for the sake of vanity. Longer procedures are simply more dangerous than shorter ones. Healthy patients fare better than people with illness. Smokers have a higher overall complication rate.

Your plastic surgeon should be part of a team of doctors caring for you. Before surgery, he may need the opinion of an internist or cardiologist, an ophthalmologist, or a psychiatrist. Active medical problems should be assessed prior to surgery, and the physicians involved should send written reports to your surgeon.

When it comes to preparing for cosmetic surgery, don't try and avoid the preoperative "hoops." I have patients who ask to not see an internist, or not to have a stress test, for example. The preoperative evaluation is necessary to fully assess the risks of surgery, and to have a team of doctors who will participate in your care before, during, and after the procedure.

All surgery has risks. In order for you to make the decision to have surgery, you need to know what the risks are. You also need to understand your specific chance of developing a complication.

When a procedure is brand new, it really is not possible to assess risk. The surgeon must make vague statements. Any procedure has short- and long-term risks. Laser skin resurfacing was not supposed to lighten the skin; it was not until years after the procedure that this problem surfaced. Similarly, the late deformities of rhinoplasty took decades to see. And breast implants started rupturing with increased frequency after many years.

Plastic surgeons work with risk numbers provided to them by manufacturers of devices or publications or at teaching courses. New York plastic surgeon Daniel Baker, M.D., tells surgeons that the actual complication rate is double what is quoted at surgical meetings!

Your Chance of a Complication Depends on Your Overall Health

If you are perfectly healthy, you have a low chance of a complication. On the other hand, if you have diabetes, heart disease, a viral disease, or another chronic or acute disease, you are more likely to have a problem after surgery. Because cosmetic surgery is performed to enhance appearance, rather than to improve health, the surgeon is responsible for not operating on patients with a particularly high chance of a complication. It's not enough that the patient is willing to accept the increased risk. The responsibility of the surgeon is similar to the fiduciary responsibility of a money manager, or the responsibility of a parent.[1]

Why Things Go Wrong in Surgery

In the field of surgery, good surgeons worry about everything. They become paranoid. They become obsessive. They must, because potential disaster lurks at every corner. Errors can occur at any stage, before, during, or after the procedure. Surgery is inherently risky, with literally thousands of individual steps that can cause problems if not planned and performed correctly. The more compulsive the surgeon, the more I respect him. From examining the patient, to planning the surgery, to ordering proper preoperative medical evaluations and laboratory tests, to performing the surgery, to assuring that surgical risks are addressed and minimized, and to providing appropriate aftercare and patient education—these are the responsibilities of the surgeon. Other than piloting an airplane or commanding an army, few situations involve more responsibility and more potential risk than surgery.

There Is No Such Thing as Minor Surgery, Only Minor Surgeons

Surgeons are fond of that saying, since they know that even the smallest surgical procedure can result in problems, even disaster. Infections and bleeding or allergic reactions can happen even with mole removal. Surgeons who try and downplay an operation to sell it to the patient are not to be trusted. On the other hand, a competent surgeon should have an air of confidence that allows a properly informed patient to be at ease.

The General Risks of Surgery

The risk of dying in *any* surgical procedure is between 1 in 250,000 and 1 in 500,000. This is similar to the risk of dying in a plane crash. In contrast, the chance of being involved in a fatal car accident was estimated at 1 in 2,000 in a study of the California highway system, and 1 in 5,000 nationwide.

Certain procedures are more risky than others. Certainly, the risk of death during eyelid surgery is much lower than the 1 in 50,000 risk of dying during liposuction. In a study of twenty-three office surgery deaths in Florida between January 2000 and May 2004, the American Society of Plastic Surgeons found that the death rate was about 1 in 16,000 procedures overall. An interesting statistic emerges when this information is critically analyzed: the death rate for non–board-certified plastic surgeons, and other doctors such as dermatologists, was 1 in about 11,000 procedures; the death rate was three times lower, about 1 in 35,000, in patients operated on by board-certified plastic surgeons.[2]

Infection

With any procedure, even one as small as a mole removal, there is a chance of infection. Bacteria, virus, and fungus can all cause infection. Humans live in a delicate balance with bacteria. We need bacteria for proper digestion and even for proper wound healing, but too many of them tip the balance and cause infection. In surgery, we clean the skin with germ killers, most commonly Betadine. Surgeons scrub their hands, wear masks that filter out germs, and use sterile gloves, drapes, and instruments. The air in the operating room is filtered, and ultraviolet lights are used to kill germs in air ducts. Patients may receive a single dose of antibiotic, through the vein or by mouth, just before the operation starts. Implants and surgical sites are often washed with antibiotics. Keeping patients warm and giving them extra oxygen during and after surgery also lowers infection rates. Even with all of these measures, it is impossible to eliminate infections after surgery. In clean cosmetic surgical procedures, "normal" infection rates are about 2 percent on the face and 4 percent on other parts of the body. Despite common practice, extended use of antibiotics before and after surgery does not lower the rate further.

Infection usually comes from bacteria already in the body. Even Betadine doesn't sterilize the skin—it simply kills the surface bacteria, leaving the bacteria in the hair follicles and sweat glands. If a patient has a urinary or respiratory infection at the time of surgery, the bacteria can spread through the blood and infect the wound. Patients who have diabetes, cancer, obesity, or another chronic disease have impaired immune systems and higher infection rates. Longer surgical procedures also have increased infection rates.

Procedures such as peels or lasers can cause the herpes virus, always hiding in the body, to activate. For this reason, antiviral drugs are prescribed before and after these procedures. In fact, the chicken pox virus (*Herpes zoster*) can reactivate from stress after any surgery and cause pain and scarring.

The wet, dark, warm environment that is created after a peel is just what fungi love, so wounds are kept clean to decrease fungal infections.

If infection does occur, additional antibiotics are necessary. They can be taken by mouth, but if the infection persists, intravenous antibiotics in the hospital may be necessary. If an implant is involved, it usually needs to be removed to cure the infection.

Bleeding

Bleeding is normal and expected in surgery. Real surgeons are trained in a variety of techniques to control bleeding during an operation. Bleeding *after* surgery is another story. Oozing is normal in many procedures, particularly liposuction and rhinoplasty. Significant bleeding that begins after surgery could become a surgical emergency, even one that is life threatening.

Many factors increase the chance of bleeding after surgery. *Aspirin* interferes with clotting and increases the likelihood of bleeding following surgery. Patients should stop taking aspirin five days before surgery. To err on the safe side, I advise my patients to stop a week prior to surgery. This is a new recommendation with excellent scientific backing, replacing the old "two-week rule." Literally hundreds of other drugs interfere with bleeding. Your surgeon will give you a list of those you should avoid. At the end of this section is a partial list of these drugs.[3]

Hypertension is notorious for causing bleeding after surgery, particularly after face-lifts. It must be well controlled prior to surgery. *Nausea* can cause bleeding when the head is lowered to vomit. Drugs are given during surgery to decrease the chance of nausea following surgery, although in about 20 percent of people nausea will occur nevertheless.

Lowering the head after facial surgery increases the chance of bleeding. Try it now and you will feel the blood rushing to your head. It is important to not bend down after surgery and to sleep on a pillow or two. Your head should be above the level of your heart.

The risk of bleeding after surgery is highest the first night after the procedure and decreases with each subsequent day. There is some risk as late as three weeks following surgery. Flying is not advised for about two weeks after surgery, since changes in air pressure have been known to cause bleeding.

Blood Clots (DVT) and Pulmonary Emboli

Clots and emboli are the dreaded complications of any surgery. Blood clots (*deep venous thrombosis—DVT*) can form in the leg or thigh during surgery. If they break off, clots can travel to the lungs, creating a potentially fatal pulmonary embolism. Factors that increase the chance of DVT are long surgery, inadequate fluids during surgery, trauma to the legs or thighs, and a history of smoking or oral contraceptive use. Birth control pills increase the chance of a clot by six times. Because of this, they should be stopped one month prior to surgery and not restarted until two weeks after surgery.

The chance of a blood clot in the leg with a surgical procedure such as a face-lift is 1 in 300. There is about a 1 in 1,000 chance that this clot will break off and travel to the lungs. Most blood clots form when patients are under general anesthesia. Intermittent compression boots keep blood from stagnating in the veins of the leg. These devices, or an injection of the blood thinner heparin, decrease the chance of DVTs. There really is no excuse for not using one of these preventive measures. Shockingly, in a survey of 273 plastic surgeons in 2001, only 20 percent appropriately used compression boots on their patients. Properly applied, they have no side effects and are inexpensive to use. *Boots, or an alternative treatment, are required for any procedure lasting over thirty minutes.* With higher-risk procedures, boots *and* heparin should be used.[4]

Nausea and Vomiting

These problems are incredibly common after general anesthesia and even after intravenous sedation. Often narcotics given during surgery are the culprits. But many different anesthetic drugs also cause nausea. During surgery most surgeons and anesthesiologists give patients a combination of a steroid called Decadron, and either Zofran or Anzemet. While the drugs are helpful, 20 percent of patients still develop nausea and vomit. Often the only cure for this type of nausea is sleep. It takes hours to rid the body of the anesthetic drugs that caused the nausea. Most nausea is gone by the morning after surgery. Nausea that persists is usually due to the continued use of narcotics such as codeine or Percocet.

Scars

Some patients think that plastic surgeons should not leave scars. If the plastic surgeon has done all the right things to maximize the cosmetic appearance of the scar, the rest is up to the genetic destiny of the patient.

Scars are the inevitable result of healing. Unlike reptiles that can regenerate body parts, humans simply patch them with scar. *There is no such thing as scarless surgery.* All procedures must create scar. During the normal course of healing, scars thicken and become red and hard for the first three months. Over the next nine months, the scars gradually soften and turn flesh colored. After a year, good scars are not particularly eye-catching from a normal speaking distance. However, all scars are visible in a magnifying mirror.

Most people heal with good-quality scars, but in less than 5 percent of people, scarring takes a different course, with a prolonged period of hard, red (*hypertrophic*) scars. This scarring can last for years and be very troublesome. In less than 1 percent of people, the scarring goes completely awry, resulting in tumors of the scar tissue, called keloids. These scars invade normal tissue, and can itch and can grow to golf-ball size. Asians and people of African descent have the highest rate of keloid formation.

Steroid injections and laser treatments can help scars mature, as can massage, pressure garments, vitamin E and other creams, and silicone and hydrogel (water-based) sheeting or topical salicylic acid or silicone gels. Most surgeons do not find these useful. But because they are harmless, many people feel they are

Plastic Surgeon's Tricks to Minimize Scarring

- Incisions placed in the best possible location and orientation
- Skin lifted to decrease tension
- Skin closed with dissolving stitches in the deep layer
- Skin closed with fine stitches or glue in the superficial layer
- Stitches removed early
- Meticulous suture technique—the signature of the plastic surgeon

better than nothing. Cryotherapy, or freezing of the scar, may help bad scars.[5] Interestingly, covering scars with paper tape seems to improve them. We don't know why this happens, but patients may benefit if they can tolerate six months of taping.[6]

Like so many other questionable treatments, scar treatments are plagued by noncontrolled studies and anecdotal evidence.

Skin Closure—Stitches, Staples, and Dermabond

The plastic surgeon's skin closure technique might include traditional stitches, or perhaps a "subcuticular" technique, where the suture is completely beneath the surface. This technique leaves no suture marks or cross tracks. Little white dots of scar are caused by stitches left in too long, and tying the stitches too tight causes tracks. Since plastic surgeons take great pride in their scars, they seldom use staples. In the scalp, however, staples actually are preferred since they kill fewer hair follicles.

A newer method of closure is Dermabond—or a competing product, Indermil. These glue-like materials eliminate cross tracks and stitch marks and do away with painful suture removal. It has been slow to catch on with plastic surgeons, perhaps because of their reluctance to give up their signature closure techniques! But as surgeons become familiar with the technique of application, Dermabond is used with increasing frequency, in everything from eyelid surgery to tummy tucks.

Smoking

It always amazes me when someone is concerned enough with their appearance to consider cosmetic surgery, but will knowingly injure their bodies and appearance by smoking tobacco. Inhaling more than four thousand toxic substances, including carbon monoxide and nicotine, increases the chance of disaster in plastic surgery. As early as 1984 it was recognized that wound-healing problems after face-lifts were nearly thirteen times more likely to occur in smokers than in nonsmokers. Most reputable surgeons will not perform face-lifts, breast lifts, or tummy tucks on active smokers. These are three procedures that lift large areas of skin, cut off excess tissue, and pull the tissues tight, sewing them in place. The skin's blood supply is dramatically reduced. Smokers who have breast reductions have complication rates three times higher than nonsmokers. Their skin can simply die in these procedures, creating a nightmarish situation. There really is no solid information on how long it takes for a former smoker to decrease her risk. Most plastic surgeons will operate if a patient stops smoking a month before surgery and doesn't restart until a month afterward.[7]

Poor Cosmetic Results

Every surgeon has good, bad, and average results. Let's hope that your surgeon has mostly good results. In fact, if a surgeon tells you he has never had bad results, choose another surgeon. He is either inexperienced and hasn't done enough surgery to see poor results, or he is dishonest.

Technically difficult procedures are more likely to have problems than simple ones. Rhinoplasty has the greatest chance of a poor result, with about 15 percent of patients requiring a revisionary procedure. Liposuction has a revision rate of 5–10 percent. Of course, skilled surgeons will have fewer poor results and poorly trained surgeons will have more bad results. Choose your surgeon wisely.

Reoperative Surgery

If the procedure doesn't turn out well, a revision may be considered. It is a difficult decision for both patient and surgeon, since secondary surgery is often more difficult than the original procedure. The tissue has been laced with scar tissue, even if it is not visible from the surface. Scar tissue is the body's response to injury, and it is a gritty material that is not pleasant to work with. In fact, scar tissue is the bane of existence for all surgeons.

Operating on a scarred nose, for instance, is like trying to separate two pieces of paper that have been cemented together. One slip and the paper rips. Surgery on scarred tissue can take hours longer than surgery on tissue that has never been violated.

When considering whether to reoperate, the patient and surgeon must again assess risks and benefits. But this time, the curve shifts. No longer is the chance of success 85–95 percent and the chance of complication 5–15 percent. If the problem is as dramatic and obvious as a large divot in the tip of the nose, the decision is easier than if the flaw is so subtle that only the patient sees it. The second procedure may have only a 50-percent chance of success, a 30 percent chance of a complication, and a 20-percent chance that it will neither improve nor worsen the situation. When the numbers shift this way, the decision to operate is very difficult. Sometimes your original surgeon shouldn't try again. You may need a surgeon who specializes in difficult, reoperative surgery. While such a choice may tarnish your plastic surgeon's ego, you need to think of yourself first.

Money becomes a factor after a bad result occurs. Many plastic surgeons will reoperate and not charge a fee. The patient is, however, responsible for the anesthesiologist and facility fees. If you find a new plastic surgeon, there would, of course, be an additional surgical fee. Unless your result is so bad that malpractice was committed, your surgeon will probably not return your fee or pay for a second surgeon. If he does agree to return your money, you may need to agree to not file a lawsuit against him. No surgeon can guarantee perfect results, and part of cosmetic surgery is the risk of a bad result and the need for a second procedure.

Rare Complications

Each surgical procedure has its own peculiar associated risks. For example, there is risk of spearing an organ with liposuction, of puncturing the lung during a breast augmentation, of blindness with eyelid lifts, and of a skull fracture with rhinoplasty. These are rare, but possible, problems. The specific risks for each procedure will be discussed in other chapters.

How to Decrease Complications with Cosmetic Surgery

A number of little details can decrease your chance of a problem with surgery. The ASPS convened a task force on patient safety in 2000, in response to a cluster of patient deaths in office operating rooms, and in 2002 published its findings.[8] Those findings, along with some of my own, include the following:

- Your surgeon should keep you warm and give you extra oxygen during surgery. If the body is allowed to cool during surgery, the heart can be affected and the infection rate is higher. Special warming techniques should be used for surgery that takes more than two hours and treats more than one-fifth of the body surface area.
- You should not have surgery outside of a hospital if more than a pint of blood loss is expected.
- No more than five kilograms (eleven pounds) of fat should be suctioned in a facility outside of a hospital.
- Large-volume liposuction should not be performed in combination with a tummy tuck.
- Only operations taking less than six hours should be performed in the office.
- If your surgery is longer than thirty minutes, you should have some sort of blood-clot prevention.

- You should receive medication to decrease nausea and vomiting following surgery.
- You should be given adequate pain-relief medication after surgery.
- You should have a thorough history and physical examination prior to surgery. The surgeon might perform a cursory exam but, let's face it, they simply will not do the job of a board-certified internist. An appropriate letter with recommendations should be sent to your plastic surgeon. I perform the exam on healthy patients under age 50 and request internist clearance on those over 50.
- Preoperative testing should include an electrocardiogram (EKG) for anyone over age 45 or younger if indicated. It is a good idea to have an EKG stress test if you are over 50.
- Laboratory testing should include a blood count and blood chemistry, and other tests depending on your individual circumstances.
- Only normal healthy patients or patients with mild disease should be operated on in an office setting. Others should be operated on in the hospital, if at all. Remember, this is cosmetic surgery. I don't believe that people with serious illness should have cosmetic surgery. Unless your risks are as low as possible, you're playing with fire.

Cosmetic Surgery in Teenagers

Rhinoplasty and otoplasty are the two most common surgical procedures in teenagers. Male breast reduction and chin augmentations are also fairly common and appropriate.

Legally, cosmetic breast augmentations cannot be performed in teenagers. The Food and Drug Administration (FDA) approves saline implants only in women older than 18, and gel implants in women older than 22. Since breast growth is often not complete until that age, medically it is inappropriate under age 18. The American Society of Plastic Surgeons advises its members not to perform breast augmentations in younger teenagers except in cases of severe asymmetry. The ASPS and most plastic surgeons and psychologists feel that women are not emotionally ready to make such important decisions until age 18.

Since the body continues to change and childhood fat deposits often disappear in the late teens, liposuction should not be routinely performed before age 18.

Often teenagers do not have a fixed body image. They may have immature goals and may want to look like celebrities. As they age, they may well outgrow these immature desires—although many adults still have them.

When I see a teenager for cosmetic surgery, I watch carefully how she interacts with me and with her parents. If she can't communicate with me before surgery, I worry that she will not be able to follow instructions after surgery, thereby increasing the chance of complications. Also, I worry about her ability to deal with a new appearance, particularly if it is not exactly what she envisioned. In the immediate postoperative period, when tissues are distorted from swelling and bruising, many teenagers become upset and even fairly mute. For all of these reasons, I often turn down teenagers and advise their parents to wait a year or two before considering cosmetic surgery.

All consultations with teenagers and children must be performed in the presence of a parent or legal guardian. Minors cannot legally consent to surgical procedures. A parent must sign the consent form.

Often, a parent will bring a child into my office and ask that a surgical procedure be performed—usually an otoplasty, a rhinoplasty, or even revision of a scar. Sometimes the child will have no interest in the procedure. In that case, I will not perform the procedure. *Cosmetic surgery is for the patient—not the parent, not the boyfriend, not the friends.* It is the patient who will have the procedure and must accept the potential risks.

Psychiatric Complications

Cosmetic surgery patients have a predisposition to body-image disorders. They are more likely to be prescribed brain-altering drugs by a psychiatrist. It is extremely important that patients be psychologically stable prior to cosmetic surgery. They must be able to tolerate complications following surgery. Any complications can be upsetting, particularly those that severely impair health. Complications that alter appearance can start a tailspin of depression. For that reason, it is important that anyone under the care of a psychiatrist or psychologist be evaluated and receive unconditional clearance for the surgery. Uncontrolled depression can lead to suicide after surgery. The risk of death from suicide is threefold higher in women who have breast augmentations. Rhinoplasty is another procedure notorious for psychiatric complications after surgery.[9]

Drugs and Foods That Must Be Stopped Before Surgery

DO NOT TAKE any medications that contain *aspirin* for seven days before your surgery or seven days after your surgery. Aspirin increases bleeding during and after surgery. Check the labels of all the medications you take; if you are uncertain whether they contain aspirin, ask your doctor or a pharmacist.

DO NOT TAKE any *vitamin E* or *herbal remedies* including *orthomolecular vitamin therapy* and even *green tea* for a week before or after surgery. These substances have unknown effects and may interfere with your anesthesia or ability to heal. You *may* take vitamin C and the B vitamins.

The following is a *partial* list of aspirin-containing products or drugs that may cause abnormal bleeding or bruising, or may have undesirable side effects. You should ask your doctor if they can be continued before surgery. Most items listed are brand names and have registered trademarks.

Acetyl salicylic acid
Acuprin
Alka-Seltzer
Amitriptyline
Amoxapine
Anacin
Anaprox
Ansaid
Aphrodyne
Argesic-SA
Arthritis Pain Formula
Arthropan
ASA
Ascriptin
Aspergum
Axotal
Azdone
B-A-C
BC (tablets and powder)
Bexophene
Buffaprin
Bufferin
Buffex
Buffinol
Buff-A Comp
Buf-Tabs
Butalbital Compound
Cama Arthritis Reliever
Carbamazepine
Carisoprodol Compound
Cheracol Capsules
Chlor-Trimeton
Clopidogrel
Codoxy
ColBenemid
Colchicine
Comtrex
Congesprin Chewable
Coricidin
Cosprin Tablets
Co-Tylenol
Coumadin
CP-2 Tablets
Damason-P
Darvon w/ASA
Darvon Compound
Darvon N w/ASA
Desipramine
Diagesic
Disalcid
Dolene Compound
Dolprin
Donnatal
Doxaphene Compound
Doxepin
Dristan
Duradyne
Durasal Tablets
Easprin
Ecotrin
Efficin
Elavil
Emagrin
Emcodeine
Empirin
Emprazil
Encaprin
Endep
Equagesic
Equazine-M
Etrafon
Excedrin
Fastin
Fenfluramine
Fiogesic
Fiorgen PF
Fioricet
Fiorinal
Fish Oil
Flagyl
Flexeril
Four Way Cold Tablets
Furazolidone
Furoxone
Gaysal-S
Gelpirin
Gemnisin
Goody's Headache Powders
Haltran

Heparin
Herbal Medicines
Imipramine
Indocin
Indomethacin
Isocarboxazid
Isollyl Improved
Lanorinal
Limbitrol DS
Liquiprin
Lodine
Lortab
Ludiomil
Magnaprin
Magsal
Marnal
Marplan
Marprotiline
Matulane
Meprobamate
Meprogesic Q
Micrainin
Midol
Mobidin
Mobigesic
Monoamine Oxidase Inhibitors
Mysteclin F
Nardil
Nicobid (gum or patch)
Norgesic
Norpramin
Nortriptyline
Novahistine
Ornade
Orphegesic
Oxycodone
Pabalate
Pamelor
Panasal
Pargyline
Parnate
Pedi-profen
Pepto-Bismol
Percodan
Perphenazine
Persantine
Phenaphen w/Codeine
Phenelzine
Phenergan
Phentermine
Phenylbutazone
Plavix
Pondimin
Presalin
Procarbazine
Propoxyphene
Protriptyline
Prozac
Pyrroxate Caps
Redux
Relefen
Rid-A-Pain w/Codeine
Robaxisal
Roxiprin
Ru-Tuss
Saint-John's-Wort
Saint Joseph's Cold Tablets
Saleto
Sine-Aid
Sine-Off
Sinequan
Sinutab
SK-65 Compound
Soma Compound
Sumycin
Supac
Surmontil
Synalgos DC
Tagamet
Talwin Compound
Tegretol
Tenuate Dospan
Tetracycline
Tofranil
Tolectin
Tolmetin
Tranylcypromine
Triaminicin
Triavil
Trigesic
Trilisate
Trimipramine
Uracel
Vanquish
Vibramycin
Vitamin E
Vivactil
Warfarin
Zorprin
Zyloprim

The following drugs must be stopped three days prior to surgery:
Advil, Aleve, Clinoril, Dolobid, Feldene, Ibuprofen, Meclomen, Medipren, Motrin, Nalfon, Naprosyn, Nuprin, Oraflex, Orudis, Ponstel, Rufen, Sulindac, Toradol, Voltaren, and other "nonsteroidal" anti-inflammatory drugs. Also Bextra, Celebrex, Vioxx, and any other Cox-2 inhibitors.

4

Computer Imaging

Modern cosmetic surgical consultations virtually require the use of computer imaging. Photos altered by the computer help patients understand how they might reasonably look following surgery. About a thousand plastic surgeons use Mirror Image, Canfield Scientific's digital imaging program. Doug Canfield, president of the company, says his programs help doctors communicate better with patients.

Computer imaging is a double-edged sword. On the one hand, the computer can allow you to better understand your surgery. In a rhinoplasty, for instance, it can simulate a narrower tip of your nose. Your image can be progressively altered, showing your appearance after different surgical maneuvers. Used appropriately, computer imaging can be a terrific aid to the consultation. On the other hand, unscrupulous doctors can use computer imaging to market their practice, promising results that are impossible. The computer can do anything, even as ridiculous as adding a third eye to your face! Trouble begins when the computer is used to sell a procedure. It should only be used to educate.

Only the surgeon who will be performing your surgery should perform the computer imaging. There is no way that a nurse, secretary, or marketer can know whether the changes made on the computer can actually be accomplished on a live patient. Only the surgeon can determine this, after listening to your medical history and examining you. Insist that he be the one to do the imaging. If someone other than the surgeon does it, take the entire experience with a big grain of salt.

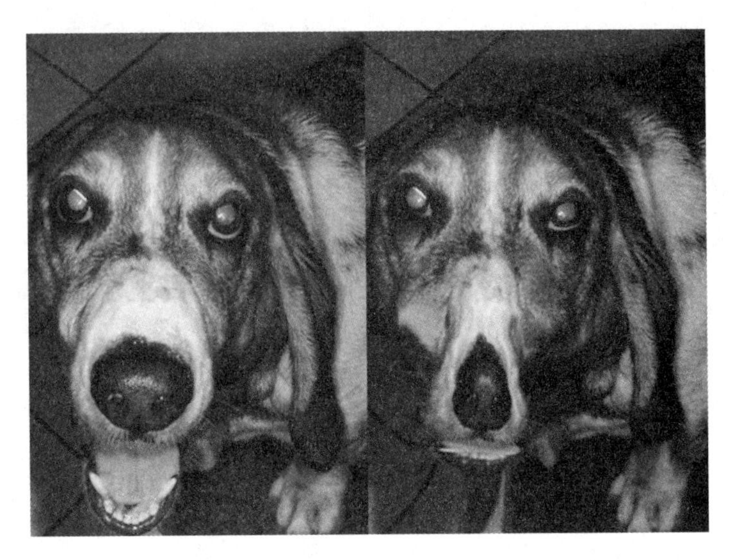

My dog has undergone an ear lift, a rhinoplasty, and a canthopexy, courtesy of computer imaging. If I can do this to Flopsy, imagine what the computer can do to you.

The Computer Can Lead to False Expectations

Most plastic surgeons are honest with computer imaging. But dishonest doctors can create computer images that are not surgically possible, which leads to false expectations and eventually to unhappy patients when the promised result is not achieved.

Be careful of the word "promise." No surgeon can promise a perfect result, or even *any* result. Every surgeon wishes he could. The differences in healing between patients make any such guarantee impossible to offer. An honest surgeon will tell the patient her chance of achieving the desired result. I might tell a patient with chubby thighs that there is a 95-percent chance that the image I create on the computer can be achieved with surgery. Why not 100 percent? Scar tissue, infection, unappreciated size of bone or muscle, or simply the fact that the surgeon is human, may all contribute to a less-than-desired result.

Surgeons Shouldn't Print Out Altered Images

Most surgeons won't. In fact, most will not even save them on the computer. Remember that imaging is an educational tool, not a guarantee of a particular result. If the altered photo is printed out, patients cling to the image and expect that result. If it is not achieved, the computer-generated image will be powerful ammunition for a lawsuit. Honest surgeons emphasize that the image is just a possibility. They may even purposely generate poor results, pointing out that these too are possibilities.

Procedures That Can Benefit from Simulation

Computer simulations are essential in some procedures and useless in others. In rhinoplasty, they allow the patient to understand how changes in one part of the nose can alter the harmony of the entire nose. They show how a chin implant makes a large nose seem smaller. The computer is also useful in liposuction, face-lifts, eyelid and eyebrow lifts, and tummy tucks. It is less useful in breast surgery and wrinkle removal.

The Future of Computer Imaging

Faster computers and higher-resolution cameras are continually improving computer imaging. New systems can assess the degree of invisible sun damage within the skin. Future technology will measure the thickness and texture of the skin, allowing accurate assessment of skin care programs. Commonly used techniques only measure in two dimensions; but Doug Canfield says that "the next generation of three-dimensional imaging techniques will allow the measurement of volume." Three-dimensional analysis will allow surgeons to accurately choose a breast implant that will create a specific cup size. Surgeons will be able to measure volume differences after liposuction, thereby adding more precision

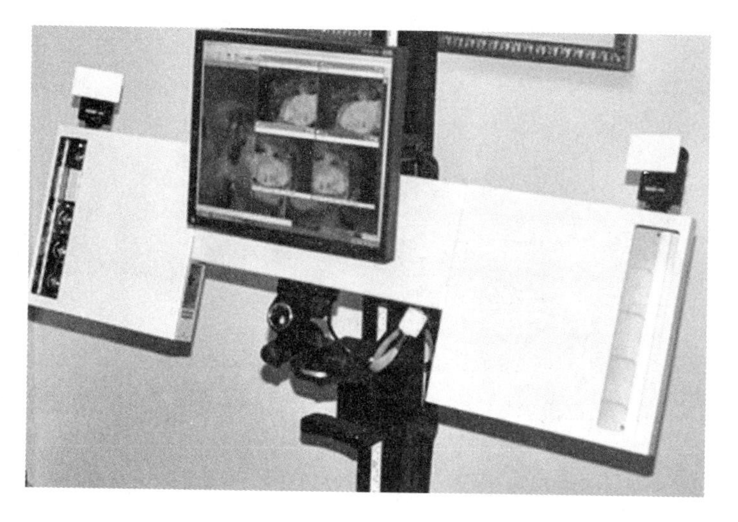

The Canfield Imaging 3D Vectra system. See color plates.

to plastic surgery planning. I have used the new generation of imaging for the past year. In a millisecond, multiple cameras snap photos from different angles. The computer blends the pictures, creating a three-dimensional image that can be turned to any angle. This technology is a quantum leap beyond current two-dimensional imaging. While various 3-D systems cost over $45,000 today, as the price drops, more surgeons will acquire them.

5

Aging

"Old age isn't so bad when you consider the alternative."
—Maurice Chevalier

All animals age. Teleology explains why things happen: we age so that we will die. Dying allows evolution to occur without overpopulating the planet. Thousands of years ago, if an animal didn't eat you, an infectious disease was likely to kill you. Well-known scourges of the time were smallpox and the plague. Viruses such as influenza and bacteria such as tuberculosis served to keep the human population in check. The average life expectancy was twenty years, so there was a good reason why puberty occurred at age 12. It was important to raise your children to an age where they could care for themselves by the time you died. If you were lucky enough to live into your 20s—or even more rarely, your 30s—more surprises were around the corner for you.

Poor vision was a death sentence before modern times. As we age, we lose the ability to focus on close objects. Presbyopia, as it is called, is a natural part of growing old. A thousand years ago, as our vision failed, we would not be able to see lions as they hunted us. But during the thirteenth century when eyeglasses were invented, we won a great battle over nature.

Advances in Medicine and Surgery

Medicines have been described since the time of Hippocrates. When people ate various flowers, herbs, and potions they noticed different reactions; these natural remedies were the predecessors of many of today's drugs. Willow bark, for instance, helped pain. Later it was found that it contained salicylic acid, the ingredient in aspirin. New medicines have altered the natural course of illnesses.

Surgery too has been practiced for thousands of years. One limiting factor was pain. In Homer's time, 800 B.C.E., opium from poppies made pain more bearable. Alcohol was the anesthetic of choice during the seventeenth century. Laughing gas, or nitrous oxide, was described in 1799 and ether in 1846. Anesthesia became progressively safer, with newer drugs and monitoring equipment. In 1977 the death rate from general anesthesia was 1 in 4,000. Now it is 1 in 250,000 to 1 in 500,000—safer than the 1 in 2,000 death rate from driving for a year on the California highways!

Antibiotics were the greatest advance of twentieth-century medicine. Simple bacterial infections in 1939 were life threatening. Death could result from cellulitis unless an amputation was performed. Five years later, a short course of antibiotics could cure the infection. Penicillin, the first antibiotic, was discovered in 1928 and, along with streptomycin, was mass-produced in 1943. By the end of the decade, dozens of antibiotics were available. There are more than ten thousand antibiotics now.

As we enter the first decade of the twenty-first century, we are again seeing the emergence of a new class of antibiotics: in this century, we are developing drugs that can fight viruses.

Modern Medicine Has Resulted in a Longer Lifespan

No longer are people dying of infection or appendicitis at age 20, as was the case in ancient Rome, or being eaten by tigers at age 40, or dying of gallbladder disease at age 50. With enormous advances in fighting heart disease owing to the understanding of aspirin, the development of drugs called statins and beta-blockers and cardiac stents and surgery, people are living longer than ever. We're beginning to learn that heart disease may not be inevitable, that it may be due to inflammation in the heart that can be controlled. As a result of this and other advances, life expectancy in Western nations is now nearly 80 years.

This increased life expectancy, teamed with painkillers and joint replacement surgery, has created a large number of physically active people in their "golden years." In the 1990s, the demand to "look as good as you feel" primed the growth of cosmetic surgery.

As Billy Crystal Said on *Saturday Night Live*, "It's More Important to Look Good Than to Feel Good"

While not entirely true, this statement is not entirely false, either. In the business and entertainment worlds, one may be able to live with pain, but wrinkles can be a career killer.

As we age, complex changes occur in the body. Women know very well when their estrogen hormone declines. After it decreases to a critical level, menstruation ceases. The timing is predetermined at birth and depends on the number of eggs in the ovary. When they run out, menopause occurs.

Other hormones drop off similarly. Thyroid hormone, growth hormone, testosterone, the adrenal androgen DHEA (dihydroepiandrosterone), and ACTH (adrenocorticotropic hormone) are just some of the hormones that diminish as we age. Both men and women have female hormones called estrogens and male hormones called androgens. Only endocrinologists and physiologists can begin to grasp the complicated mechanisms and interactions of hormones. Like the discovery of new planets around stars, new hormones are found at regular intervals. We know more are out there, but they haven't yet been seen or measured. This adds to the mystery of aging.

Without blood tests, we can't tell when most hormones decline. But along with their decline are the obvious signs of aging. Hair turns gray, wrinkles appear, and skin sags. Muscles shrink and fat collects. The integrity of muscles, cartilage, and bone decreases. We get fractures, joint and back pain, and arthritis. We begin to forget more; we take longer to solve problems. And our creativity declines. Together, these are called cognitive changes.

Aging Spurts™

Just as growth in children occurs in spurts, aging also occurs in spurts. For years my patients would tell me that they developed wrinkles over the course of a few weeks. At first I did not take them seriously. I thought that they noticed the aging when looking into the mirror for the first time in a while. But too many patients told me the same story. And we have all heard of a person whose hair turned white over a few weeks after physical or emotional trauma. I call these "aging spurts." These spurts are the result of thousands of complex hormonal changes in the body. When the body is exposed to physical or emotional stress, the adrenal glands produce extra steroids such as cortisol. Stress leads to elevated levels of a chemical called interleukin-6. And this may injure DNA. Memory loss and diabetes are just two of the aging symptoms that may result.[1]

Even sleep plays a role in aging. The quality declines as we age. Sleep turns out to control many hormone functions: growth hormone is largely produced during sleep; steroid levels increase with sleep deprivation and are associated with an increased rate of aging. And so, improved sleep may stall aging.

The complex science of aging is not well understood. Certainly, your genes predetermine much of what will happen to you. But not everything. The environment strongly influences what diseases you will develop and how rapidly you will age. According to Caleb E. Finch, Ph.D., professor and director of the Gerontology Research Institute at the University of Southern California, many factors play a role in aging. These include lifestyle, exposure to infections, stress, and diet, as well as chance variations during development.

New York City plastic surgeon Darrick E. Antell, M.D., F.A.C.S., has studied aging changes in a series of twins. He has shown that sun damage and cigarette smoking cause enormous changes in appearance. One would expect twins to age at the same rate but the photos seen here demonstrate that environment triumphs over genetics.[2]

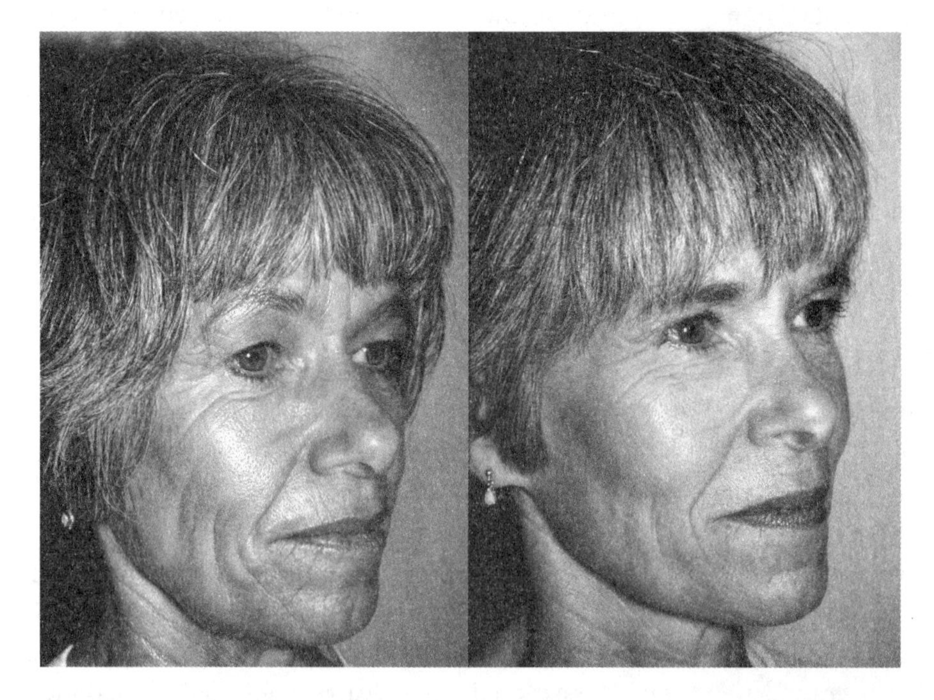

The twin on the left has had significant sun exposure; her sister, on the right, has had much less exposure to the sun. Neither has had any surgery. Photos courtesy of New York City plastic surgeon Dr. Darrick E. Antell. See color plates.

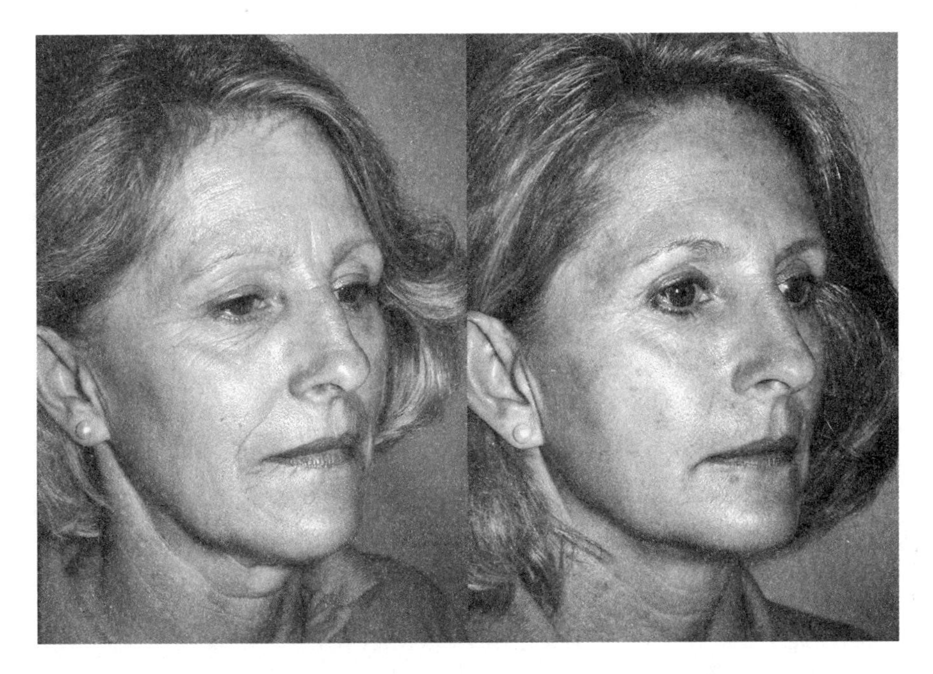

The twin on the left has smoked for thirty years; the twin on the right does not smoke. This environmental influence on aging is reason enough to "kick the habit." Photos courtesy of New York City plastic surgeon Dr. Darrick E. Antell. See color plates.

The precise cause of aging is not known. Possibly it is one too many mutations in our DNA, or changes in the sugars and proteins of the body. A favorite theory is that free radicals progressively damage cells. In actuality, aging may result from all of these processes.

The RealAge System

Michael Roizen, M.D., in his *RealAge* books, says that certain environmental factors can be altered to increase lifespan.[3]

To stall aging, he suggests that you—

Take the following medications:

- Daily aspirin
- The following vitamins and minerals: folic acid, vitamins B6 and B12, calcium, vitamin D, magnesium, vitamins C and E, flavenoids, and potassium
- Any physician-prescribed medications
- Vaccinations against common illness

- Medications to keep your blood pressure and cholesterol low
- Do NOT take hormones

Eat the following foods:

- Coffee, green tea, tomato sauce, and a well-balanced diet that is low in calories and high in nutrients
- Unsaturated, nontrans fat
- Five servings of fruit *and* four to five servings of vegetables a day
- Nonfried fish three times a week
- One half to one alcoholic drink per day and no more than three drinks
- Daily breakfast
- Whole grains and fiber

Make these lifestyle changes:

- Floss and brush teeth daily and see a dentist every six months
- Wear a seatbelt when driving and a helmet when biking
- Get ten to twenty minutes of sun each day
- Avoid secondhand smoke
- Have safe sex—the more, the better
- Laugh a lot and keep a positive attitude
- Stimulate your brain
- Sleep seven to eight hours a night
- Exercise enough to burn about 500 calories a day, weight-train, and build stamina
- Avoid air pollution and toxins
- Reduce stress and have a stress relief plan
- Don't smoke
- Own and walk a dog
- Maintain your weight in an appropriate range

If you follow this advice, you will live longer. Some of these suggestions are easier to follow than others. Many more factors are listed in Dr. Roizen's *RealAge* books, but these are a start.

Facial Aging

The changes of aging in the face are predictable enough to allow carnival entertainers to accurately guess your age. The first signs occur by age 30. Early skin creases appear around the eyes. The muscles pull on the overlying skin and crin-

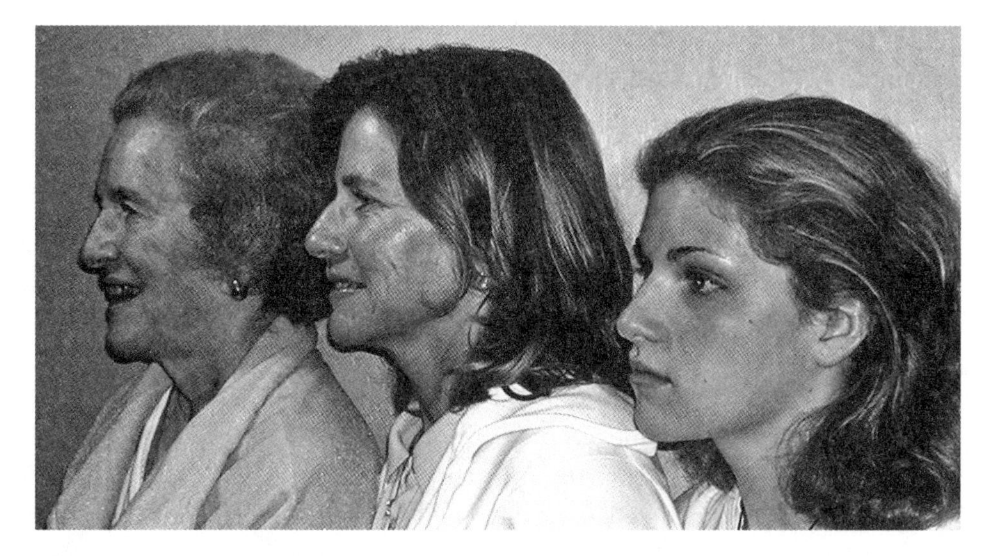

Three generations of the same family: the 74-year-old grandmother is on the left, the 48-year-old mother in the middle, and the 15-year-old daughter on the right. Note the progressive descent of the eyebrow, the upper eyelid skin, and the development of the nasolabial folds, jowls, and excess skin of the neck over the decades. See color plates.

kle it. The wrinkles form at right angles to the direction of the muscle pull and are actually *stress fractures of the dermis.* Plastic surgeons call these "dynamic wrinkles." They are different from the "static wrinkles" caused by weathering of the skin. The distinction between the two is important, since dynamic wrinkles return rapidly after treatment unless the underlying muscle is paralyzed or removed. Static wrinkles, on the other hand, can be treated without regard for underlying muscle.

The nasolabial folds begin to deepen by the late 30s, causing considerable dismay. By the 40s, other wrinkles appear on the face. Eyelid skin begins to droop and fat protrudes from the lids. The "tear trough" appears as the cheek fat heads south. Fat often begins to collect under the chin. By the 50s, sun damage has caused splotchy brown pigmentation to collect as wrinkles spread to the full face and begin to deepen. Excess skin in the neck invites remarks such as the "turkey gobbler," and the jowls rear their ugly heads. The eyebrows head south along with the earlobes. The tip of the nose and the ears increase in size, since cartilage can continue to grow throughout adulthood. By the 80s, the soft tissue of the face has significantly drooped, giving the elderly a skeleton-like appearance.

Dr. Scott Bartlett, a plastic surgeon at the University of Pennsylvania, has observed that while the height of the face decreases with age, the width increases. But unless the teeth fall out, the face does not shorten. In people who have lost their teeth, dentures and dental implants preserve the height of the face.[4]

Plastic Surgeons Help to Preserve a Youthful Appearance

Methods to turn back the clock include lifestyle changes such as exercise and diet modifications, medicines and hormone treatment, drugs for the skin, and surgery. All of these topics will be discussed in detail in the course of this book.

A *diet rich in antioxidants* may improve the quality of the skin. Skin water content and fat levels increase (the more the better in the skin). The antioxidants lutein, alpha lipoic acid, vitamins C and E, carotenoids, and polyphenols improve the skin, whether they are eaten or applied in cream form. When the antioxidant vitamins C and E are included in the diet, there is less sun damage after ultraviolet light exposure. Of course, for the antioxidants to help the heart, brain, eyes, and other organs, they must be eaten.[5]

Exercise is crucial in aging. According to Dr. Caleb Finch and many other scientists, exercise can actually reverse many of the changes of aging. It improves muscle power, endurance, and reaction time. It increases bone density and improves posture. Exercise is a natural antidepressant and mood elevator.

To be effective, exercise must increase the heart rate to about 50 percent above normal. Vigorous exercise increases the heart rate to 85 percent over normal. You must exercise for at least twenty minutes, three times a week. To stimulate muscle growth, the exercise must tire the muscle. Any exercise will work.

Before beginning an exercise program, you should be examined by an internist. You're less likely to be injured if a physical therapist or personal trainer supervises your exercise. And you will progress faster.

A *skin care program* is the foundation for healthy skin. Chapter 6 will elaborate on this subject.

Cosmetic surgery can be performed to improve appearance or to decrease signs of aging. *The goal should be to look as if you never had surgery.* The telltale signs of facial distortion, odd facial movements, or noticeable scars are the hallmarks of poor surgery. As long as the patient and I agree that the result is satisfactory, I like it when I'm told that no one noticed she had surgery. Comments like "You look great" or "You look rested" are what I want. Not "Who did your surgery?"

The goal of the next century is to cure killers such as heart disease and cancer, and neurological diseases such as multiple sclerosis and Parkinson's disease, and then to further improve the quality of life by developing better drug and surgical treatments for the cosmetic signs of aging.

Plastic surgeons spend much of their time battling the physical signs of aging. The following chapters explore the measures you can take to help look younger.

II

Facial Surgery

6

Skin Care: The Foundation for a Healthy Appearance

The skin care industry is enormous. The spa and salon industry shares this popularity with plastic surgeons and dermatologists. Tens of thousands of products are produced, some of which may not work, despite their claims. People pay an enormous amount of money for skin care products, often even without scientific studies that demonstrate their effectiveness. Many women get advice from attractive young saleswomen whose goal is to maximize their own commissions by selling the most expensive products.

Paula Begoun's books *Don't Go to the Cosmetics Counter Without Me* and *The Beauty Bible* dissect out the ingredients in skin care products and dispel their myths. In a skin care consultation, patients tell me what products they use. After researching them, I often inform the patients that they have spent their money on worthless products. Price seems to have little relevance: lower-priced products may work better than their more expensive counterparts.[1]

In general, products purchased in stores have very low levels of active ingredients. Glycolic acid, for instance, requires a 7-percent concentration to have a noticeable effect. Most glycolic products sold in stores have less. Concentrations high enough to *actually* do something to the skin can cause problems such as irritation and pigment changes. When that happens, medical advice is needed.

For this reason, manufacturers generally reserve products with higher concentrations of active agents for sale in physicians' offices. If there is a problem, you can simply call the doctor.

The Double-Edged Sword of Doctors Selling Products

Twenty-five years ago, this concept was unheard of. In fact, it would have been considered "cheap" and even unethical. Now, *most* plastic surgeons and dermatologists sell products.

In my office, I carry only competitively priced products that I feel are useful to my patients. By understanding the chemistry of the products, I can protect patients from wasting their money at the cosmetics counter. The cosmetics industry standard is to recommend a retail price at a 100-percent markup. As in a store, this markup pays for the time required to order, stock, and sell the product, as well as for floor space. Wastage and returns also account for some of this markup. In the end, most physicians do not make a lot of money on the products: they are providing a service.

There are several reasons why doctors sell products. Ideally, they research the products and offer only those that deliver the promised results. Cheaper products should be available if they provide the same result as a more expensive item.

Physicians also provide products so that patients do not have to search in retail stores. After surgery or peels, most people use cover-up cosmetics, which are more opaque than traditional makeup and hard to find in stores. Another item plastic surgeons carry is broad spectrum sunblock containing zinc oxide, also difficult to find in stores.

Unfortunately, doctors also sell products to make up for lost income in a world of declining insurance reimbursement. But doctors can take advantage of their patients' trust; naïve patients may buy unnecessary costly products from unscrupulous doctors. When a person goes to the cosmetics counter in a store, she realizes that the salesperson will try to sell her the most expensive product. She doesn't expect that in a doctor's office. That is the difference. The concept there should be to *offer* products, not to *sell* them. To help prevent abuse, doctors should not award sales commissions to their employees.

Skin Care and Cosmeceuticals

In the 1960s, skin care meant washing the face with soap and applying a moisturizer such as "cold cream" at night. By the 1970s, exfoliation (removal of the upper layer of dead skin) became popular. By the late 1980s, an explosion of skin care had begun, with the rapid introduction of hundreds of new drugs for the skin. No longer were these substances "cosmetics," but rather cosmetic pharmaceuticals, or "cosmeceuticals."

Women have been using cosmetics for five thousand years. Makeup hides imperfections and makes skin look healthier. It enhances sexuality. Cosmetics were originally made from clay, rocks, or plants. The Egyptian queen Cleopatra used black kohl and green copper pigments around her eyes. Every culture since has used cosmetics. And for two thousand years women have used chemicals to decrease wrinkles. Some cosmetics were safe, but those that contained lead, arsenic, and mercury eventually ate away the skin. By the mid-nineteenth century, safer cosmetics with bismuth, vegetable rouge, zinc oxide, talc, and starch were used. Soaps and regular bathing became popular. Near the end of the century, the use of lipstick became widespread and beauty salons appeared. Modern cosmetics such as eyebrow pencil, mascara, and rouge were first manufactured in the early part of the twentieth century.

The Federal Food, Drug, and Cosmetic Act

This law was passed in 1938, banning toxic cosmetics such as eyelash dye in the United States. Additional laws passed in 1966 and 1977 strengthened the powers of the FDA. Other nations' governments often mirror its findings. The FDA defines cosmetics as "articles intended to be rubbed, poured, sprinkled, or sprayed on . . . for cleansing, beautifying, promoting attractiveness, or altering the appearance." It defines drugs as "articles intended for use in the diagnosis, cure, mitigation, treatment, or prevention of disease . . . and intended to affect the *structure or any function* of the body of man or other animals." Some products, such as flouride toothpaste, are both drugs and cosmetics. The FDA does not define a cosmeceutical.

Cosmetics don't have to be FDA approved to be sold, and registering them is voluntary. There aren't even regulations on manufacturing practices. *Cosmetics need not be effective, just safe, to be marketed.* Without their safety having been demonstrated, a warning must be placed on their labels. The FDA has the authority to halt production of cosmetics and criminally prosecute companies. Drugs, on the other hand, must be approved and recognized as *safe and effective* prior to sale. Nonprescription drugs marketed before 1972 may continue to be sold without further review; they are considered "grandfathered."

The Line Between Cosmetics and Drugs Is Blurring

Forty years ago, this distinction was easy. Makeup was a cosmetic and penicillin was a drug. Now, with the addition to skin creams of many "natural" substances that behave like drugs, the line between cosmetics and drugs is often hard to decipher. As a result, a variety of creams and liquids sold over the counter *actu-*

ally do something. This chapter will explore some of the more common skin drugs.

Legally, a cosmetic manufacturer could put topsoil in a product and market it without prior FDA approval. The FDA does have a list of prohibited ingredients, but it is short. Only when a company claims that its product affects the structure or function of the body is it then classified as a drug.

If skin creams actually did what they claim—wrinkle and stretch mark reduction, stimulation of collagen formation, fat destruction, and increased testosterone, for example—they would be classified as drugs. The FDA points out that, under these circumstances, they are not recognized as safe and effective. They therefore qualify as new drugs and require a "New Drug Application" in order to be legally marketed.

One may wonder how the FDA allows skin drugs to be sold without the protections afforded to either prescription or over-the-counter drugs. The answer is not that these are benign creams that do not work. The answer probably has more to do with the priorities of the FDA, and its regulatory responsibility for life-saving drugs, medical devices such as cardiac stents, and the entire food industry. Cosmetics simply are less important.

That may all be changing. In January 2005 the FDA warned two companies about making unsubstantiated claims. Basic Research LLC of Salt Lake City, Utah, is the manufacturer of a variety of skin creams, including Strivectin-SD products. Strivectin claims to improve stretch marks and wrinkles and increase collagen production. The company's Dermalin-APg product claims to decrease fat. Its Mamralin-ARa product is supposed to protect a woman's breasts from sagging and shrinkage. Its TestroGel is supposed to improve sexual performance and raise testosterone levels. The FDA warned this company that its claims would define the products as drugs. It told Basic Research that the products are not recognized as safe and effective and therefore are considered "new drugs" that may not be legally marketed in the United States without FDA approval.

University Medical Products USA, of Irvine, California, is the maker of FACE LIFT Collagen 5 products (Cell Regeneration Cream, FACE LIFT, BODY LIFT, and others). These products claim to increase collagen production, diminish wrinkles, decrease cellulite, reduce weight, and eliminate bloating. The FDA again warned the manufacturer that these claims classify its products as "new drugs" that could not be marketed without approval.

We can only hope that the FDA's actions will end the unsubstantiated, sometimes ridiculous claims made for skin products. If they do work, they should be

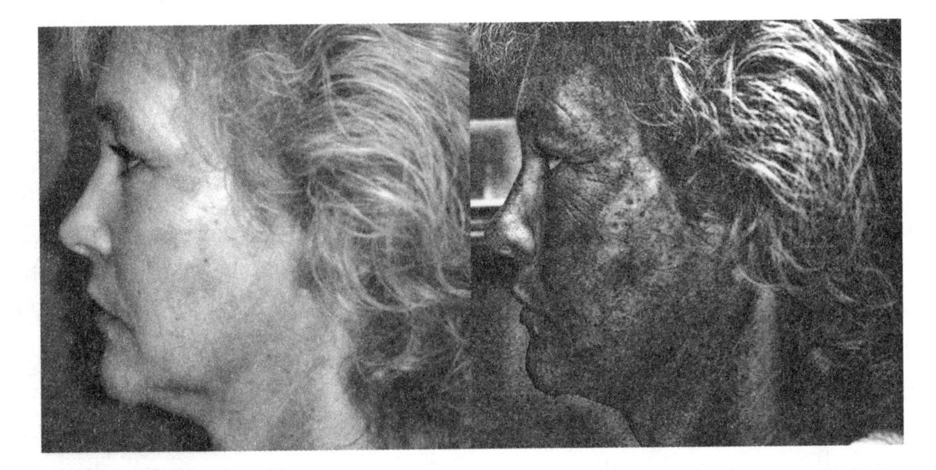

The use of special ultraviolet filters allows the camera to see sun damage years before it is visible to the naked eye. On the left, a 50-year-old woman; on the right, her appearance using ultraviolet filters. Photos such as these are taken prior to beginning a skin care program and at regular intervals thereafter to chart progress. See color plates.

subject to independent review and demonstrate efficacy in controlled clinical trials. The FDA owes consumers no less.

Drugs for the Skin

The skin care industry is now a *$6-billion industry* in the United States alone. By 2006 over a hundred additives claimed some sort of anti-aging property (see list on pp. 73–76). How is the consumer supposed to determine which cream to use?[2]

Gertrude Stein's saying, "A difference, to be a difference, must make a difference," couldn't be more apropos than when analyzing cosmetics. These creams are touted to do something special, but unless they perform they really are useless. In order to determine whether a cream actually does something, we need to compare it to one without the active ingredient. When a drug is studied in this manner, we can begin to trust the results. If the science is acceptable, the study will be published in a peer-reviewed journal, not a "throwaway" journal, where just about anything can appear.

Perhaps the best example of a questionable product is thermal spring water. Five ounces cost $18, but you do get five gauze pads with it. No study has ever shown this product to be superior to tap water, which is free.

Most skin cream studies rely on the perceptions of the patients ("my skin feels better using the cream") or subjective opinions by examiners ("her skin appears smoother, less wrinkled, less mottled"). These types of evaluations are not particularly scientific.

We are just entering an era when skin texture and wrinkles can be *objectively* assessed. A material somewhat like Silly Putty can be pressed against the skin, placed under a scanning electron microscope, and analyzed by a computer. This procedure evaluates the effects of various creams. Better yet, new high-resolution cameras that can distinguish between skin marks 5 millimeters apart are now available. These computer-aided digital cameras can analyze brown pigment, texture, and pores in the skin.

If your doctor recommends a particular cream for your skin, he should have read the literature and be able to quote studies demonstrating the effectiveness of the drug. Let us hope he has not just read about the cream in *Vogue,* or is selling it in his office because it will add to his profit.

Classes of Drugs for the Skin

The major classes include the alpha hydroxy acids (AHAs), also called fruit acids; vitamin C; vitamin A as tretinoin (Retin-A); and the newest drugs, the peptides. In addition, there are pigment reducers, exfoliants, moisturizers, toners, and cleansers.

Some cosmeceuticals actually do work. Some do not. Many creams do not work because they simply cannot penetrate the protective outer layer of the skin. Even though a cream may work on cells in a test tube, that does not mean it will work clinically and make skin healthier and more attractive. Small molecules, such as AHAs and vitamins, have a chance at penetrating the upper layers of skin and acting on the lower layers. Large molecules, such as growth hormones, proteins, collagen, and hyaluronic acid, really have little chance of penetrating the skin. Whether peptides, strings of amino acids, can penetrate the skin is also questionable.

The skin can be viewed as a chain-link fence. Ping-Pong balls and marbles fit easily through the holes. Tennis balls may be able to squeeze through, with a little help. Anything larger, such as a baseball, won't pass through. Even though some creams may be capable of astounding marvels in a test tube, if they can't get through the fence, they can't play ball. They will be useless. In a nutshell, that's the problem with many large-molecule creams. Botox is injected through the fence; glycolic acid, vitamin C, and tretinoin pass right through the holes. But proteins, collagen, and other large compounds just sit outside.

Honest skin chemists call this the "500 Dalton Rule." A Dalton is a chemi-

The 500 Dalton Rule

Skin cream	Weight	Can it penetrate skin?
Glycolic acid	76	Yes
L-ascorbic acid (vitamin C)	176	Yes
Tretinoin (vitamin A)	300	Yes
Hexapeptides	680	No
Hyaluronic acid	From 10,000 to 10,000,000	No
Collagen	300,000	No

cal's unit of weight. No chemical with a molecular weight over 500 Daltons can penetrate the skin.[3] Interestingly, an average peptide containing just four amino acids weighs 540; hexapeptides contain 6 and weigh on average 800.

It is impossible for consumers and even physicians to independently know whether a cream is effective. Or whether two creams are equivalent, or if using two or more creams together is better. Companies add their interesting-sounding chemicals to these products, but only the AHAs and retinoids, vitamin C, and a handful of other small molecules actually have proven clinical effects. I wonder how companies can justify the high cost of new cosmeceuticals without proof they work.

Some cosmeceuticals contain human proteins. This opens up a whole new problem, since dangerous infectious diseases potentially can be transmitted.

Tretinoin (Retin-A) and Other Vitamin A Creams

The public knows tretinoin as Retin-A even though similar drugs called tazarotene (Avage) and adapalene (Differin) and generic versions are available commercially. Tri-Luma combines tretinoin with the pigment reducer hydroquinone and an anti-inflammatory steroid.

Tretinoin is one of the most useful drugs for the skin. Without vitamin A, the skin becomes dry and the hair and nails become brittle. Vitamin A and tretinoin act on DNA. Tretinoin lessens fine wrinkles, decreases brown pigmentation, diminishes acne and visible pore size, increases collagen and elastin and hyaluronic acid, and hydrates the skin. In addition, it can decrease premalignant lesions and even prevent skin cancer. Blood flow to the skin increases. The immune system is enhanced. The skin becomes firmer and more elastic. Healing is faster. Dark circles around the eyes lessen. Tretinoin is a first-line drug for acne, rosacea, and seborrheic dermatitis—almost a wonder drug![4]

Because vitamin A drugs are destroyed by light, they must be applied at night. Tretinoin comes in a cream, gel, or liquid and in a variety of concentrations. I start my patients on a low dose (0.025 percent) and increase the concentration every month until they reach the highest concentration (0.1 percent). After several days, patients develop a red, dry skin that intensifies over the first few weeks. Many patients give up during this phase. However, eventually the skin clears up and looks better. The maximum effect is achieved at about six months. The higher the concentration, the greater the effect. I tell my patients to continue tretinoin forever.

The drug should be applied generously after washing with a mild soap. Some think that the pH (measure of acidity) of the skin should then be restored to normal with toner, to permit better penetration of the tretinoin. You shouldn't exercise after applying tretinoin because sweating increases the irritation.

The over-the-counter retinol-type drugs appear to have effects similar to tretinoin. Other vitamin A drugs like retinyl propionate, retinyl palmitate, and retinaldehyde also work even at low concentrations.[5]

Alpha Hydroxy Acids (AHAs)

Alpha hydroxy acids, also called fruit acids, were really the first cosmeceuticals and today are the dominant drugs in the skin care industry. Described by two dermatologists in 1974, now about fifteen hundred products contain AHAs. The smallest molecule in this class is glycolic acid, a chemical found in sugar cane. Other naturally occurring AHAs are citric acid from fruit, lactic acid from milk, and malic acid from apples. Other AHAs are made synthetically in laboratories.[6] AHAs stay on the skin for four hours after application unless they are deactivated (neutralized).

Glycolic acid comes in varying concentrations. The stronger the acid, the more effective it is. But this is not without side effects. Glycolic acid causes shedding of the upper layers of the skin (*stratum corneum*). This smoothes the skin and decreases acne. By exfoliating, superficial brown pigmentation decreases. The chemical spurs collagen production, calms inflammation, and is an antioxidant. But it does not improve scarring and probably only minimally helps wrinkles. Glycolic acid does not prevent skin cancer and is not as effective as retinoic acid and salicylic acid in repairing aging skin. Its effects remain for two to three weeks after use is discontinued.[7]

Opinions differ concerning the benefits of AHAs. The Beverly Hills dermatologist Zein Obagi, M.D., claims that concentrations over 12 percent may actually *irritate* the skin.

Glycolic acid is used in skin creams, washes, lotions, gels, and foundations and also as a "lunch-hour peel" (Chapter 10). Concentrations of 4 to 15 percent are commonly found in creams. Peels start at 10 percent and increase to 70 percent.

Phytic acid is a mild acid that acts as an exfoliant. It may cause less irritation than glycolic acid.

Sunscreen must be used with AHAs. The manufacturers don't like to tell you that AHAs increase the skin's sensitivity to ultraviolet B radiation. This effect lasts for a week after the drug is stopped, increasing the chance of sunburn and possibly skin cancer. Fortunately, the damage is reversible. The FDA issued guidance to the AHA industry in 2005. AHAs can also cause burning, rashes, swelling, pigment changes, blisters, and skin peeling.[8]

Polyhydroxy Acids (PHAs)

These chemicals are billed as the next generation of AHAs. They have many of the same benefits but are less irritating. As antioxidants and moisturizers, they improve the appearance of sun-damaged aging skin. PHAs can be used in combination with retinoic acid for acne and with hydroquinones to lighten the skin. PHAs do not increase sensitivity to UV light, and they provide AHA-type antiaging benefits as well as moisturization and protection from free radical damage.[9]

Beta Hydroxy Acids

Salicylic acid, the ingredient in aspirin, is a beta hydroxy acid. It dissolves the upper layers of epidermis, thereby causing a smoothing of the skin. It is also useful in acne, since it helps clear debris from the pores.

Vitamin C

Vitamin C (L-ascorbic acid) is the major antioxidant in the body. The virtues of vitamin C have been extolled since the days of Linus Pauling. In the skin, vitamin C increases collagen production. With a vitamin C deficiency, called scurvy, collagen cannot be produced and the skin breaks down. There is a limit on the amount of vitamin C that can be eaten in a day. Anything over about 1,300 milligrams will simply be urinated out. Vitamin C placed on the skin (topically) boosts the skin's vitamin C concentration twenty-fold.

Like vitamin A, vitamin C is rapidly destroyed by light. For this reason, it must be constantly replaced for maximal effect. A concentration of at least 5 percent is needed to work on the skin. It is commercially available in concentrations up to about 20 percent.

Vitamin C can protect against sunburn, reduce sun-induced wrinkling, and decrease skin cancers. A powerful antioxidant, it decreases dangerous oxygen free radicals and inflammation and decreases sunburn after sun exposure. It is more effective when used with a sunblock. It can decrease the redness associated with acne rosacea after just three weeks of use. In addition, vitamin C slows the formation of brown pigmentation in the skin.[10]

Vitamin C is one of the medicines for the skin that makes sense and is backed by strong scientific evidence. In addition to its long-term effects, it irritates the skin enough to cause swelling that temporarily decreases wrinkles—a side effect that most women like. Interestingly, three months of use fail to improve the texture of the skin.[11]

Vitamin E

This is the most abundant naturally occurring antioxidant in the skin. It is the first line of defense against environmental damage. The oils of the skin carry the vitamin E to the surface, where it neutralizes free radicals in the cells. It also moisturizes the skin.

It makes sense that vitamin E helps the skin. However, since we can increase its level in the skin by eating more vitamin E, it really doesn't have to be a component of skin creams. There isn't much evidence that if it is placed on the skin, it has any additional benefit.

Antioxidants

Plants produce antioxidants as their own sunblock because they can't move out of the sun. We can take advantage of the plant's natural sunblock by using their many-colored antioxidants. Alpha lipoic acid, Coenzyme Q10, and Idebenone (Prevage) are antioxidants that improve fine lines and roughness and decrease brown pigment.

Prevage is a new cream, made by the Botox producer, Allergan. To fully understand how this drug works, it helps to be a chemist. The main ingredient, the antioxidant idebenone, fights mitochondrial disease, at least in test tubes. (The theory is that aging is a mitochondrial disease.) It is very similar to the antioxidant Coenzyme Q10. The drug allegedly reduces wrinkles and roughness and hydrates the skin. Although the company may have its own data, as of this writing no controlled studies have been published that demonstrate its effectiveness. Prevage is marketed by Elizabeth Arden or sold in physician's offices.[12]

Other antioxidants found in skin creams contain flavonoids from soy, catechins from green tea, beta-carotene, lutein, lycopene, and grapeseed extract. While these do have a scientific basis, there really is scant hard science supporting their use.

Hydroquinones

This class of drugs stops the production of brown pigment in the skin (*melanin*). Although commonly referred to as "bleaching creams," the drugs are not like Clorox—they do not immediately remove pigment. The brown in the skin gradually wears off over months.

Hydroquinones come as lotions, liquids, creams, or gels. Freckles, the mask of pregnancy (*melasma*), and brown pigmentation after inflammation can be treated with this twice-daily drug.

Hydroquinones are more effective when used in combination with tretinoin or AHAs. They are less effective when used with skin steroids such as hydrocortisone. Among the hydroquinone creams on the market, some contain tretinoin and some contain steroids and sunscreens. The skin must be washed before application and the cream is applied liberally. The effects are short term; they are gone after two days.

Rashes and stinging are common side effects. Because the protective melanin pigment is reduced with this cream, it is critical that sunblock with zinc oxide be used to prevent skin cancers and wrinkles. Interestingly, in dark-skinned people, the use of hydroquinones can cause an irreversible *darkening* and thickening of the skin. More important, hydroquinones may cause leukemia and other cancers and may impair fertility.

Because of this, the FDA plans to ban hydroquinones from all nonprescription products and classify all skin bleaching products as new drugs requiring their approval. The European Union took this action in 2001. Currently, hydroquinones with a concentration above 2 percent require a prescription. Lower concentrations are commonly found in over-the-counter creams.

Another hydroquinone-like drug is called kojic acid. It is less effective than hydroquinone and is sometimes used in people who are allergic to hydroquinones.

Glabridin, an ingredient of licorice extract, is a natural skin lightener. While it does not work as effectively as hydroquinone, it does not appear to have long-term side effects. Glabridin shows promise as a skin lightener but needs more studies, and like so many "naturally occurring herbs," there is little standardization of purity of glabridin and little agreement on the needed concentration.

Emblica, a family of four naturally occurring chemicals that lighten the skin, is now being used in cosmetics. Only one of the four is small enough to penetrate the skin. Emblica is also an antioxidant and stops the breakdown of collagen. There is no evidence of any side effects from this herbal remedy and so it may replace hydroquinone in over-the-counter skin lighteners.

Copper Peptide

Copper peptide (glycyl-L-histidyl-L-lysine-Cu—whew!) is actually a protein growth factor that has been isolated from human plasma. In 1994 it was shown to improve wound healing. It increases collagen formation better than tretinoin, melatonin, and vitamin C. The collagen stimulation is due to the complex molecule and not to the copper or the peptide alone. Copper peptide increases the growth of the cells that make collagen (fibroblasts) and causes them to make more growth factors. These in turn have enormous effects on the skin. Cells exposed to radiation therapy improve when exposed also to copper peptide.[13]

Unfortunately, various cosmeceutical manufacturers may not have read this research carefully and have begun putting copper into their creams. Copper alone will not have any effect. It must be part of the peptide compound. The copper peptide evidence is compelling in test tubes; the big question is whether this large molecule can actually penetrate the skin and cause real change in actual humans.

Retinol has been added to some copper peptide products. This teaming with anti-aging drugs that are known to be effective hints that either the copper peptide alone may not be clinically effective or the vitamin A drug may enhance the copper peptide.

Kinerase

Kinerase (kinetin) is a growth hormone from plants. In test-tube-type studies, it delays aging of cells. It can also make fruit flies live longer. However, few scientific studies in humans demonstrate that this drug actually does something helpful for the skin.

Peptides

In 1999 four proteins in fermented yeast extract were found to speed wound healing. These proteins defend against protein damage. The C6 peptide, as one is called, may improve wound healing and stimulate collagen formation in test tube studies. However, there are no clinical studies that show its benefits and justify the high price of these products, which can be as much as $570 a bottle.[14]

Products like Regenerist, Strivectin-SD, Wrinkle Relax, and Reclaim all contain these types of drugs. Many products contain substances called pentapeptides, hexapeptides, and oligopeptides. These are combined with fatty acids to allegedly help the peptide soak into deeper layers of the skin. Some of these peptides are billed as stimulators of collagen production. Others are supposed to block a protein that allows muscle to contract, giving a Botox-like effect.

The data supporting the claims of the peptides are scant, at best. In one study twenty women had significant improvement in their wrinkles when a hexapeptide was used twice a day for a month. Peptides increased the production of collagen and other skin proteins more rapidly than vitamin C. Most physicians are skeptical, believing that proteins placed on the surface of the skin cannot actually penetrate the skin and perform the marvels that the companies describe. On the other hand, the placebo effect may be proportional to the amount of money spent on the cream.[15]

Often, these wonder creams combine their products with retinoids and AHAs. Nicer skin results from the older, proven drugs and the new "wonder drug" distinguish the cream from other products and help boost the price of the new product. Hmmm.

New Wrinkle Creams

I call them swellers. These chemicals decrease wrinkles by causing skin swelling. Any drug that promises *immediate* results achieves them by causing swelling. The swelling temporarily decreases wrinkles. While this may sound fine, I worry about long-term use of these drugs. Because swelling stretches the skin, repeated use can actually *cause* wrinkles and extra skin. The result is similar to the medical condition known as *blepharochalasia,* where allergies cause eyelid skin to swell repeatedly. The skin stretches and eventually needs to be removed.

Women who use these swellers on their lips acknowledge that their lips look larger for a few hours. However, they also say that their lips feel strange. They are conscious of them when speaking, eating, and kissing.

Again, no studies show that these drugs are effective and, more important, safe.

Botox-Like Creams

Certain new creams claim to paralyze the underlying muscles, resulting in a Botox-like effect. These creams contain peptides called argireline or acetyl hexapeptide-3. In test tubes these drugs stop formation of the nerve chemicals that control muscle. It may be fantasy to think that this type of drug can safely penetrate an eighth to a quarter of an inch of skin and reach muscle. No controlled studies show these drugs to be effective.

Another new ingredient is Boswelox, an herb. This drug is an antioxidant and may improve immune function. L'Oréal's Wrinkle De-Crease contains Boswelox and claims to reduce wrinkles. Again, no peer-reviewed controlled studies in the medical literature support the claim.[16]

Freeze 24/7

This is a concoction containing GABA (gamma amino butyric acid) and eugenol, a clove derivative. The GABA is supposed to relax muscles, and when sprayed on the face, it seems to limit movement, temporarily lessening wrinkles. There are no good scientific studies, however, that show its effectiveness.

Stretch Mark Creams

Striae Stretch Mark Crème with Regenetrol Complex allegedly improves the appearance of stretch marks, presumably by an increase in collagen and elastin production that thickens the skin. Unfortunately, the studies were not published in peer-reviewed journals and therefore may not be reliable.

TNS Recovery Complex

This new cream contains a variety of growth factors and other chemicals. It is supposed to reduce wrinkles, increase elasticity of the skin, and decrease age spots. The initial reports are promising, but again there is a paucity of controlled studies that demonstrate its effectiveness. Since growth factors are giant molecules, it is asking a lot of them to be able to pass through the protective skin. A variety of growth factors have been incorporated into other cosmeceuticals.

Photoprotective Iron Chelator Technology

This brand-new approach to skin aging is not even on the market yet. Powerful antioxidant drugs treat a particularly aggressive free radical, iron ions. The Cheladerm Company, which produces them, is a joint venture between Procter & Gamble and the University of Florida.

Exfoliants

Exfoliants remove dead cells from the skin. Exfoliation can be done mechanically by washing with a washcloth or a loofah pad, or it can be done chemically by a variety of creams. There is nothing magical about exfoliants; they simply remove the upper layer of dead epidermis, resulting in cleaner skin with less acne. Too much exfoliation can result in irritation.

Sunblock

Part of the sunlight that reaches earth is called ultraviolet (UV) light. The two commonly known forms are A and B. A third type, ultraviolet C, does not occur naturally on the earth's surface. Both UVA and UVB can cause sunburn and skin cancer. Tanning salons commonly use UVA and falsely state that it is safe. *There is no safe ultraviolet light.*

Sunblock mechanically or chemically blocks ultraviolet light. The sun protection factor (SPF) is a common measure of the strength of sunblock. The higher the SPF, the longer you can stay in the sun without looking like a lobster.

Chemical sunblocks do not completely protect from all ultraviolet light. Different chemicals are blended to block various wavelengths of light. UVB sunblocks are readily available, but UVA blocks have been troublesome to develop. Some chemical sunblocks last only an hour or so in direct sunlight. Zinc oxide, a mechanical sunblock, in a concentration of at least 3.5 percent, can sufficiently block ultraviolet light for the average person. The zinc in zinc oxide has the added advantage of being an antioxidant, reducing free-radical damage in the skin. Even titanium dioxide does not completely block UVA.

The FDA classifies sunblocks as over-the-counter drugs, not cosmetics. Because of the high cost of bringing a new drug to market, and the relatively low cost of sunblock, new sunblocks are slow to come to market. The first new class of sunblock in the United States since 1988 was approved by the FDA in 2006. L'Oréal's Anthelios SX, known as Mexoryl SX in Europe, contains two chemicals in addition to the UVA blocker ecamsule. Sunblock with zinc oxide or Mexoryl that completely blocks ultraviolet light and minimizes sun damage to the skin should be used by most people.[17]

Glass is commonly—and incorrectly—thought to screen out all ultraviolet radiation. Because UVA goes through window glass, people get sun damage on their left cheek from driving and on the side of the body that faces their office window.

Sunblock is the single most important step in long-term skin care, because aging is accelerated by sun. Sunblock allows the skin to repair some of the damage that has already occurred.

Other Skin Products

Hydrocortisone is a common steroid that is available without a prescription in concentrations less than 1 percent. It has anti-inflammatory properties and can calm rashes and irritation caused by other skin creams. Ointments are more potent than creams. However, it is often the ointment base (petrolatum) that patients are allergic to, so I usually prescribe a cream for rashes. If a more rapid response is needed, much more potent steroid creams can be prescribed.

Toners help clean the skin and restore its pH. While many women like the feel of toners, if a mild cleanser is used, there isn't much evidence that toners are necessary. The older types of toners contained alcohol, which irritates the skin.

The public's need for moisturizers may be one of the great myths in skin care. Most people do not need moisturizers, except in cold, arid weather. The skin makes its own oils and provides its own moisture. This means the upper layer of skin, the stratum corneum, contains more than 10 percent water, keeping it soft and supple. Many moisturizers contain oils such as petrolatum (Vaseline), lanolin (sheep-wool oil), and mineral oil. They block water evaporation and can actually clog pores and increase acne. They can interfere with the use of drugs such as tretinoin and AHAs. Because dry skin reflects more ultraviolet light than hydrated skin, some dermatologists feel that moisturizers actually accelerate skin aging.

There's a sucker born every minute, they say. And purchasers of oxygenated water fit the bill. This product adds extra oxygen to ordinary water and is supposed to enhance sports and cardiovascular and muscular performance. Websites that promote its use imply that it can also help varicose and spider veins, hair loss, wrinkles, dry skin, and a host of other conditions. A simple study proved that this product does not work. In fact, a single breath of air contains more oxygen than a whole bottle of oxygenated water. Even if the water could increase the oxygen in the body, it would do so for only a few seconds. And this product costs only $2.50 per 12-ounce bottle![18]

Plain and simple, there is no evidence of any kind that extra oxygen blown onto the skin does anything. Extra oxygen does improve wound healing. But wound treatment has little in common with chronic wrinkle management. Some plastic surgeons offer oxygen therapy for the skin. Ask them to show you evidence that this treatment does anything at all.

Additives in Skin Creams and Their Possible Benefits

Additive	Possible benefit
Acerola	Tropical fruit containing vitamin C
Algae extract	Moisturizer
Allantoin	Anti-irritant
Aloe vera	Anti-inflammatory (5% necessary for effect) and antibiotic?
Alpha lipoic acid	Antioxidant, anti-inflammatory
Amino acids	Skin growth factors
Aminophylline	Asthma drug—unproven as cellulite reducer
Arbutin	Skin lightener in hydroquinone family
Arnica extract	Decrease bruising (unproven)
Avocado	Wrinkle reduction, increase elasticity
Barley extract	Antioxidant, anti-inflammatory
Beech tree bud extract	Increases activity of skin cells, antiseptic
Biotin	B vitamin, necessary for many enzyme steps
Bisabol	Anti-inflammatory
Black Cohosh	Anti-inflammatory, wrinkle reduction
Borage extract (Borago officinalis, herb)	Anti-inflammatory, moisturizer
Bromelain	Protein digesting enzyme
Calendula extract	From marigold flowers; antiseptic, anti-itching, and anti-inflammatory, moisturizer, styptic
Carotenoids (lutein, lycopene, beta-carotene)	Antioxidants, protect against sun damage
Centella asiatica	Antioxidant, stimulates collagen production
Cerebrosides	Glycolipid from animal brains—unproven moisturizer and skin smoother
Chitosan moisturizing factor	Moisturizer from shellfish
Citric acid	Astringent and antioxidant
Cocamidopropyl betaine	Moisturizer
Co-enzyme A	Activates fatty acids, breaks down amino acids

Additives in Skin Creams and Their Possible Benefits *(continued)*

Additive	Possible benefit
Co-enzyme Q10	Stops collagen breakdown
Collagen	Protein from connective tissue; does not penetrate skin
Collagen peptides	Stimulates collagen production
Cucumber	Moisturizer, anti-inflammatory
Date palm	Anti-inflammatory, wrinkle reduction
DMAE (dimethylaminoethanol)	Stabilizes cell membranes, activates muscle contraction
Echinacea	Antibacterial, anti-inflammatory
Elastin peptide	Complex changes between cells
Elhibin	Inhibitor of elastase and tryptase enzymes
Emblica	Skin lightener
Estrogen	Wrinkle reduction, thickens skin
Eyebright extract	Anti-inflammatory
Flavonoids	Antioxidant
Fruit acids	Exfoliation, decreases brown splotchiness, increases collagen
Gingko biloba	Astringent, antioxidant
Ginseng	Moisturizer, enhances immune function, improves blood supply
Glycolic acid	Exfoliant, moisturizer, improves wrinkles
Glycosaminoglycans	Moisturizer
Green tea (catechin)	Antioxidant
Herbal extracts of artichoke, burdock, walnut and orange	Antioxidant, moisturizer
Hexyl nicotinate	Increases blood flow by dilating blood vessels
Hohn sugar extract	Moisturizer, exfoliant
Human placenta and amniotic fluid	Protein with no medical claims
Hyaluronic acid	Sugar-like substance that constitutes and moisturizes the skin

Additives in Skin Creams and Their Possible Benefits *(continued)*

Additive	Possible benefit
Hydrocotyl extract (Centella asiatica)	Antibacterial, anti-itching agent
Hydrolysed wheat protein	Moisturizer
Hydrolysed yeast protein	Moisturizer
Hydroquinone	Reduces brown pigment production
Jojoba oil	Moisturizer
Kojic acid	Inhibits production of melanin pigment
Licorice extract (Glycyrrhiza glabra, Glabridin)	Anti-oxidant, antibacterial, skin lightener
Lipophilic fruit acid	Skin lightener
Lyophilized collagen	Moisturizer and reduces irritation
Mannitol	Antioxidant
MDI complex	Collagenase inhibitor, antioxidant
Menthol (50% peppermint oil)	Stimulates blood circulation
Mimosa bark	Anti-oxidant, antiseptic
Niacinamide (vitamin B3)	Stimulates collagen and leptin, skin lightener, wrinkle reducer
Panthenol (pro-vitamin B5)	Moisturizer, stimulates cell proliferation
Papain	Exfoliant
Peanut oil	Moisturizer
Plant extracts of mulberry, grape, saxifrage and scutellaria	Skin lightener, antioxidant
Pomegranate	Anti-inflammatory, antioxidant
Protein	Moisturizer
Pyridoxine (vitamin B6)	Moisturizer
Retinoids	Vitamin A type drugs
Saccharides	Moisturizer
Salicylic acid	Anti-inflammatory, exfoliant, antiseptic
Seaweed extract and protein	Anti-inflammatory, antioxidant, antiseptic
Selenium	Element needed by antioxidant enzymes

Additives in Skin Creams and Their Possible Benefits *(continued)*

Additive	Possible benefit
Shea butter	Moisturizer
Silk amino acids	Moisturizers
Sodium PCA (pyrrolidone carbonic acid)	Moisturizer
Soybean germ	Moisturizer
Soybean oil	Stimulates collagen, elastin, and glycoprotein production, antioxidant
Spirulina	Moisturizer
Squalene	Moisturizer
Stinging nettle extract	Anti-inflammatory, astringent and bactericidal agent, contains Vitamin E
Superoxide dismutase enzyme	Antioxidant
Tea tree oil	Antibacterial, helps seborrhea, psoriasis, eczema, dermatitis, and acne
Theophylline	Similar to caffeine, increases blood flow, may reduce fat
Titanium dioxide	Sunblock
Tyrosine peptide	Pigment reducer
Vitamin A	Vitamin necessary for skin health
Vitamin C	Major antioxidant in body, stimulates collagen production, skin lightener
Vitamin D	Necessary for skin metabolism, made by skin after exposure to UVB
Vitamin E	Antioxidant, free-radical scavenger
Wheat germ protein	Acts like soap and moisturizer
Yeast extract	Normalizes sebum production, moisturizer
Zinc oxide	Mechanical sunblock

Note: The benefits listed are derived from various reports. They are not necessarily scientifically proven.

Your Individualized Skin Care Program

Depending on your precise issues—whether you have wrinkles, oiliness, brown pigmentation, visible pores, acne, or loose or thin skin—your physician will prescribe a regimen that includes one or more of the above creams. If you are serious about your skin, you will have to make some changes.

A skin care program begins with clean skin. You should wash thoroughly in the morning and the evening. Your cleanser should be mild and not leave a residue. Creams need to be applied in the morning and evening. Common products include tretinoin, vitamin C, hydroquinone, hydrocortisone, an AHA as well as a sunblock, and maybe a toner. Other treatments may be prescribed; we will look at them later.

7

Growths on the Face

I am always amazed when people can come into my office and point to some subtle abnormality of their lips, for instance, but are not bothered by that giant mole in the middle of their face. When a person has a mole on her face, it creates what my wife the psychologist likes to call cognitive dissonance. The face is, by nature, symmetrical. Two ears, two eyes, and the midline nose and mouth. A mole throws off that balance.

Some people view moles as beautiful. Consider Marilyn Monroe, who probably drew in her famous left-cheek mole. This mole was not present when Marilyn was a brunette, and it seems to have moved to various locations on her face later in life.

Marks on the face are commonly called moles or birthmarks. Plastic surgeons term these marks lesions. A veritable zoo of different objects can grow on the face, from benign to malignant. Treatment depends on the diagnosis. Sometimes the diagnosis is so obvious that treatment can begin without a biopsy. However, if there is any question about what a lesion is, a biopsy is in order. A biopsy takes a piece of the specific growth and sends it to a pathology laboratory, where a pathologist looks at it under the microscope. Once a diagnosis is made, the lesion is given a proper name and treatment can begin without guessing.

Types of Things That Grow on the Face

Freckles are collections of brown pigment that typically occur in children, morphing into age-spots in adults. Kids like freckles. Adults don't. Freckles are

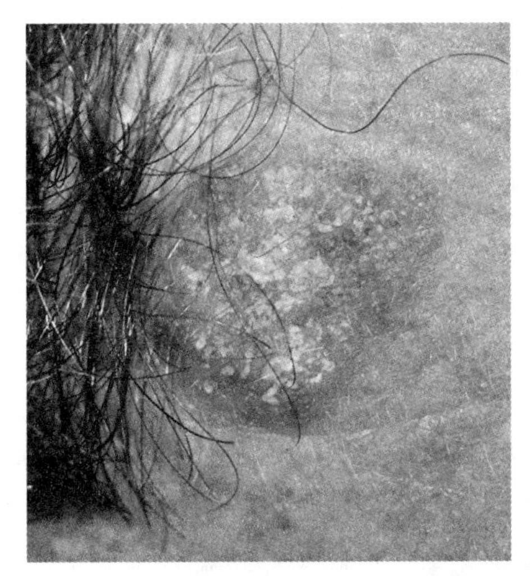

Seborrheic keratosis.

treated with fading creams, the hydroquinones, or varying levels of chemical peels. If the plastic surgeon needs to make a payment on his boat, he'll (over)use the laser.

When brown spots begin to raise up, they may be benign *seborrheic keratoses.* Plastic surgeons like to shave these off. They can be frozen off, but this procedure often leaves a white mark on the skin. Also, when the lesion is shaved, a specimen is available to look at under the microscope. Not so with freezing. The importance of pathologically examining the lesion under the microscope cannot be overemphasized. In a study done at Harvard Medical School, more than nine thousand specimens were sent to the pathologist with the presumed diagnosis of seborrheic keratosis. Under the microscope, sixty-one turned out to be melanomas, the most lethal type of skin cancer. The lesson is: Unless the specimens are looked at under the microscope, cancers will be missed.[1]

Warts are common on the face. They are caused by a virus and have an entirely different appearance than warts on the feet or hands. Most people are surprised when I tell them they have a wart on their face. Unlike other warts, I usually excise those on the face. They rarely recur.

Skin tags are overgrowths of skin and are particularly common around the eyelids. They also occur on the neck and underarms, and around the breasts. They get snipped off, with or without anesthesia.

Moles are the most common brown growth on the face. Everyone has dozens of moles on their bodies. Cancer occurs in 1 in 10,000 moles. Most are benign for life. However, if your mole is larger than the eraser on a pencil, has more than one color, has color that extends beyond the raised border of the mole, bleeds,

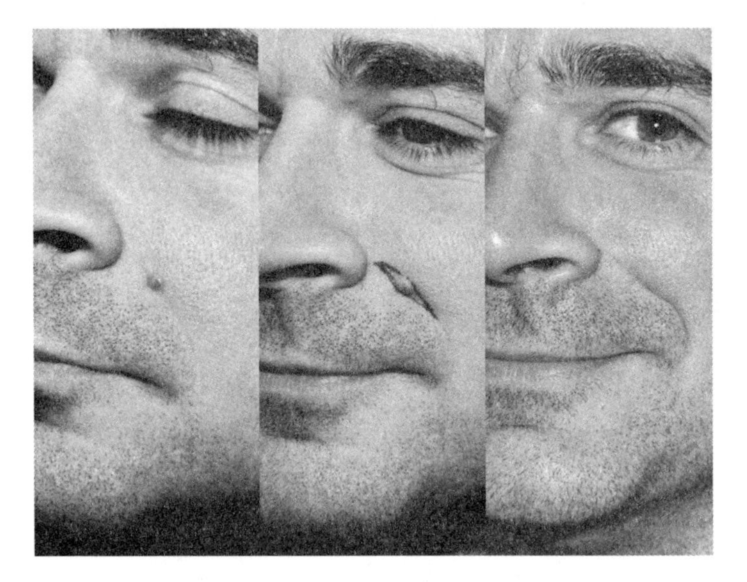

This man had a mole on his cheek that was distracting. The middle photo demonstrates the method of removal, by placing a scar along the natural fold of the face. On the right, his appearance a year later. See color plates.

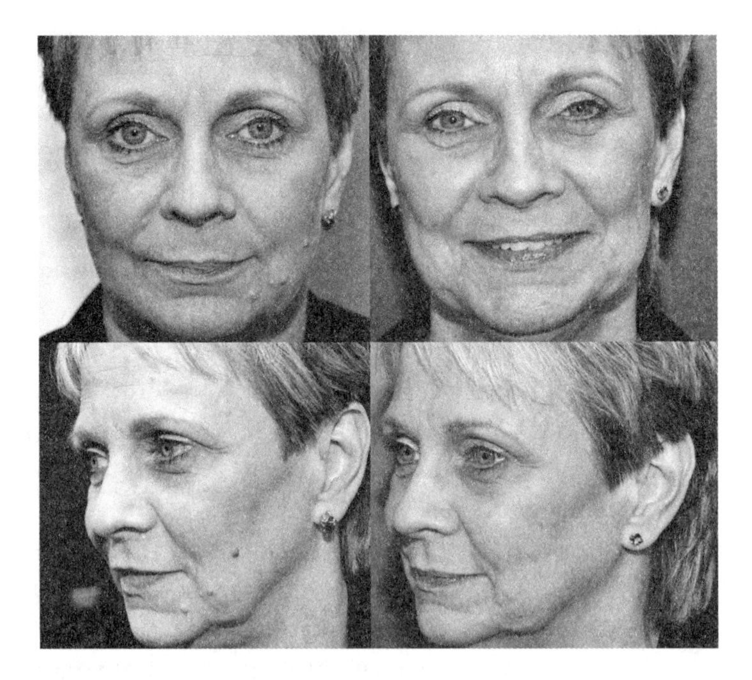

This woman had four facial moles removed, significantly improving her appearance. See color plates.

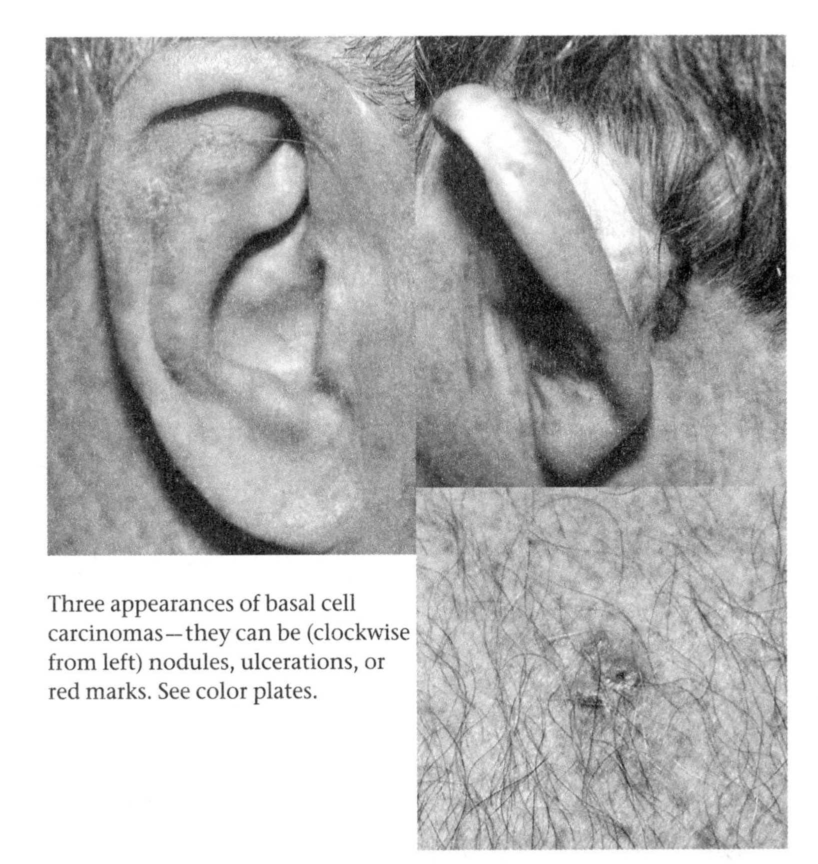

Three appearances of basal cell carcinomas—they can be (clockwise from left) nodules, ulcerations, or red marks. See color plates.

itches, or has irregular borders, then you should show it to your dermatologist or plastic surgeon. These moles would likely be removed.

Of the many different types of skin cancers, the most common are *basal cell carcinomas.*

These slow-growing cancers can be raised pink nodules, ulcerations, or just red areas. Typically, they break open and bleed, then heal. They do not spread beyond the local area. Left untreated, they can become huge. They can get so large that they have been called rodent ulcers, conjuring visions of rodents chewing away at your skin. To avoid this unpleasant scenario, have the skin cancer taken off early. Basal cell carcinomas are usually biopsied to establish the diagnosis. The tissue is immediately examined under the microscope (called a frozen section) to assure that all of the cancer is removed. Alternatively, some dermatologists perform a Moh's excision, a removal of the skin cancer with immediate examination of the tissue under the microscope to assure complete removal. A plastic surgeon will then repair the wound. Recurrence is possible, so it is im-

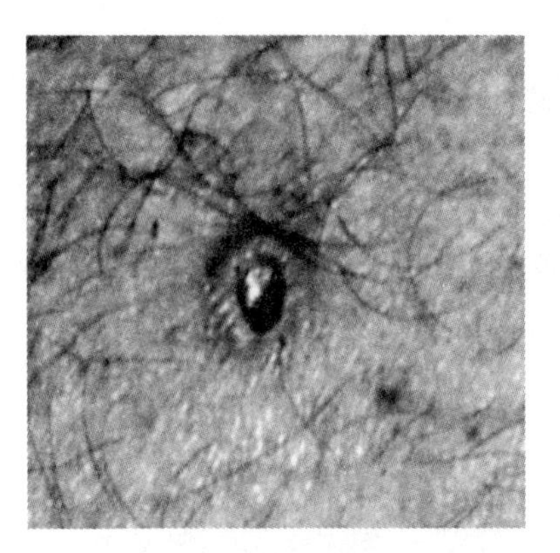

Squamous cell carcinoma. See color plates.

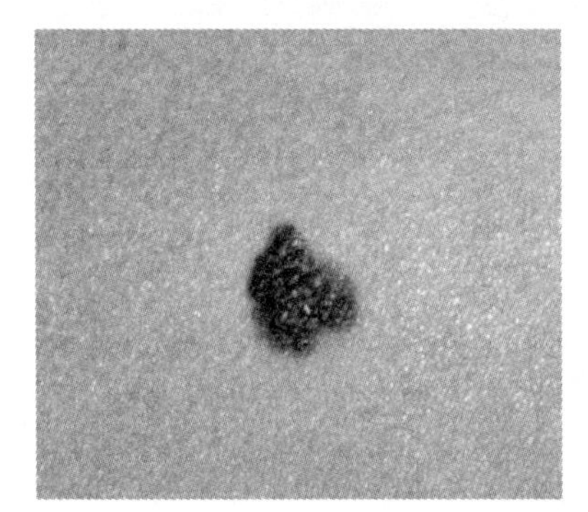

Melanoma. See color plates.

portant to follow up with your doctor every three months for the first year after excision and every six months for the next two years.

Squamous cell carcinomas are the next most common skin cancer. Like basal cell carcinomas, they occur in sun-exposed areas such as the face, in particular the nose, eyelids, and ears. Squamous cell carcinomas can grow rapidly, and can even spread throughout the body if allowed to become large. They should be removed within a few weeks of diagnosis to prevent growth.

Melanomas are the most ominous type of skin cancer. They can grow spontaneously or can arise from moles. They should be surgically removed without delay. The prognosis is largely related to the thickness: thin melanomas are readily curable; intermediate-thickness melanomas may require removal of the local lymph nodes; thick melanomas can spread throughout the body and have a very poor prognosis. Melanomas usually are pigmented, but some are flesh colored.

Lipomas are benign fatty tumors that can occur anywhere on the body. On the face, they are common on the forehead. They occur randomly, but some say they can occur after trauma. Removal is performed through a small incision, allowing the lipoma to be forcefully expelled.

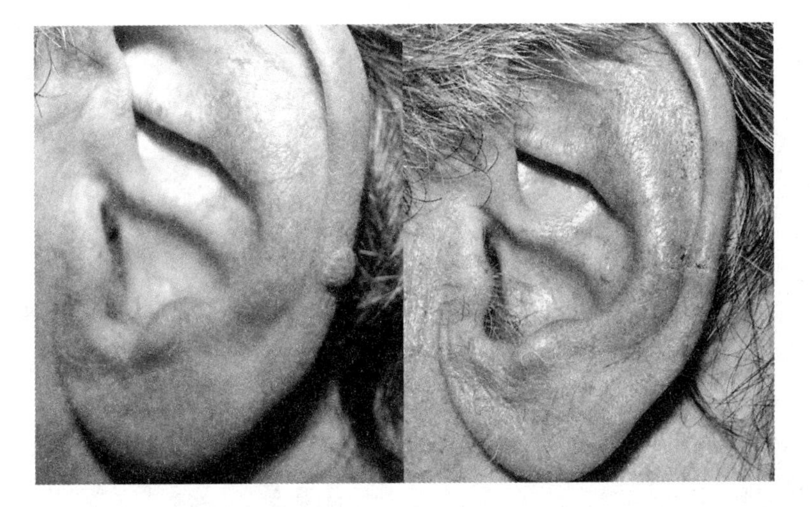

This man had a flesh-colored mole on his ear for three years. It turned out to be a melanoma. On the right is the ear after removal. More skin had to be removed later. See color plates.

Cysts are trapped collections of skin caused by ingrown hairs or pimples. Since cysts eventually get infected, they are removed by excising the skin along with the underlying hardened material.

The Technique

Most skin lesions are removed in the office under local anesthesia. If frozen sections are required, the procedure may be performed in the hospital. A plastic surgeon removes the lesion using principles that maximize the cosmetic result (Chapter 3).

The lesion is excised in the shape of an ellipse, which looks like a football. If the growth were to be cut out and allowed to heal, which some dermatologists prefer, there will be a white, depressed scar. Plastic surgeons like to make better scars. To do this, they follow certain important principles. The ellipse and resulting scar are oriented along an existing or future wrinkle line. Any line that joins sections of the face, such as between the upper lip and cheek, will do. The skin is cut down to the fat. The skin edges are lifted in a step called undermining. This lift takes the tension off the skin and allows it to be stretched. Two layers of stitches are used. The lower layer is made of a dissolving material that will be gone in weeks to months. The skin is then closed with either fine stitches or glue.

Sometimes lesions are too big to close with a line scar. In this case, the plastic surgeon may lift a flap of skin from an adjacent area. This is never completely detached from the body. Or a skin graft may be needed to close a large wound.

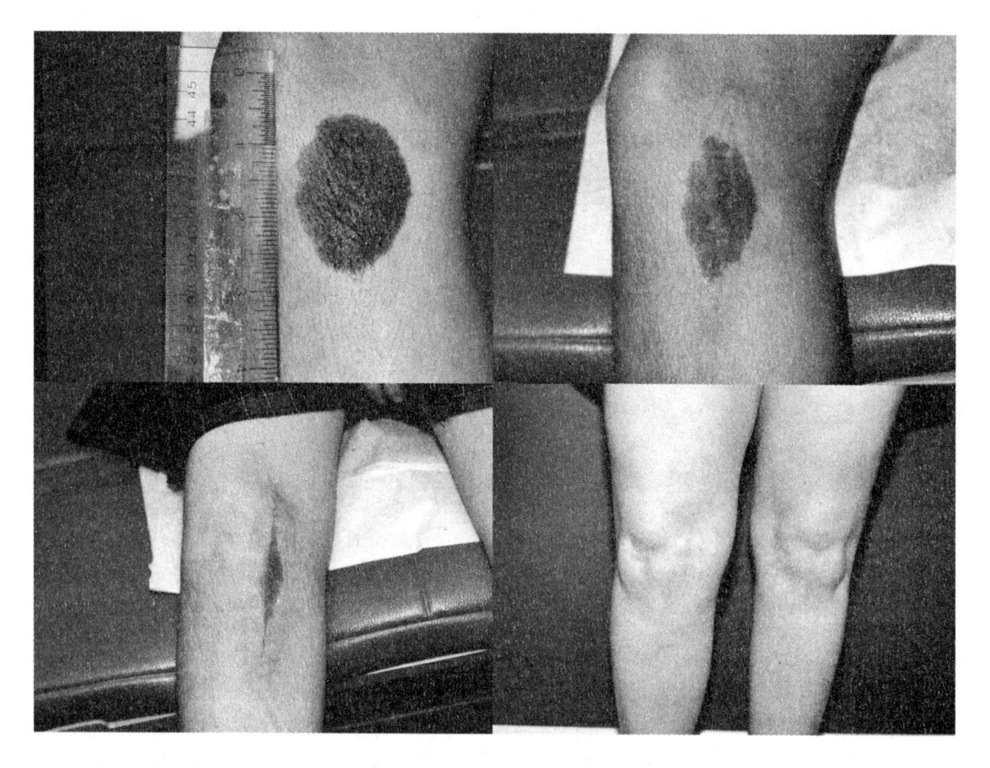

This 12-year-old girl was born with a giant mole. It was removed in a series of three operations allowing for the stretching and growth of skin between excisions. The various stages are shown, with her appearance at age 18 at lower right.

The top layer of skin from another area is shaved off with a machine like a cheese slicer, being completely separated from the body before being sewn onto the wound. Or thicker grafts are "borrowed" from behind the ears and sewn onto the face.

Other techniques include serial excision of the mole, which involves cutting out a piece of the mole, letting the wound heal, allowing nature to stretch the skin, and repeating the process in a few months. More complicated techniques include tissue expansion, where a gradually inflated balloon-type device causes the skin to grow. Finally, microsurgical transfer of tissue from one area to another, connecting blood vessels and nerves, is sometimes necessary.

The Costs

The cost associated with the removal of skin growths is infinitely variable. Skin tags may be removed for a few hundred dollars, whereas giant moles may cost thousands to excise. Removal of an average facial mole will be in the $700–$1,000 range.

8

Tackling Wrinkles

Your skin is the first thing people see when they look at you; scars, wrinkles, and acne color their perception. As the largest organ in the body, the skin helps regulate the body's temperature. It helps fight infection and it makes vitamin D.

Anatomy of the Skin

People speak about the many layers of the skin. But for practical purposes, the skin consists of just two layers. The top layer is the *epidermis,* the waterproofing layer. It keeps everything on the outside from entering the body and it keeps you from drying out. It contains fatty acids that are natural antibiotics, killing bacteria and preventing them from invading the body. It is also a barrier to viruses and fungi. The outer layer of the epidermis is made up of dead cells that mechanically protect the skin. This layer thickens in areas that are chronically irritated, such as the hands of a construction worker. When the epidermis is broken, as in a scrape or a burn, liquid oozes out and quickly clots into a scab. This scab is a mechanical barrier that keeps the tissue from drying out. It also helps keep germs from getting into the body and causing infection. Epidermis contains the brown pigment melanin, a natural sunblock. Melanin absorbs ultraviolet light and prevents it from injuring the DNA.

The lower layer of the skin is called the *dermis.* This is the structural layer, the leather of the skin. It holds you together. The dermis is made up mostly of *collagen* and about 4 percent *elastin* fibers. Collagen provides the strength of the skin, while elastin allows it to stretch and snap back.

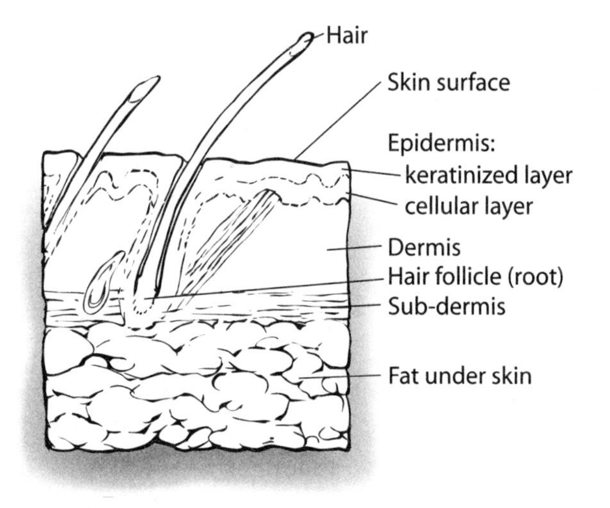

A cross-section of the skin.

Collagen and elastin fibers are held together in a pool of very complex sugars and proteins. Hyaluronic acid is one of the sugars of the skin. It is important when we consider moisturizers and fillers of the skin, since it can absorb a thousand times its weight in water.[1]

Within both layers of the skin are hair follicles and sweat glands. Hair helps trap warm air around the skin. Nerves at the base of hair follicles trigger reflexes when stimulated. When a bug crawls on your skin, it is these nerves that give you an early warning before it bites. Sweat cools the skin by evaporating when you are overheated. Sweat also increases the ability of the dry palm of the hand and the sole of the foot to grip objects. If you didn't have sweat glands, you would have difficulty holding onto things. Sweat also kills some bacteria.

Beneath the skin is a layer of fat. This fat insulates the body and keeps it warm. This nutritional storage zone also absorbs impacts during sports and trauma. It allows the skin to glide over the muscles and bones.

Aging Skin

As skin ages, the collagen layer thins and becomes unorganized when viewed under the microscope. Sun exposure causes these changes as well. Elastin also decreases as we age. In children, stretched skin quickly snaps back into position; but in the elderly, the skin first stays put and then gradually sinks back.

Remember your grandmother with the "see-through" skin. As the collagen thins, you can see more blood vessels beneath the skin. Aging skin thins, becomes less elastic, develops wrinkles, and becomes dryer as oil production decreases. The two layers of skin become less tightly attached to each other, allowing easier injuries.

Sun damage causes an increase in elastic material and oil production. Collagen decreases dramatically. Skin cancers and other growths, such as seborrheic keratoses and age-spots, increase in sun-exposed skin.[2]

Wrinkles Are Actually Fractures in the Dermis!

In most people, the first wrinkles do not appear until their late 30s. However, certain conditions accelerate the formation of wrinkles.

The most significant of these is sun exposure. Simply put, the more sun exposure you have had, even as a child, the earlier your wrinkles will appear. I have seen patients in their early 20s with wrinkles. These are the beach babies —they spend hours each day working on their tans, and their future wrinkles. And today's freckles become tomorrow's age-spots. It is the ultraviolet light (UV) contained in sunlight that is responsible for much of the aging of the skin. Chronic UV exposure causes wrinkles, splotchy pigmentation, visible capillaries (*telangiectasia*) and loose, rough skin. UV causes damaging substances called free radicals to be produced. UV also depresses the natural antioxidants in the skin, contributing to more damage. Ultraviolet light damages your genetic material (DNA) every time you are in the sun. To convince yourself, compare the skin on the back of your hands with the skin on the inside of your upper arm. The upper arm looks better in everyone.

Smoking accelerates wrinkling. A single puff on a cigarette releases toxins that constrict blood vessels for nearly four hours. That's four hours of decreased nutrition and oxygen to the skin. It's no wonder that smokers have unhealthy, gray-appearing skin. They look older than their age. We may joke that it is better to look good than to feel good, but smoking causes you to look bad *and* feel bad.

Acne scarring accentuates wrinkles later in life. It creates mini "facial crumple zones™" throughout the face, particularly just below the corners of the mouth. The bacteria responsible for acne actually eat away at the dermis. Decades later, when the skin thins, the weakened dermis crumples in old acne-scarred areas.

There Is No Wrinkle Cure

Treatment starts with prevention. You need to limit your sun exposure to twenty minutes a day, or religiously use a zinc-oxide–containing sunblock if you must be in the sun. Elimination of smoking and its toxins will help maximize your genetic potential. Drugs such as tretinoin, alpha hydroxy acids (AHAs), and topical vitamin C stimulate the formation of collagen and elastin and help prevent your dermis from deteriorating. The dermis can actually thicken with long-term use of these drugs.

Once you have wrinkles, prevention is a moot point. You can use skin creams to limit ongoing damage, but only more aggressive treatments will reverse them. Wrinkles are treated by filling them, by sanding, lasering, peeling, or shrinking them, by stretching them out, or by stopping the constant pull on the skin that continually recreates them.

9

Filling Wrinkles

Facial wrinkles are a new phenomenon. Until a few centuries ago, humans didn't live long enough to become wrinkled. In fact, the presence of wrinkles foretells aging. A healthy, youthful appearance wanes with advancing years. Every glance into the mirror reminds us of our mortality.

Attempts to correct wrinkles by inflation with fillers began as early as 1902. Paraffin (wax), silk, celluloid, latex, and Vaseline were injected into the face. These procedures were abandoned twenty years later because of disastrous complications. One surgeon actually combined various ingredients in a vegetable grinder and injected them into the face.[1]

Surgeons have long searched for the ideal filling material—one that was simple to use, inexpensive, lasted a long time, and had no complications. Until Restylane was introduced in Europe in 1996 only one wrinkle filler was in common use. For two decades collagen was the only choice. And until 2003, it was the sole filler in the United States. Now many fillers are in use around the world and more are on the way. Only time will tell which are useful and which are dangerous. As Peter McKinney, M.D., C.M., professor of plastic surgery at both Northwestern and Rush medical schools, says, "The one that lasts longer with fewer problems wins."[2]

In 2005, there were 40,000 Radiesse injections, 221,000 collagen injections, 1,194,000 hyaluronic acid injections, and 35,000 Sculptra injections in the United States alone.

Some fillers are designed to be placed in the skin and some belong underneath the skin, in the fat. Fine-wrinkle filling requires different materials than deep folds. After deciding on the depth of the material, we consider how long the material lasts and its cost.

While no surgeon wants to be the last to perform a new procedure, it is not wise to be the first, either. New fillers need to last for a long enough time to be cost-effective but not cause short- or long-term problems. Consumers should be sure that their physician purchases fillers from approved sources within his own country. If a product is brought in from outside the country, there is no way to be sure the product is genuine.

Fillers are not just used for filling wrinkles or folds. Innovative plastic surgeons are now using fillers to reshape noses, chins, jawlines, and virtually every area of the face.

Wrinkle Fillers

Collagen

Collagen is a protein made from cow skin. It is processed, mixed with the anesthetic lidocaine, and made into a paste that is injected into fine wrinkles. Collagen was first used in the late 1970s but did not become popular until it was commercially available in 1981. As a foreign protein, collagen is attacked and digested by the human body. The body can become confused, making antibodies against the cow collagen and even the body's own collagen. The result may be an *autoimmune disease,* such as dermatomyositis, rheumatoid arthritis, lupus, or scleroderma, in which one's own collagen is attacked.[3]

Collagen lasts between two weeks and six months, after which the injection must be repeated to maintain the effect. Under the microscope, we can see that the body digests the collagen within six months. Allergic reactions may result; therefore, the patient's skin must be tested prior to the injection. One, or even two, tests are usually performed a month beforehand. Collagen is marketed by the Allergan Corporation under the names Zyderm and Zyplast. The average cost in the United States for an injection is about $400.

Cosmoderm and Cosmoplast

Cosmoderm is human collagen. It was introduced in 2003, in an attempt to decrease the allergic reactions to cow collagen. Made from human skin tissue that is grown in a laboratory, Cosmoderm, strengthened by the chemical glutaraldehyde, is not as thick as Cosmoplast. My major objection to Cosmoderm is that it is made from dead humans. There is no way to sugarcoat that fact. And there is no way to assure with total certainty that a disease will not be passed along from the dead donor to the recipient of the collagen. Still, infectious diseases such as AIDS or hepatitis, West Nile virus, or Cruetzfeld-Jacob disease (the human mad-cow disease) have never resulted from a Cosmoderm injection.

On the other hand, every type of tissue ever transplanted from one human to another has eventually been linked to infectious disease transmission. Hepatitis C has been passed along, even when the organ donor tested negative! Diseases that we can't even test for yet might be in the tissue. There has been one death and at least two dozen serious infections from the transplantation of other types of contaminated tissue from one person to another. In 2003 the U.S. Centers for Disease Control and Prevention reported that a tissue donor had transmitted the hepatitis C virus to eight people. And there is no way to test for Cruetzfeld-Jacob disease. I worry about human cadaver materials causing disease in patients who simply want to look better.[4] Diseases can be transmitted even after the most stringent testing.

In 2005 bone, skin, and tendons were illegally taken by a New Jersey tissue bank from New York funeral homes and sold to companies for implantation into other humans. Lifecell Corporation, producer of the collagen implants Alloderm and Cymetra, was one of these companies. These tissues were *never* tested for infectious diseases. Scary enough that I never use a human product unless the problem is life threatening.

Cosmoderm and Cosmoplast cost about $100 more than cow collagen.

Evolence

This is a brand-new injectible made from pig collagen. The manufacturers have tinkered with the molecule and linked it to a sugar. This appears to make it a "super-collagen" capable of lasting far longer than standard collagen. Because it is made from collagen, allergies can occur and so skin tests are necessary. The company says it can last over fifteen months. It is available in Europe, Canada, Japan, and Israel, but not yet in the United States.

Cymetra

This material contains collagen, elastin, and other proteins from human cadavers. It is the injectable form of Alloderm. While infectious diseases have never been transmitted with Cymetra, the very possibility makes many plastic surgeons and patients nervous.

Transplanted Cymetra lasts about two months. It should not be used around the eyes. It is not a particularly popular filler.

Fascian

This is basically human collagen that is taken from fascia (connective tissue) of the calf muscle. The material is used to fill wrinkles, scars, nasolabial folds, and depressions. Fascian may be replaced with your own collagen at some point, so there is some chance of permanence. At $85 per syringe, Fascian is much cheaper than other fillers. But I do not believe that we should choose a filler material solely on the basis of price. As with other cadaver-derived products, I get nervous since this product is taken from a human.[5]

Isolagen

In this interesting approach to wrinkle filling, the Isolagen Corporation grows pure cells in the laboratory. A small piece of skin is taken from a patient and sent to the Isolagen lab where, in six weeks, millions of cells are produced. The cells, called fibroblasts, make collagen. They are sent back to the doctor and injected into the patient. Presumably they will then "live long and prosper," growing new collagen within the patient's skin. This procedure is currently under study in the United States.

Restylane, Perlane, Hyalaform, Juvederm, and Captique

The early 2000s saw the race for a better wrinkle filler. Half a dozen companies waged a furious battle to gain early market share for what they hoped would be the next blockbuster cosmetic drug—the next Botox.

Restylane was approved in Europe in 1996 and in the United States in 2003. It is made of hyaluronic acid, created by bacteria through the wonders of genetic engineering. Because no animals are used to create this product, there is no chance of contracting infectious disease. And because this sugar-like substance is already normally present in our bodies, skin testing for allergy is not required. Although the FDA says that Restylane is comparable to Zyplast collagen at six months, many physicians find it to be superior. Under the microscope, this material is seen to last nine months. Bruising, swelling, and pain on injection are common and depend on the skills of the doctor.[6]

Restylane fills wrinkles nicely and clinically lasts between six and twelve months, although this duration is not guaranteed. Some patients need a second touch-up injection a few weeks after the first because swelling occurs immediately, making precise injections impossible.

Restylane is not destroyed by the body. It simply diffuses into the skin, like a drop of ink diffusing into a glass of water. Gradually it settles into the skin, and the wrinkles are exposed again.

The company that produces Restylane also produces lower- and higher-viscosity (thickness) products. Perlane is Restylane's thicker cousin. It will be used to correct deep folds and to plump lips. Restylane Fine Line is a thinner material that will be used in fine wrinkles, and Restylane SubQ uses larger particle sizes to fill deeper folds and to build up the cheekbones and the chin. These products will serve to confuse physicians and consumers since their indications overlap.

Juvederm, approved in 2006, is Allergan's answer to hyaluronic acid. Juvederm comes in three different viscosities. Like the Restylane products, the thicker materials are used for deeper folds; the thinner fillers are used for more superficial wrinkles. Juvederm claims to have the highest concentration of cross-linked hyaluronic acid, allowing the chemical to stay in the body longer than competing products. Because Allergan also makes Botox, the co-marketing of these two substances will propel Juvederm's popularity.

Restylane is today's "injectable of choice." It has caught on rapidly. Not only is it used for fine wrinkles, but creative uses keep being described. It fills folds, scars, and various depressions. It is being used to plump up the eyebrows, giving a lifting effect. Used in the lower eyelid, it pushes up the lid, shaping it and filling in the "tear trough." It can highlight the bones in the upper eyelid and increase the size of lips. It has even been injected into sun-damaged cleavages!

Because Restylane is a gel, it cannot be mixed with local anesthetics. Without lidocaine, the injections hurt and so some sort of anesthetic is usually used.

Outside the United States, different types of Restylane are available. One, an experimental hyaluronic acid mixed with the anesthetic lidocaine, is currently in development. Early reports indicate that injection of this material hurts less than other hyaluronic acids.

Restylane is most commonly injected into the fine wrinkles around the mouth and face. It is less effective for the deeper wrinkles and folds. It can also fill depressions of the face and nose.

Other brands of hyaluronic acid are also available. Hyalaform is made from roosters' combs, as opposed to the test-tube production of Restylane. Plastic surgeons say that Hyalaform may disappear sooner than Restylane—perhaps after just three months. Restylane packs four times as much hyaluronic acid into its product as its competitors, which may explain its better clinical results. There may be a higher allergic reaction rate with Hyalaform than Restylane. The same

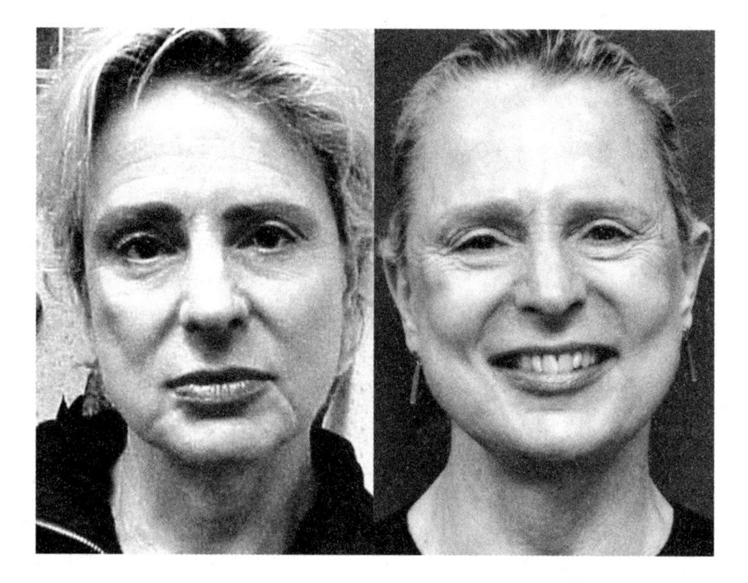

Another surgeon excised cysts below this woman's mouth and filled in the scars with too much fat. I re-excised her depressed scars and filled them with Restylane (on right). See color plates.

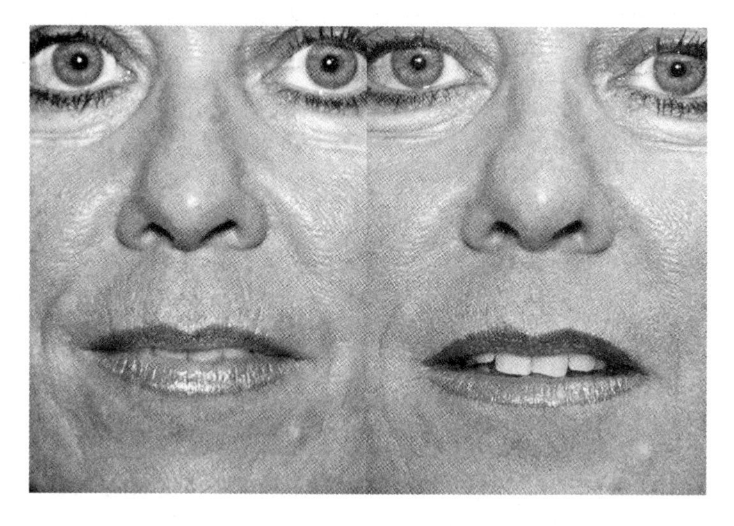

This woman had Restylane injected into the wrinkles around the mouth. On the right, two months after injection. See color plates.

company that makes Hyalaform also has a test-tube-created product called Captique.

Restylane is an excellent product, with a very good safety profile. As its use increases, we are sure to see problems, however. There have already been reports of immediate and late allergic reactions.[7]

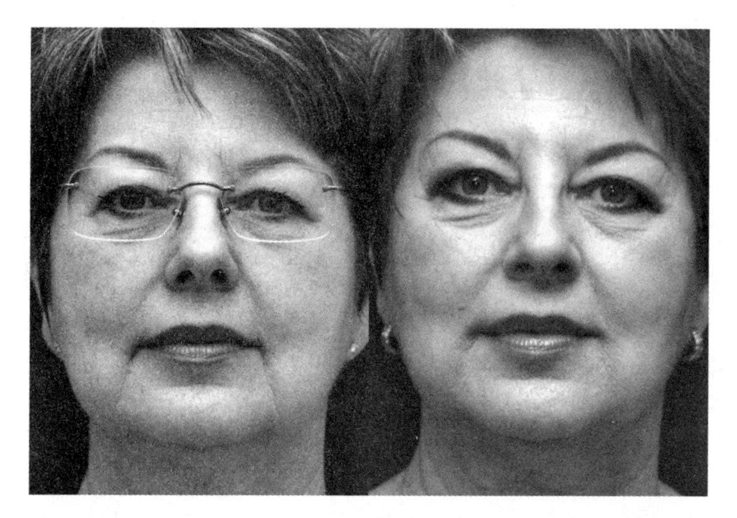

This woman had a combination of Radiesse injected deep into the nasolabial folds and Restylane injected more superficially. Left, before; right, following her second Restylane syringe a week later. See color plates.

At the surgeon's cost of $240 per cc, Restylane is more than ten times as expensive as gold. Hyalaform is less expensive.The average cost to a patient for a Restylane injection is $500.

Puragen

This is a souped-up hyaluronic acid. The Mentor Corporation has altered the molecule to make it more stable. That means it takes longer to fade away. Puragen Plus also has the anesthetic lidocaine added to the gel, making injections less painful. Puragen is already approved in Europe and will most likely be approved in the United States.

Radiesse

When a product changes its name, it reminds me of a *Dragnet* episode: "The names have been changed to protect the innocent." This injectable product, formerly called Radiance, is now called Radiesse (apparently, the original name was too close to other medical products). Radiesse is made of *calcium hydroxylapatite,* the building block of bone and teeth. Other chemicals in the mix have long been allowed as the carriers for injectable medications. Radiesse is completely synthetic; no animal products are used in its production. That gives assurance that infectious diseases cannot be contracted through its use. The nontoxic chemical degrades over time, broken down into calcium and phosphate, normally present in the body. Collagen (or scar) surrounds the degraded Radiesse particles, allowing the filler to last a long time.[8]

Radiesse is not used for fine wrinkles but for deeper depressions, such as the nasolabial folds. Approved by the FDA for tissue filling in 2006, its list of uses has been steadily growing. Not only has it been used to fill the marionette lines between the corner of the mouth and chin, but it has also corrected sunken cheeks, filled dimples, and rejuvenated the depression between the jowl and the chin. Radiesse is also being used as a nonsurgical implant to make the chin and cheekbones larger.

Radiesse is increasingly popular for correcting minor deformities of the nose after fractures or rhinoplasty. Almost magically, when Radiesse contacts bone, *new* bone is formed. When not touching bone, it creates bulk and scar. Part of the improvement may be permanent, although the manufacturer states that this filler lasts between one and two years. Some surgeons claim it lasts longer.

The placement of Radiesse is truly an art, and results will vary with the skills of the surgeon. Its injection is really surgery: surgeons use the injecting needle to break attachments of skin to underlying tissue before the injection, and they use the material like steel rebar in concrete—crossing folds to reinforce them and hopefully stall recurrence.

No allergic reactions have been reported to date, and no skin test is needed. Radiesse cannot be used to make lips larger, however. In a third of patients it clumps and forms lumps in the lips. If these lumps occur, the lip must be punctured and the material squeezed out. This usually solves the problem.

Radiesse costs the surgeon $295 for a 1.3 cc syringe. The average cost for a Radiesse injection is $900.

Sculptra

Sculptra was approved as a filler in 2004. It has been marketed outside the United States, as NewFill, since 1999. Chemically similar to suture material, Sculptra (*poly-L-lactic acid*) currently is approved only for filling facial fat loss (lipoatrophy) in HIV-positive patients. Injected into the deep dermis, it may last as long as two years, although one study showed it had disappeared in four months. It may last longer, because it stimulates the body to make a collagen scar around the material. As with Radiesse, lumps may appear when it is injected into the lips. But there have been troubling cases of stubborn lumps in other parts of the face, prompting a warning about the long-term effects. I expect that this material will soon be approved for cosmetic purposes. It will then be used to fill nasolabial folds, marionette lines, and other facial hollows. Unlike Restylane,

Sculptra can be mixed with lidocaine to lessen the pain of the injections. New, creative uses for Sculptra and other fillers are described almost daily. Sculptra has even been used, off-label, by podiatrists to plump up unattractive feet![9]

The average cost for a Sculptra injection is $1,000.

Autologen

This novel product is collagen made from a patient's own skin. A piece of skin is processed in a laboratory and returned to the patient's doctor for injection as collagen. Despite the hope that the collagen would be permanent, since it is the patient's own tissue, it only lasts as long as cow collagen.

Plasmagel

This substance, not particularly popular, is made from the patient's own blood and vitamin C. It lasts about three months. I do not know any plastic surgeons who use it.

Silicone

Silicone was used extensively in the 1970s. I remember "Imus in the Morning" stating that the greatest achievement of a Playboy model was enduring the pain of silicone injections into her breasts. In fact, silicone liquid has been injected into breasts and faces for more than thirty years. It sounds like a good idea, in that it does not break down over time. However, most authentic plastic surgeons have seen disasters from these injections. The FDA says silicone can cause swelling, reddening of the skin, lumpiness, and soft-tissue tumors. The scar around the silicone can be very thick, creating nodules, or it can be thin, allowing the material to drift through the tissues, ending up via gravity in the most southern location.

Once injected, the game is over. If the area becomes infected, the infected tissue needs to be cut out. Countless women have undergone mastectomies or removal of whole sections of the face because of infected silicone. The American Society of Plastic Surgeons says injectable silicone can harden, migrate, cause inflammation, and kill skin. As a result, the FDA banned it in 1991. The next year, it enforced its ban by having three noted New York dermatologists sign an agreement to stop injecting liquid silicone. At the time, FDA commissioner David Kessler, M.D., said, "People who undergo these injections are exposing themselves to unknown, potentially dangerous risks." Again in 1993, the FDA took action against a Miami plastic surgeon who had injected liquid silicone.

The next year, silicone oil was approved by the FDA for treating retinal de-

> The use of silicone violates a respected principle in plastic surgery: when a foreign substance is implanted in the body, it must either eventually dissolve or be retrievable.

tachment in AIDS patients. This created an avenue for physicians to once again inject silicone into wrinkles, arguing that its use would be off-label and not illegal. In 2004, highly purified silicone oil was injected into the atrophied faces of AIDS patients, restoring contour without side effects. Long-term problems were not assessed, however. In 2006, an unlicensed nurse injected silicone into a patient's buttocks. Some of the silicone traveled through the bloodstream into the lungs, causing respiratory failure.[10]

I know that there are still physicians who inject silicone. I wonder why. As a patient, I would select a physician who practices "safe surgery" and not "what I can get away with" surgery.

Artecoll and ArteFill

ArteFill consists of tiny balls of plastic (polymethyl-methacrylate) mixed with the anesthetic lidocaine in a collagen base. (Hard contact lenses are made out of this material.) Once injected, ArteFill becomes surrounded by scar. A plastic, it never breaks down. Like liquid silicone, it violates that basic principle of plastic surgery: *any foreign substance that doesn't dissolve must be removable.* ArteFill can't ever be removed. A 2004 study extolled the virtues of Artecoll, another filler, showing clinical improvement in facial wrinkles for up to a year. But plastic surgeons who fail to study the history of silicone are doomed to repeat its problems.

Artecoll and ArteFill sound good, since they permanently fill in scars and wrinkles and plump lips. However, like silicone, they cause lumps and areas of inflammatory scarring called granulomas. These problems prompted the Swiss Society for Dermatology, the Swiss Society for Plastic Reconstructive and Aesthetic Surgery, and the Swiss Society for Aesthetic Medicine to advise against the use of Artecoll and silicone in the face. In addition, it can migrate to unwanted areas. Artecoll has been used on four hundred thousand patients in Europe and around the world since 1994 and was just approved in the United States. A skin test is required before injection because the filler contains highly allergenic collagen.[11]

Gore-Tex

While not technically injectable, tiny pieces of Gore-Tex (expanded polytetrafluoroethylene, e-PTFE) are threaded into fine wrinkles. Poor results are frequent. Scar is made around the Gore-Tex and is often visible through the skin. A foreign material, it can become infected and extrude through the skin. I have removed Gore-Tex from many patients. A fine material for dental floss and shoes, it is a poor wrinkle filler. The material is supposed to be inert and not biodegradable, but in 2006 the Environmental Protection Agency found that Teflon (the closely related chemical polytetrafluoroethylene) is "a likely carcinogen." In November 2006, the company voluntarily withdrew all Gore-Tex products from the plastic surgery market.

Surgisis Soft-Tissue Graft

An interesting material, not technically an "injectable," Surgisis is a sheet of collagen taken from pig intestines. All the cells are removed and the material is sterilized. Tissue grows into it when implanted into humans. The material is many times stronger than human tissue and therefore can strengthen the abdominal wall in tummy tucks. In the nose, it can smooth bumps. Surgeons are using it to help pull up tissues in face-lifts and to make lips larger. Time will tell how satisfactory this material is in the long run.

The Technique

Fillers are injected into either fine wrinkles or deeper folds or other tissues. Collagen and Restylane are the most common agents for fine wrinkles; Radiesse, Sculptra, and fat, for the deeper folds. A very fine needle is used to inject the filler directly into the wrinkles. Collagen numbs as it is injected, but Restylane and Radiesse do not contain anesthetic. For this reason, when I fill wrinkles in the center of the face or make lips larger I first numb the nerves to the cheek and chin. Through four injections, about 80 percent of the area around the mouth can be numbed. Sometimes, I numb the skin with EMLA anesthetic cream.

After filler injections there are little red pinpoints and bruising from the needles. These spots can be immediately covered with makeup. It may take up to two weeks for all of the bruising to resolve. Makeup can be applied the next day, but the swelling will last about two days.

When filler is injected into fine wrinkles, it is deposited in ribbons and blobs under the skin. The surgeon flattens the material with massage. If lumps are felt later, the patient can smush the gel with her finger. This step is important, because the blobs and ribbons will be camouflaged by swelling for a few days. Since hyaluronic acid absorbs water, wrinkles actually further improve in the days following injection. A typical hyaluronic-acid wrinkle-filling session takes thirty to forty-five minutes. The immediate result usually generates a "wow" from the patient.

The Risks

While wildly popular, these injections are not without risk. Collagen injections have a high chance of causing allergic reactions. Skin tests are required prior to injection. Repeated collagen injections may lead to the development of diseases similar to rheumatoid arthritis. Allergan recommends that no more than an ounce be injected per year. Allergic reactions are much less likely with the newer fillers.

More important, the semisolid wrinkle fillers can be accidentally injected into blood vessels. When this happens, disaster looms. Around the eyes, nose, forehead, nasolabial folds, and even the lips, blood vessels link the skin with the eyeball and even the brain. If filler gets in these blood vessels, the result is immediate and permanent blindness. At least forty-three cases of blindness have been reported from injections around the eyes. If the material floats into the brain, a stroke can result. In other areas of the face, the consequences are not as severe, but loss of skin can result. The most common area of skin loss is between the eyebrows: the material mechanically blocks the blood vessels, resulting in the death of tissue in the affected area. Liquids such as Botox cannot occlude the blood vessels and therefore will not result in this problem.[12]

Surgeons take special precautions to avoid injecting into blood vessels. Ice and anesthetics with epinephrine shrink the blood vessels, and the doctor injects only while withdrawing the needle. Some physicians make a tiny stab in the skin and use a blunt needle to place the filler, further limiting the chance of injecting into a vessel. The fillers are injected using minimal force. *Many doctors simply will not take the risk and will not inject around the eyes.* As more injections are performed by less-trained people, disasters will be more common.

Any injectable material can activate herpes infections. Patients who have herpes sores should receive an antiviral such as Valtrex before and after the injection. And injections should not be performed during an acne breakout.

When fillers are injected, they eventually break down and are absorbed by the body. No one knows what happens if they are exposed to different types of lasers or radiofrequency energy. We may be in for some surprises. Until we know better, I recommend not lasering areas that have been filled. Laser first, then fill.

The Bottom Line

So many fillers are available that it can be confusing for both doctors and patients. Patients should be sure that their surgeon injects only materials approved by the FDA or an equivalent agency. My own treatment regimen is fairly simple. When wrinkles are above the midnose, I usually use Botox or peeling or lifting procedures. I inject fillers around the eyes only sparingly. In the nasolabial folds, I prefer fat. If the patient wants a filler from a bottle, I choose Radiesse for the deep folds and Restylane for the superficial wrinkles. For other wrinkles on the face, I prefer Restylane. Stay tuned, however; the field is changing rapidly.

10

Peeling and Lasering Wrinkles

In 1958 two renegade plastic surgeons in Miami, Florida, noticed that people with flash burns to their faces healed with fewer wrinkles. At the same time, they learned of women in Hollywood salons who painted the faces of their clients with a chemical that erased wrinkles. The young physicians experimented with deep chemical peeling on animals and then on themselves. Twenty-nine years later, when I was Dr. Thomas Baker's cosmetic surgery fellow, he still illustrated that story by showing the spot on his arm where he bravely performed the first deep chemical peel.

The results were so dramatic and so revolutionary that other plastic surgeons refused to believe them. It wasn't until the doctors brought a patient to the annual meeting of the American Society of Plastic Surgeons that other plastic surgeons accepted the technique. Baker and Dr. Howard Gordon opened up an entire field of plastic surgery that culminated in 556,000 chemical peels in the United States in 2005!

Peels and Lasers Are Methods of Fooling the Skin into Healing

Virtually all methods of nonsurgical wrinkle control have one thing in common: they wound the skin. This starts a complex series of events, which ultimately leads to healing. The skin heals by making new epidermis and shrinking the dermis. The body races to close the wound before bacteria can enter and cause infection. Shrinking skin, however, can lead to deformities. But the body values life before function. When surgeons injure the skin to decrease wrinkles, it responds by pumping out new collagen and elastin. Virtually every known chemical and almost every known electronic device that decreases wrinkling injures the skin and fools it into healing and shrinking. And so, the science of wrinkle reduction is the science of wound healing.

Deep Chemical Peels

Deep chemical peels use a solution containing phenol (the substance in Chloraseptic sore-throat spray) mixed with water, the caustic agent croton oil, and soap. The mixture is applied to the face slowly, because rapid peeling can cause irregular heart rhythms.

The peel is painless because the phenol numbs the skin in much the same way that Chloraseptic soothes a sore throat. Phenol is actually carbolic acid, an antiseptic that was first used by Joseph Lister. It causes the upper layers of skin to peel off. The face swells like a pumpkin for about a week. But when the skin finally heals, it is remarkably smoother.

Not only does this peel shrink the skin, it also causes the body to make new, healthy-looking collagen. For nearly three decades, the Baker-Gordon deep chemical peel was the best method to lessen wrinkles. As an added benefit, the peel decreases the chance of future skin cancers by removing sun-damaged cells. Profound skin lightening makes this peel less useful on people with any degree of brown color in their skin.[1]

The terminology is a little confusing, because skilled doctors can alter the depth of the peel. Deep peels can be performed lightly and, as you'll learn, light peels can be performed deeply.

Deep Chemical Peels Are the "Two-Week Peels"

Before the peel, the skin is treated with tretinoin to make the skin more uniform, and with hydroquinone to decrease brown pigment. Peels are performed in the operating room, with an anesthesiologist present to administer sedation or general anesthesia.

After the skin is cleaned with acetone, the peel is applied slowly. The skin immediately turns a stark white color, a process called "frosting" by surgeons. A thick layer of moisturizer is placed on the skin. If the peel is performed while the patient is awake, it is not necessary to numb the skin.

While the deep peel does not hurt when it is applied, it does cause enormous swelling of the face. Within minutes, the skin starts oozing blister fluid. Repeated cleaning of the skin, with either a hydrogen peroxide solution or a weak vinegar (acetic acid) solution, is necessary to dissolve the crusts and keep the skin soft. If crusts are allowed to form, the patient's misery level will be even greater. The crusts crack when she is speaking or eating, causing bleeding and delaying healing. After cleaning, a thick layer of a moisturizer is applied. I prefer Crisco shortening, made with all-vegetable oil. It is gentle on the skin, inexpensive, and has a low chance of allergic reactions. (And if there is any left over after you heal, you can bake with it!) Alternatively, a lighter moisturizer, squalene, made from olives, can be used.

The face heals in seven to fourteen days, leaving the skin beet red. Until the skin is healed, there is a significant amount of pain, requiring narcotics such as codeine. Makeup can be applied very gently after healing has occurred. The healed skin is dull, while skin that has not yet healed is shiny. The redness of the face lasts for nearly three months, requiring a relatively opaque makeup. For this reason, deep peels are uncommon in men.

When healing is complete, the skin tone is more uniform. Brown pigmentation and wrinkles are significantly reduced. Fine and even coarse wrinkles are immediately decreased; further improvement is seen in the first three months, as the skin continues to shrink.

Everyone's skin lightens after a deep peel. Depending on your skin color, this might be a major issue. And a line may be visible where the peel ends at the jaw, despite "feathering" of the peel. The neck cannot be peeled because there is too high a chance of scarring. The face can also scar, but this is rare unless infection occurs.

The deep chemical peel, while not extinct, is certainly much less popular than it was two decades ago. Even Dr. Baker, its inventor, says, "I still do an occasional phenol peel . . . but public demand is for the laser." The more sophisticated and expensive laser peels have many drawbacks, however, and the phenol peel seems to be making a comeback.[2]

Laser Peels Are the Modern Day Peel

The *carbon dioxide laser* became popular in 1994 as a high-tech method of wrinkle reduction. Computers shape the high-intensity light energy to patterns of squares, lines, or circles. Like an artist paints with different brushes, the surgeon uses these varying shapes to vaporize different zones of the face. The initial results were dramatic and the laser soared in popularity, but as with so many new technologies, problems and complications arose.

Pigment changes and scarring are common with this laser, and some of these changes do not appear for months or years. The laser was expected to cause fewer pigment problems than deep chemical peeling because the pigment-

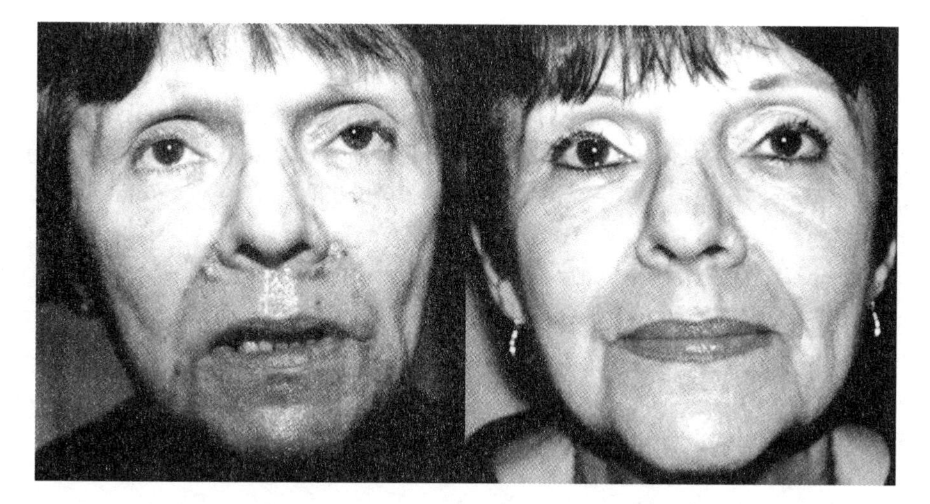

On the left, lasered skin at one week postop; on the right, at one month. See color plates.

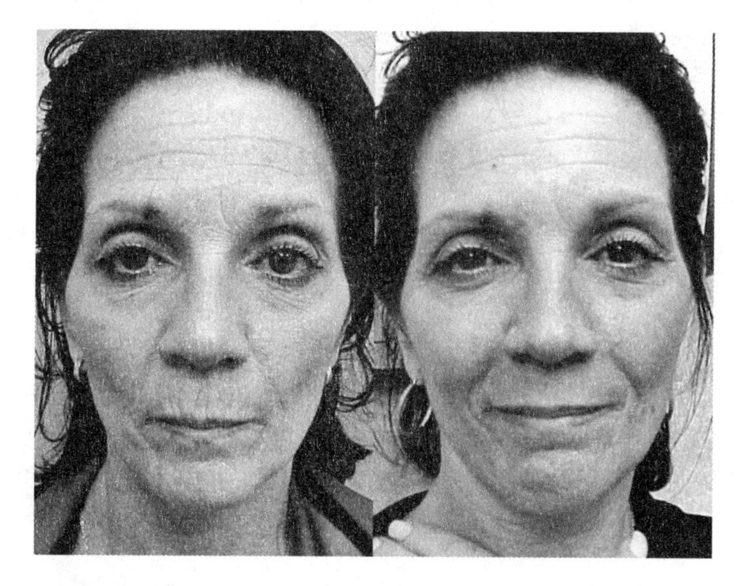

This 49-year-old woman spends a lot of time in the sun, working on her wrinkles. She has undergone a full-face CO_2 laser resurfacing. See color plates.

producing cells (*melanocytes*) were thought to be preserved. Yet over half of patients have lightened skin after the laser.[3]

Because of scarring and pigment problems, a "lighter" laser, the *Erbium,* was introduced in 1999. It causes less heat damage than the carbon dioxide laser but is less effective in treating wrinkles. There seems to be a "no pain, no gain" principle here. Heat injury apparently contributes to some of the wrinkle reduction. The Erbium laser does remove the weathered upper layers of skin, allowing new skin to be formed and causing some shrinkage of the skin during healing.

Preparation is identical to that for a deep chemical peel. Laser peels are performed under local anesthesia with intravenous sedation or with general anesthesia. The eyes are protected with steel contact lenses.

The pulsed carbon dioxide and the Erbium laser both sequentially remove layers of skin. With the CO_2 laser, the skin is blasted two or three times. The first pass removes the epidermis. Nurses in the operating room are very impressed with the second pass, as the skin visibly shrinks before their eyes. The third pass is usually reserved for the deeper wrinkles. Recovery and aftercare are similar to that for the deep chemical peel.

Problems with Laser Peels

Many complications can occur with lasers. The laser can set hair and surgical drapes on fire, making for an unpleasant day. It can ignite the oxygen being delivered to the patient, causing a potentially deadly explosion.

In the early days of the laser, corneal injuries and eyeball perforations were frequent. Even today, scarring and pigment changes are common. Brown pigment streaking and lines demarcating where the laser stopped can occur. This hyperpigmentation can be extremely troublesome and may require skin creams such as tretinoin and hydroquinone, peels, microdermabrasion, and even repeat lasering. It may even be permanent. The skin can scar over the years, stiffening and lightening. The cells that make pigment are destroyed by the heat of the laser, making it dangerous to be in the sun without complete sunblock.

Infections from laser peels can cause scarring. If a herpes infection occurs, it can create chicken pox–type scars. Antiviral drugs such as Zovirax and Valtrex can prevent this complication and should be used in all patients undergoing a laser peel.

About 10 percent of patients do not achieve much wrinkle reduction. A second laser a few months later usually further decreases wrinkles.

Thick red scars or even keloids are possible after lasering. Accutane or steroid use prior to lasering will delay healing for up to a year, which can cause intense scarring.

Since lasering really does work and really does shrink the skin, it can cause the eyelids to shrink to the point that they do not close. An experienced surgeon will know how deeply to laser and whether the lids require a tightening procedure such as a canthopexy (Chapter 12) to prevent this complication.

The laser is safe only on the face, since the skin of the neck and the rest of the body heals differently. Scars are more common in other areas.

The laser will permanently damage teeth. They are protected with a gauze pad or mouthguard during the procedures.

After the skin has regrown, it is bright red for months, even up to a year. A skin reaction called milia, or "whiteheads," often occurs a few months after the peel; it is helped by tretinoin and exfoliants.

The deep laser peel is declining in popularity, mostly because of the high chance of skin lightening and the uncomfortable, lengthy recovery. In 1996, there were 46,000 deep laser peels performed in the United States. After a peak in popularity around the year 2000, by 2005 there were 58,000 deep lasers performed. Because it so effectively lessens wrinkles, the laser peel is still selectively used despite its problems. On the other hand, 418,000 so-called nonablative laser peels (see Chapter 38) were performed the same year. In these, the upper layers of skin are not removed.

To take advantage of the aggressive skin smoothing effects of the carbon dioxide laser but to limit healing problems, lower power levels are now being used. The skin is left intact and acts as a dressing. The "light CO_2 laser," as it is called, does decrease wrinkles, but not as aggressively as the more powerful version. It can be performed with only EMLA cream anesthesia and patients recover within a week without oozing. It doesn't appear to cause skin lightening.

Dermabrasion

In 2005, there were 42,000 dermabrasion procedures in the United States, a 34 percent decrease from the prior year. Only 2,800 were performed in men.

Preparation for dermabrasion is the same as for deep chemical peels. Again, local anesthesia with intravenous sedation or general anesthesia is used. Dermabrasion is performed with a device similar to a Black & Decker sander, which physically removes the top layers of skin. There is less heat injury with dermabrasion than with the laser, so there is less lightening of the skin. But this older technique is more difficult for the surgeon to learn than the laser. It is also very risky. Eyelids have been ripped off with the dermabrader. I prefer to reserve dermabrasion for the upper and lower lips, where it is great at decreasing the "lipstick bleed" wrinkles. I also use dermabrasion for sanding acne scars off the face. (For this problem, the deep chemical peel has no role.) Dermabrasion is equivalent to the laser in effectiveness, but the laser is probably safer.

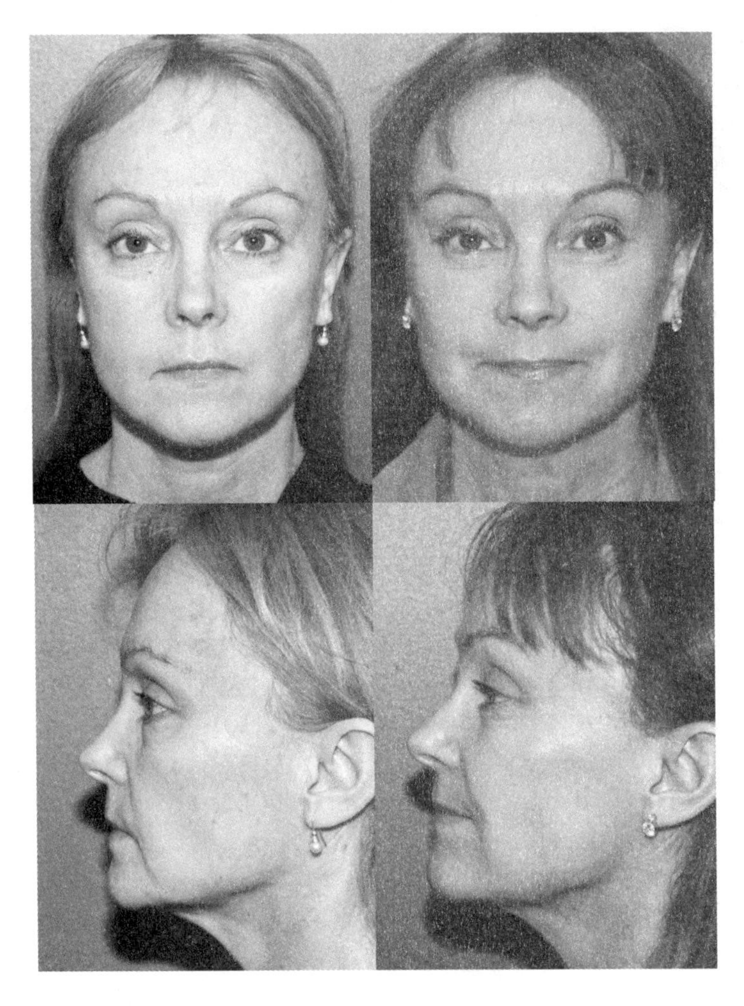

This woman had scattered brown pigmentation of the face and fine wrinkling. Note the improvement in pigmentation and wrinkles on the right, after the TCA peel. See color plates.

Light Chemical Peels

Many people don't relish the idea of two weeks in hiding while waiting for their new skin. Less drastic types of peels were introduced in the 1980s, using *trichloroacetic acid (TCA).* This procedure hurts when performed, but peels more superficially than when phenol is used. The epidermis and superficial dermis peel, healing in five to eight days. The more superficial the peel, the fewer wrinkles are removed. TCA concentrations range from 15 percent to 50 percent, alone or in combination with other chemicals. Higher concentrations result in deeper peels. The TCA peel is useful for pigmentation irregularities, such as sun-

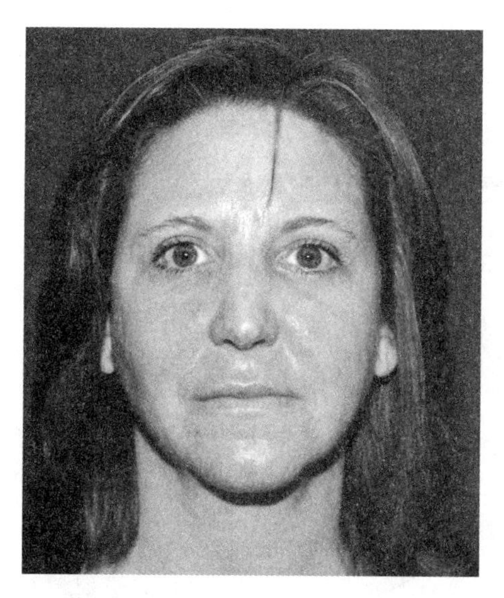

This degree of redness of the face is expected one week after a TCA peel. See color plates.

damaged skin and discoloration from pregnancy. It helps wrinkles and other sun damage. Acne and large pores are improved. Even some premalignant conditions can be corrected.

Literally dozens of factors can be altered in the TCA procedure, resulting in different levels of peeling. The thickness of the skin, the pre-peel use of creams, the type of cleansing the day of the peel, the TCA concentration, how hard the chemical is rubbed in, how many strokes the surgeon uses to apply the peel, and how the peel is cared for afterward all affect the result. *The TCA peel is part art and part science.*

To try and achieve better results, TCA has been combined with dry-ice peels, Jessner's solution (resorcinol, salicylic acid, lactic acid, and ethanol), methylsalicylate, and glycolic acid. None of the combinations is better than TCA alone.

In the 1990s, Zein Obagi, M.D., the Beverly Hills dermatologist, added a blue dye, glycerin, a thickening base, and soap to the TCA. He could better judge the depth of this "blue peel." In the hands of experienced doctors, the results became more predictable.

Many people do not want *any* downtime. Not only does it take nearly a week for the TCA peel to heal, but there can be two weeks of redness afterward.

TCA Peels Are the "Long Weekend Peels"

TCA peels require extensive preparation. Patients must "reprogram their DNA" with at least two weeks of tretinoin in order to achieve an even peel. In addition to the tretinoin, two weeks of a pigment reducer (hydroquinone) is helpful. You

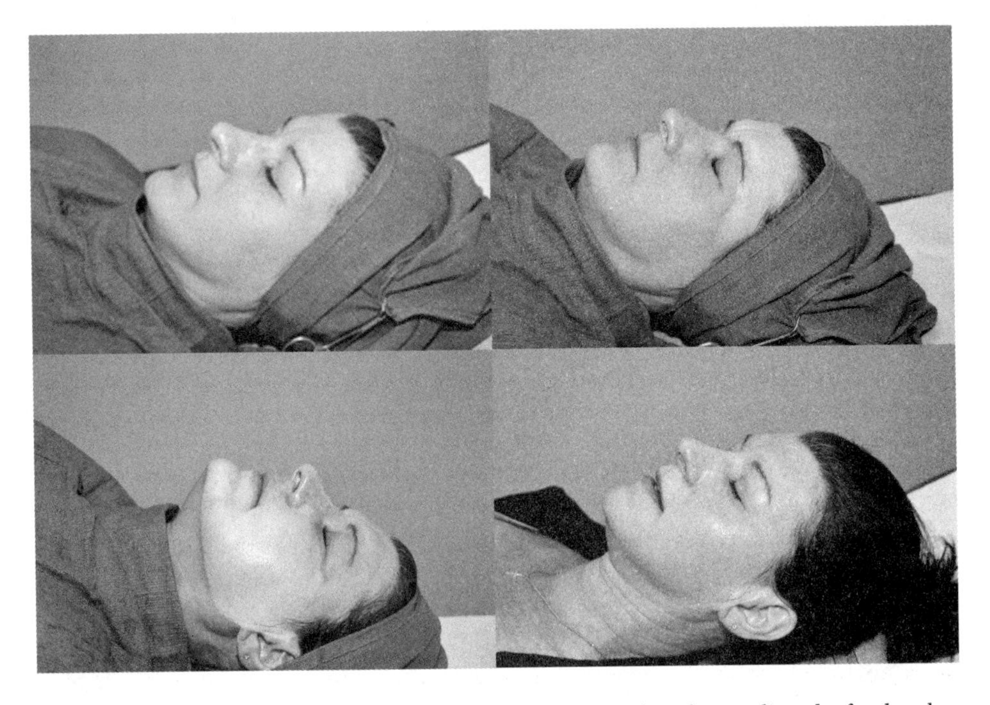

A TCA peel in progress. Upper left, before the peel. Upper right, after peeling the forehead; frosting is visible. Lower left, after peeling the mid and lower face, with the forehead already recovering. Lower right, face recovering. See color plates.

won't actually see any lightening of the skin, but the drug turns off brown pigment production in the skin. Without it, the pigment-producing cells can go wild after the peel, leading to splotchiness. These two drugs remind me of the chemicals used on lawns to get the grass all growing the same way!

The actual peel hurts. My wife usually swears at me during the peel and makes some weak threats. I often prescribe a mild narcotic such as codeine, and the anxiety drug Valium, to be taken an hour before the procedure. Prior to the peel, the face is washed with soap. I then remove the oils from the skin with nailpolish remover (acetone). I apply the TCA with a fine gauze pad, stroking the skin repeatedly until the chemical soaks in. I carefully observe the skin as the chemical causes frosting—a reaction of the acid with keratin, a protein in the skin.

I peel the forehead first, then allow the patient a few minutes until the pain lessens. Next, I sequentially peel the cheeks, the nose, the lips, and the chin. Last are the eyelids. The neck can be peeled with a lower concentration of TCA. Cool saline compresses are applied to the skin for about ten to fifteen minutes, until the frosting subsides. Once this happens, there is usually no more pain.

Now comes the fun part. The first evening after the peel, the skin looks reddened and sometimes gray. The face is washed twice a day and a moisturizer is applied. The next day, it looks dry; by the third day, the skin looks strikingly like that of a molting reptile.

By the fourth day after the peel, the skin begins to flake off, in much the same way a bad sunburn does. By the fifth day, the peeling is usually complete and makeup can be applied. Older people usually look red rather than gray. If the peel is purposely deepened, as is required for the treatment of wrinkles, healing can take as long as fourteen days.

The redness of the skin will persist for two to six weeks. After the skin has peeled, the patient is again given tretinoin and hydroquinone. The darker the skin color, the longer hydroquinone is used after the peel. I encourage the long-term use of tretinoin.

Any peel can cause scarring, if performed too deeply. And any peel can injure the cornea if splashed into the eye. A nurse has a bottle of eyewash in her hand during the procedure, in the event of any such accidental spillage.

Though TCA peels are superficial, bad results can still occur. Brown pigment can return after unprotected sun exposure (hyperpigmentation). I have seen this effect in olive-complected people, particularly if they scratch their skin during the weeks following the peel. Sometimes the peel will not fully remove brown pigment. In this case, it may be repeated in six to eight weeks.

I've seen more than one allergic reaction to various skin creams after the peel. The protective layer of the skin has been thinned, and creams that were previously tolerated may cause problems now. Itching is common. It is treated with Benadryl tablets and hydrocortisone cream.

Scratching is the curse of the peel. It can rip off the delicate new skin and lead to infection. Infection can cause scarring. Trim those long fingernails after a peel!

The effects last between one and two years, depending on the type of skin care that is used and the amount of sun damage sustained. If desired, the peel can then be repeated.

New Peels

Chemical peels are constantly being tinkered with. In one new peel, TCA is blended with glycolic acid. Other peels use a whole chemistry set. I'm sure there are many possible peeling agents and many combinations that really do work. But I'm wary of "new" peels that have not been thoroughly studied, for both short-term effects and long-term complications.

If You Can't Hide for a Week, Consider a Glycolic Acid Peel

With glycolic acid, a series of progressively stronger peels, applied for increasing amounts of time, can improve skin tone and color. The skin gradually acclimates to the acid, allowing its beneficial effects to occur without actual peeling. It might be called the no-peel peel! It gives the skin a reddish hue, similar to that simulated with makeup. Adult-type occasional acne is reduced.

With glycolic acid peels, you can literally go back to work the same day. They have subtle results at best, when used in low concentrations. In one study, patients who underwent six 20-percent glycolic peels on one side of the face and microdermabrasion on the other side thought the glycolic acid side looked better, but their doctors could not see a difference between the sides. In another study, the skin's appearance improved after five monthly 30-percent glycolic peels, but not quite as much as with one 35-percent TCA peel. Elasticity and water content increased with both treatments, but the TCA peel showed more improvement of wrinkles. After the study, patients chose to continue with glycolic acid peels, as opposed to TCA peels, presumably because less pain and downtime was involved.[4]

Glycolic peels are the "lunch-hour peels." But they are not really peels—true peeling would be considered a complication. During this office procedure, the skin is prepared with two weeks of home glycolic acid products.

No peel can be performed if the patient has used the drug Accutane or steroids in the past year. (Some people think the time frame should be two years.) Both drugs impair healing. Peels should be performed with great care, or not at all, on anyone with healing problems. Women who get herpes infections should take an antiviral such as Valtrex from the day before the peel until the skin has healed. Severe scarring will result if a herpes outbreak occurs before the skin has healed.

The peel begins after all traces of makeup have been removed. Some doctors first scrape the skin with a blade to remove the dead upper layers of skin. Then a 20-percent to 70-percent solution of glycolic acid is applied. In salons, no more than 10 percent should ever be used by aestheticians. The acid is applied for two to seven minutes. The actual peel stings a little. Since the goal of this type of peel is no downtime, I usually start with a low concentration of acid, applied for a short time. The peels are performed every two to four weeks, increasing either the concentration or the amount of time the acid is applied. The face is carefully observed while the acid is on the skin. If any swelling or blistering occurs, the peel is immediately stopped. While the goal is no peeling, blistering or flaking can occur.

This peel is a well-thought-out procedure. While the acid is relatively strong, small amounts are applied. The peel is stopped by spraying a weak base on the face, so weak that it could even be sprayed into the eyes without harm. When the acid is neutralized, a small amount of heat is produced, along with water and oxygen, causing the acid to foam. The face is sprayed with the neutralizer until there is no more foaming. A moisturizer is then used.

The face is slightly red for several hours, after which makeup can be applied. The skin occasionally blisters, although a skilled physician *gradually* increases the strength or duration of the peel to avoid this result. I've had a few patients who developed allergic reactions to the moisturizer used after the peel. Since the upper layers of epidermis have been removed, any chemical can better penetrate the skin: substances that were well tolerated with an intact skin can now cause allergies. My worst complication was in a woman who scratched her forehead after the peel had healed, causing a staphylococcus infection and scarring.

The latest twist on the glycolic peel is citrate peel boosters. Solutions containing some combination of citric acid, mandelic acid, the amino acid arginine, and the lighteners kojic acid and arbutin, pretreat the skin prior to the peel. At an American Academy of Dermatology meeting, texture, pigmentation, and acne improved after booster peels. The chemicals may indeed improve the peels, but the results have yet to be published.

Salicylic Acid Peels

In 1998, 30-percent salicylic acid peels were developed. This is a beta hydroxy acid dissolved in alcohol and water. These peels cause fading of pigment spots, decreased surface roughness, and reduction of fine lines. They are particularly effective when patients have acne, as salicylic acid is the chemical in many acne scrubs. However, the peels have never been particularly popular.[5]

Topical Aminolevulinic Acid

This drug is a relatively new addition to the skin care world. A solution of 20-percent aminolevulinic acid is painted onto the skin. The chemical is absorbed into the skin cells; precancerous and cancer cells absorb more of the chemical than normal cells. A specific type of light (such as intense pulsed light, Chapter 38) is shined on the face, purposely causing the production of free radicals that destroy the cells that are growing fastest—namely the bad ones. For about five days the skin is swollen and red. Afterward the skin looks better and has fewer precancerous growths.

Usually patients have two treatments, spaced a month apart. When compared with intense pulsed light alone, the addition of topical aminolevulenic acid significantly improves skin rejuvenation. Side effects include swelling, redness and flakiness of the skin.

The Cost

The average cost of a chemical peel is $850. Of course, this takes into account glycolic peels, which cost about $200 each, as well as TCA peels, at $900, and phenol peels, at $2,000. Laser peels cost $2,500 on average. Dermabrasion runs $1,400.

11

Botox

More than any other procedure, botulinum toxin type A, more commonly known as Botox, has come to represent the public's fascination with cosmetic surgery.

Originally identified as a toxin from *Clostridium botulinum* bacteria, Botox was first used in 1977 to treat strabismus, a problem with the eye muscles. The Food and Drug Administration approved it in 1989 for this purpose. When it was later used to treat uncontrolled muscle spasms in the face, patients had fewer wrinkles in the skin overlying the treated muscle. Doctors started using Botox for cosmetic reasons in 1996, and the FDA approved this use in 2002.

Botox Is Big Business

Sixty-five thousand Botox injections were performed in 1997. By 2005, there were 3.3 *million* injections in the United States alone. The manufacturer, Allergan, sold $831 million worth in 2005. This meteoric rise is due to a number of factors. First and foremost, the drug works. It does what it is supposed to do, with a minimum of associated problems. Combine this fact with the public's increasing desire to look young, the increase in lifespan, and a highly competitive business environment, and you have the formula for Botox's success.

Botox is the most common of the poisons produced by botulinum bacteria. It uses the type A toxin, while other companies market similar products with different toxins. Since this bacterium produces a soup of different toxins, and animals such as snakes also produce paralyzing toxins, expect competing products in the future to promise better or longer-lasting results. Sounds a little like snake oil!

Botulinum type B toxin, marketed as Myoblock, begins to work in three days, more quickly than Botox. However, it usually lasts less than ten weeks, whereas Botox lasts three to four months. Myoblock might be useful in people who are resistant to Botox or when *blended with Botox* to take advantage of both.

The Medicis Corporation will soon market Reloxin in the United States, Canada, Japan, and Europe. Called Dysport outside the United States, this drug will compete head-on with Botox. Perhaps then Allergan will lower the ever-increasing price of Botox.

Botox Is Most Commonly Used on the Wrinkles Between the Eyebrows

This area is called the *glabella.* The muscles between the eyebrows are called the *corrugators* and are useful for only one purpose: to create a fearful expression. Called vestigial muscles, they were at one time important because the scowling appearance scared away enemies, telegraphing an imminent attack. Let us hope that humans don't need these muscles anymore.

Botox is also used to eradicate the horizontal lines of the forehead. The injection must be placed in the middle portion of the forehead; otherwise, drooping of the brow will occur. Creative Botoxing can elevate eyebrows, evening out asymmetry. The crow's-feet area, to the side of the eye, is also commonly treated. Newer locations include the lines on the sides of the nose (called bunny lines) and dimpled chins.

Brave surgeons may inject the muscles around the mouth, to decrease the wrinkles of the lips and even to lower the upper lip in the case of a "gummy smile." But sixteen muscles around the mouth all contribute to the smile: the chance of injecting them exactly symmetrically is not high. Asymmetrical injections will make it look as if you have had a stroke. And overdoing the injections will cause speech problems, even drooling. It will affect kissing and playing the trumpet. I do not advise these injections. Some surgeons have begun to inject the neck to reduce the vertical bands. I do so only in select patients, women with early bands and relatively good skin quality. If done too aggressively, the injection can cause problems with swallowing. But if it works, it can stall the need for a neck lift for many years. In addition, some surgeons inject the muscles of the chin to smooth out a rumpled appearance.

When Botox is used in the upper portion of the face, it works in two dimensions. The forehead and eye injections are on the bone. The farther down the face

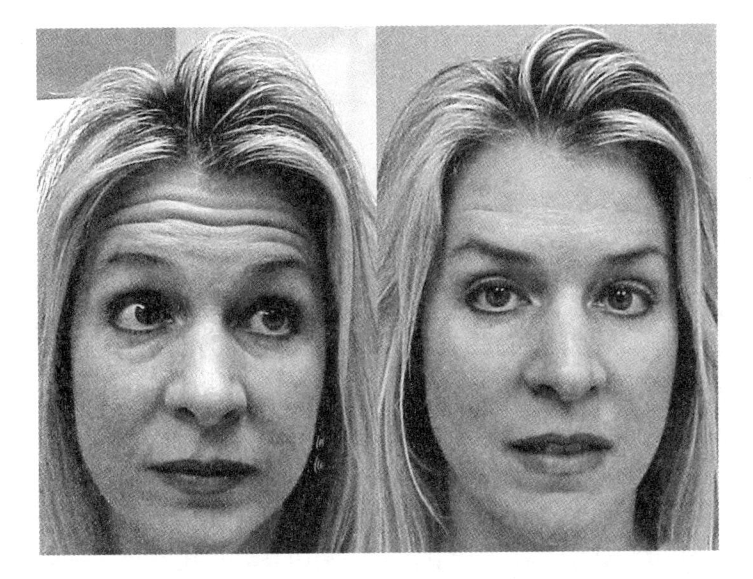

A 43-year-old woman attempting to raise her forehead before (left) and two weeks after Botox treatment (right). Note that her brows still work near the temples but are paralyzed in the central forehead.

it is used, the deeper it penetrates. It soaks nearly half an inch in all directions and when it spreads deep, its effects become three-dimensional—an interesting twist! Injection around the mouth and in the neck is very dicey.

A recent use of Botox is to balance the face in cases of facial paralysis. After face-lifting mishaps, Botox can be a lifesaver by balancing the facial muscles while waiting for function to return.

Botox Is Really a "Chemical Knife"

It takes a few days to see the "cut." Injection of Botox into the muscles causes paralysis. Wherever it is injected, the junctions between the nerves and the muscles are destroyed. When the muscles are not used, the wrinkles are not continually being re-created. Eventually the wrinkles actually "heal": the dermis gradually thickens and the wrinkles lessen.

To understand how this happens, consider stroke victims. The day after a stroke, the patient has all of her wrinkles but can't move one side of her face. A year later, the wrinkles on that side have softened and decreased. Botox does the same. Many people are disappointed when the wrinkles don't disappear overnight. Remember that we are talking about a biological process, similar to plants growing. But the body battles Botox and begins to repair itself soon after injection. Between two and six months later, the muscles begin to work again. It's time for another injection.

Long-Term Botox Use

The goal of long-term Botox is to keep muscles from repairing themselves between injections. After a nerve injury, a physical therapist's job is to keep muscles moving and electrically stimulated. The goal is to prevent atrophy of the muscles while awaiting regrowth of nerves. The muscles will begin to work when the nerves connect. It is the opposite with Botox. We want to beat down the muscles to the point where they can no longer recover.

After a few years of continuous use, the Botox effect persists longer in many people. By repeatedly hurting the muscles, and assuring that they don't recover between injections, we can destroy them. Again, this concept is the opposite of what physical therapists do for patients with nerve injuries. They electrically stimulate the muscles to try and keep as much function as possible so that when nerve regrowth occurs, the muscles haven't wasted away.

On the other hand, some patients who have used Botox for a long time become resistant to its effects, because of antibodies that effectively neutralize the Botox.

The Procedure

As for any surgical procedure, prior to Botox you should not use drugs that can increase bruising. Review the list of medications that you should avoid (Chapter 3). You should not have Botox if you are pregnant or if you have any neuromuscular disease, such as amyotrophic lateral sclerosis (ALS—Lou Gehrig's disease), Lambert-Eaton syndrome, or myasthenia gravis.

The Technique

Prepare yourself for an assault with six to thirty needle sticks. To do this, you may want to use a cream that numbs the skin. Several are available. I like EMLA cream, which is a mixture of the anesthetics lidocaine and prilocaine. Apply the cream thickly, like cream cheese, onto the skin. Do not let it dry out. To achieve this, you can cover the cream with Saran wrap. (You will look quite attractive when doing this.) The cream must be applied one hour prior to your procedure in order for it to work. Many of my patients come into the office and have my nurse apply the cream; they then sit alone and read until their procedure.

The surgeon removes the EMLA cream and cleans your skin with alcohol.

I use gel ice packs to cool the skin, providing a little more pain relief and also shrinking the blood vessels to decrease bruising.

The surgeon administers the Botox. A series of carefully placed injections deliver the toxin directly to the muscle. The zone of action is nearly half an inch on either side of the injection. If placed too low on the forehead, it can spread to the eye muscles and cause double vision and problems moving the eyelids and face. *Do not let a technician or nurse inject the drug.* Botox is a "chemoknife," destroying anything it touches. If your surgeon is too greedy to spend ten minutes to inject the Botox himself, find another doctor.

Proper injection of Botox is a true art form. Some patients require complete paralysis of a particular muscle while others just want a weakening of the muscle. Surgeons can "Botox" portions of a muscle, not necessarily the whole thing. They can look at stronger lines on one side of the face and weaken the muscles on that side of the face to create better symmetry.

The end result depends on many factors, including the injected dose. Botox can be used in concentrated or diluted solutions. It is not the amount of solution or the number of injections that is important—it is the number of *units* used. Botox is ridiculously expensive: I pay more than $500 for a vial of a hundred units. Unscrupulous doctors might use very dilute Botox to stretch out that quantity. Your result will be less profound if a lower dose is used. Beware.

After Your Surgery

Following the injection, you should stay still for about ten minutes to allow the liquid to be absorbed by the muscle. Then go home and be tranquil for the next two hours. Botox will have adhered to the muscle by then and should not spread. You can apply light makeup and go about your business. But don't exercise on the day of the injection.

You won't see immediate results. It takes between two and fourteen days for the Botox to work. I do have patients who have immediate results and are paralyzed before they leave the office. I hope they are not reading this, because they will learn that I believe instant paralysis to be physically impossible; it must be psychological.

Once the Botox takes effect, it lasts two to six months. You should allow the first injection to wear off before having another. That way you will know how long it lasts in *you.* After that first injection, don't let the effect fully wear off. The goal is to keep the muscle from working until it atrophies. We *want* to kill the muscle. We don't want it to recover. After about two years of continuous Botox use, the muscle takes progressively longer to recover. You can begin to space out your injections. I have patients who have two years of paralysis between injections! Eventually, it may be possible to stop the Botox when you have permanently destroyed the muscle.

The Risks

While Botox is incredibly safe, there are potential side effects. Bruising is the most common nuisance. Paralysis of adjacent muscles can occur, but is largely dependent on the skills of the doctor. If it occurs, double vision or drooping eyelids or brows may result. Headache, an influenza-like syndrome, and nausea are reported to be relatively common side effects, although I have rarely seen them.

Allergic reactions are exceedingly rare, but always possible. One patient died from an anaphylactic reaction to Botox mixed with lidocaine, underscoring the fact that Botox is a real drug with real potential complications. Its use should not be considered lightly, and it should be administered with the precautions necessary for all injected drugs.[1]

Botox can lose its effectiveness after multiple injections because some people develop antibodies to the drug. The Botox just seems to stop working, although no one is sure why.[2]

The most common problem is an uneven cosmetic result. A touch-up injection, done artistically, can resolve the issue. A tiny touch of Botox can make a huge difference.

In about 7 percent of people, the Botox doesn't work. Repeat injections in these individuals, performed a month later, usually do the job.

In 2004, four people who received botulinum toxin developed symptoms of botulism. When this story broke, plastic surgeons knew something strange had happened. It didn't make sense. Indeed, it turned out that the unlicensed osteopathic doctor who injected himself, his wife, and two friends had used industrial grade botulinum toxin—intended for research and not for clinical practice. The bottle that was injected contained ten million units of toxin, not the one hundred units contained in a bottle of Botox. One unit of Botox can *kill* a mouse. The lethal dose of Botox in humans is three thousand units, which is thirty times the recommended maximum dose. I guess that doctor missed the pharmacology lecture in medical school! He'll have plenty of time to study this topic now, since he was sentenced to three years in a federal prison.

In 2005 an Oregon physician and his nurse were indicted for using unapproved botulinum toxin in eight hundred patients. The only reason any doctor would use this unapproved drug is to save money. Veterinary-, research-, or foreign-grade materials are cheaper. About once a month, I used to receive a fax offering research-grade botulinum toxin. I always threw away the ad, wondering why the company would solicit a practicing surgeon rather than a researcher. I now know that the company banked on finding unethical, greedy doctors. I wonder how many used this toxin instead of Botox. The moral of this story is this: Choose your surgeon wisely—character matters in politics *and* in plastic surgery. For added safety, if you're unsure of your surgeon or if the fee is suspiciously low, ask to see the bottle before the Botox is drawn into the syringe.

Years of Botox Actually Costs Less than Muscle Removal Surgery

The major drawback of Botox is its short duration. Botox of the glabella costs about $500 and must be repeated three times a year, so over seven years a patient would spend $10,500. Some might think that surgery would be cheaper. However, a brow-lift with excision of the corrugator muscles costs about the same as the seven years of Botox. But if invested at 5 percent interest, the amount spent on the surgery would yield $500 per year, extending the use of Botox by another two or so years. By the nine years of Botox that you could afford with your surgical fee, your muscles might have regenerated and wrinkles might well have returned. For this reason I believe Botox is better than surgery. In Chapter 13 I will discuss this topic further.

New Uses

Hollywood will never be the same, now that Botox has been found to decrease underarm sweating. About a dozen injections will keep the underarms dry for several months. Sweating decreases by over 80 percent. Similarly, sweating on the palms of the hand and the soles of the feet can be lessened with Botox. The treatment is painful and fairly expensive.

People who grind their teeth may develop large muscles, called *masseters*. Injection of Botox into these muscles can decrease their size.[3]

Botox is currently in use for hundreds of other medical problems, including the treatment of migraine headaches and overactive bladders. It has truly become a chemical knife for many problems.

And now Botox has been used to lift breasts! The fact that this procedure is totally illogical doesn't stop some people from trying. It has even been injected into the wrinkles above the knees. While you may not be able to walk properly afterward, your knees will look beautiful . . .

The Ethics of Botox

Plastic surgeons, dermatologists, ophthalmologists, and otolaryngologists can perform Botox injections. For those doctors, as long as they have a thorough understanding of the anatomy of the facial muscles and have an artistic sense, Botox is easy. Other physicians, such as family doctors, nurses, physician assistants, dentists, and even podiatrists, have begun injecting Botox. I have even heard of cosmetologists injecting Botox. Let the buyer beware! These individuals have not operated on the muscles they are treating, and probably have not seen those muscles since anatomy lab, if ever. Besides, they have no formal training in facial aesthetics. I worry about poor results and complications caused by amateurs.

Taking advantage of the current fascination with Botox, unscrupulous physicians have even held parties with alcohol, food, and . . . Botox! Intimate groups of women gather at a neighborhood house, hotel room, or office and receive Botox from an enterprising doctor. There are even Botox bridal parties. The phenomenon of Botox parties shows how far some physicians will go to make money.

Medical procedures are private, not public, affairs. Procedures are performed for medical reasons, not entertainment. Botox should be injected in a medical facility, not a home or a hotel room. Very occasionally, patients develop fainting reactions. These can take place with blood drawing and do sometimes occur during Botox injections. Without proper medical equipment, this type of reaction can be life threatening. Doctors' offices are regulated by state licensing boards for good reasons. Rules for cleanliness, safety, record keeping, and privacy are necessary; they are moot when medicine is practiced outside of the office. In a group session, privacy is impossible.

Steer clear of grandstanding doctors who offer Botox in a home or a salon. Medicine and surgery belong in appropriate settings.

The Cost

Nationally, Botox averages $380 per injection per area. Larger areas such as the forehead cost more than smaller areas such as the crow's feet. Typical fees in the northeast United States are $500 for the crow's feet, glabella, or forehead, $400 for the bunny lines and over $1,000 for the neck. Remember that your surgeon is an artist. You can have a picture painted by an amateur or by a professional. In the end, you have two very different pictures. You get what you pay for.

12

Eyelid Lifts

Facial aging occurs not just in the skin, but in all the soft tissues. The skin loses its elasticity, and the connective tissues and muscles elongate and sag. As this happens, the skin actually grows and fat shifts—it appears in areas like the neck but disappears from the midcheek. The end result is sagging. The eyebrows head south, the upper eyelids droop, sometimes interfering with vision, and the lower lids drift downward, sometimes even pulling away from the eyeball. And as this happens, troughs appear below the eyes, and the skin of the midcheek mounds up between the nose and the mouth, creating nasolabial folds and marionette lines. And further down the cheek, tissue that once resided high in the face sinks down to form the jowls. In the neck, the loose skin fills with fat and becomes reminiscent of poultry as two stark bands become visible.

The constellation of sagging and stretching of the tissues is treated differently than the fine wrinkling in the skin. Here, we need bigger procedures that are designed to lift tissue and remove excess skin and fat.

The eyelids are the first area of the face to show signs of aging. By the time a person reaches 40 years old, she usually sees the first fine wrinkles around the eyes. With further aging, these wrinkles sag. The upper lids eventually droop over the eyelashes. Fat begins to protrude from both the upper and lower lids, creating what I call the "Mario Cuomo syndrome," after the former New York governor. Finally, by her 60s, the lower lid loosens and droops, causing the white of the eyeball to become visible and interfering with tear drainage.

Excess eyelid skin has been nipped and tucked for over two hundred years. In fact, the formal name for the procedure, *blepharoplasty* (meaning "change of shape of the eyelid") was coined in 1818. It wasn't until 1951 that fat removal was added to skin contouring. By removing extra skin and fat, the eyelids can be dramatically rejuvenated. Heavy upper eyelid skin, along with bags under the eyes, give the impression of a tired, old person. While the body can be covered up with clothing, the face cannot. No amount of makeup or hair can disguise aging eyelids.

In 2005 there were 231,000 blepharoplasties in the United States, making it the third most common cosmetic surgical operation. One in every seven people undergoing this procedure was a man, making it the second most common procedure for males. The potential benefits are great and the risks are relatively low. Almost all eyelid surgery can be performed under local anesthesia.

The Consultation

During your consultation, the doctor will note how much extra upper lid skin and fat you have. Some people also have fat behind the muscle of the eyelid. Called ROOF (*retroorbicularis oculi fat*), this fat was not recognized until 1990.[1] It is responsible for the fullness of the outer upper eyelid found in many people, particularly those who are overweight. When this fat is trimmed, the bones around the eye become more visible and beautiful.

The position of the eyelid on the eyeball needs to be assessed. If it is too low, a problem called ptosis exists. The eyelid should rest in a position halfway between the top of the iris (the pigmented area around the pupil) and the pupil (the center of the eyeball). When the muscles stretch out, causing ptosis, the lid sags and can interfere with vision. The surgical treatment involves shortening the muscle. Some plastic surgeons perform this type of surgery; many work together with a special type of ophthalmologist called an oculoplastic surgeon.

The lower lids are evaluated not only for skin and fat, but for elasticity. If the tone and support to the lid is poor, the position of the lid may drop. Patients with poor tone have lower lids that can pull far away from the eyeball. If not dealt with at the time of the lid lift, a pull-down of the lower lid, called an ectropion, may result. It looks bad, but, more important, can lead to a drying of the eyeball and infection, possibly even blindness.

Think of the eyelid as a pair of pants. If your pants have creases, you simply pull down on the legs to smooth them out. But if your belt is loose, your pants will fall down. Similarly, if you have wrinkles in your eyelid and you pull on the lid, the result can be a pull-down. By tightening the belt of the eyelid, we stabilize the lid and we can pull more aggressively. This type of belt tightening, called

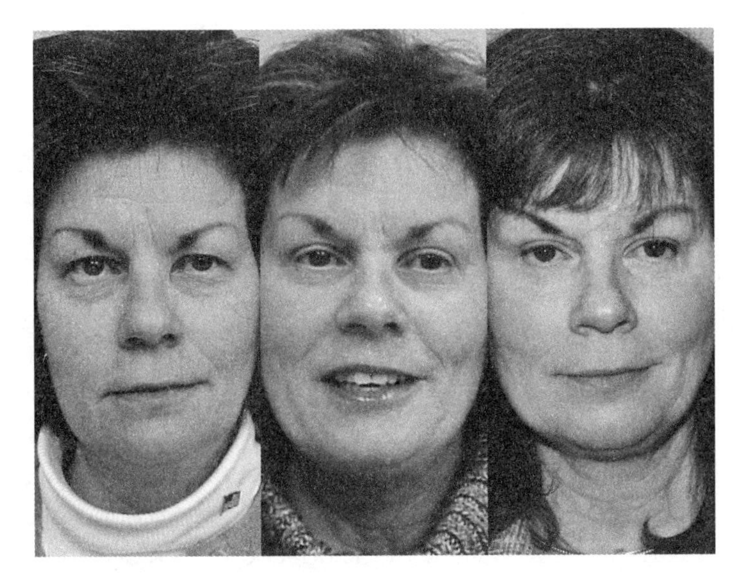

This 51-year-old woman underwent an upper and lower eyelid lift. On the left, before surgery she had heavy skin and ROOF fat. The middle photo shows her several months after her procedure. On the right she has undergone several years of Botox between her brows, decreasing vertical wrinkles.

a canthopexy, attaches the muscles of the lower eyelid to their original, higher position on the bones of the outer eye.

The importance of the lower lid belt has been appreciated only in the last decade. Ten years ago I rarely performed canthopexies; now I perform them on about half of my patients. The canthopexy allows the removal of more skin from the lower lid and decreases the chance of a pull-down.

During your consultation, the plastic surgeon will examine your lids and make these observations. He will take photographs and perform digital simulations.

An ophthalmologist, a medical doctor specializing in eyes, should do a complete examination prior to surgery. He'll evaluate your teardrop production and quality. If not enough tears are produced, dry eye problems are likely to follow the surgery. In this case, more conservative surgery may be performed. In fact, patients with dry eyes can undergo careful blepharoplasty.[2]

The Surgery

Blepharoplasties are usually performed under local anesthesia with intravenous sedation. Some are performed under straight local anesthesia, if the patient has

> In plastic surgery, it is often how much tissue is left in, rather than how much is removed, that determines whether there will be a good aesthetic result.

a high pain threshold. Some are performed under general anesthesia, if the patient wants to hear and see nothing.

In the preop area just before surgery, the surgeon marks the excess skin and fat of the eyelids. He may measure the distance from the eyelashes to the new crease. He will mark with the patient in both the lying and sitting positions.

The lids are numbed, steel contact lenses are put in place for protection, and the operation is performed. In an upper lid lift, there is not much debate about the procedure. An incision is made in the upper eyelid crease. The premarked crescent of skin is removed, along with a narrow strip of muscle.

The membrane that holds in the fat that surrounds the eye is then penetrated, exposing the two fat pads of the upper lid. This fat completely surrounds the eyeball and cushions it during normal activities and trauma. As we age, its encasing membrane weakens and the fat pushes outward. A judicious amount of this fat is now removed. This is the part of the operation that requires a sense of artistry. The most common error I see when evaluating the work of non-plastic surgeons is overremoval of this fat. Fat close to the nose is usually removed, but removal of too much fat from the middle of the eye can create a gaunt appearance. Once the fat around the eyeball is removed, ROOF fat—present in the outer eyelid, under the muscle but over the bone—is removed.

There are several techniques that are used in lower blepharoplasty.

Traditional Blepharoplasty

In this procedure, an incision is made in the skin just below the eyelashes. The skin is lifted, exposing the membrane that holds the three lower-lid fat pads. This membrane is penetrated, exposing the fat pads. Precisely the right amount of fat is removed from the lids to allow for a flat lower lid. We do not want bulging fat, nor do we want a sunken lid. Again, here is where artistry joins medicine. Finally, a 2–4 millimeter strip of excess skin is removed.

The Pinch Blepharoplasty

This older, simple technique of lower eyelid lifting is poised to make a comeback because of a 2005 paper in the plastic surgery literature. Fat is first removed through an incision *inside* the eyelid. A pinch of skin is then removed from the lower lid, without invading any deeper tissues or removing muscle. This technique has a lower chance of complications such as a pull-down of the lids. However, patients with very lax lids may still need a canthopexy.[3]

Laser Blepharoplasty

Some surgeons use the laser to perform this procedure. The laser is a tool, like a knife or a scissors. It can be used to perform nearly any operation, but most honest surgeons agree that it slows operations and adds risk. Lasers cause a heat injury that results in increased healing time. And when used in a blepharoplasty, it could perforate the eyeball. Surgeons who do laser blepharoplasties will admit in the locker room that they use it as a marketing ploy for their practice.

Transconjunctival Blepharoplasty

Instead of a skin incision, the cut is made inside the lower eyelid, usually with an electrocautery. A portion of the fat pads is removed. No visible scars and possibly a lower chance of lid pull-down are the advantages, but wrinkles and sags aren't helped. In my experience, this procedure is useful only for young patients with bulging fat and no extra skin.

The Canthopexy

A canthopexy is a simple procedure that tightens the lower lid and elevates its position. Usually one or two dissolvable or permanent stitches attach the muscle of the lower lid to a slightly higher position on the outer bony rim. No additional scars are created. If the lid is very loose, it may be necessary to perform a more radical tightening procedure, perhaps one that removes a wedge of eyelid skin or one that detaches and repositions the lower lid muscle from the bone.

On average, a simple blepharoplasty takes between two and two and a half hours. A canthopexy will add about half an hour to the procedure.

After the Surgery

The patient is watched for two hours in the recovery room. Her vision is checked repeatedly since the majority of bleeding complications and sudden blindness occur during this early period. For this reason, patients are kept tranquil with the head above the level of the heart. After this interval, the patient can be driven home. Ice packs help control swelling and bruising. Since bleeding can occur as late as three weeks following surgery, I recommend elevation of the head and not sleeping on the eyelids during that period.

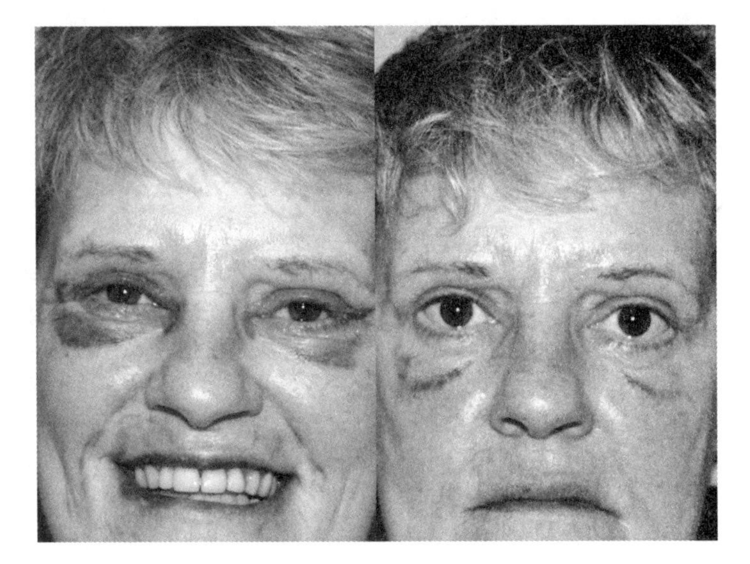

On the left, one week following an upper and lower eyelid lift, canthopexy, browpexy, and Restylane injections to lips and wrinkles around the mouth. On the right, two weeks postop. See color plates.

Oozing normally occurs during the first twenty-four hours following surgery. You'll have enough bruising to look as if you went nine rounds in a prizefight. The bruising and swelling peak between twenty-four and forty-eight hours and then rapidly settle down.

You will be seen one or two days after surgery. During this visit, the lids are cleaned with hydrogen peroxide, and ophthalmic-grade antibiotic ointment is placed on the incisions. Often you will be given a thick night ointment to protect your eyes until they can fully close. When I perform a canthopexy, I give the patient steroid eyedrops to keep swelling to a minimum. Sometimes temporary stitches are placed between the outer portion of the upper and lower eyelids for a few days to control swelling.

By ten days, most of the swelling and bruising has settled down and you can go back to work. If it is critical that no one know about the surgery, you might want to wait a full two weeks before returning to the public eye.

Festoons

Although they sound like something that might float along a canal in Venice, festoons actually are the secondary bags of skin just below the actual eyelids. They present a real problem in blepharoplasty. None of the dozens of techniques

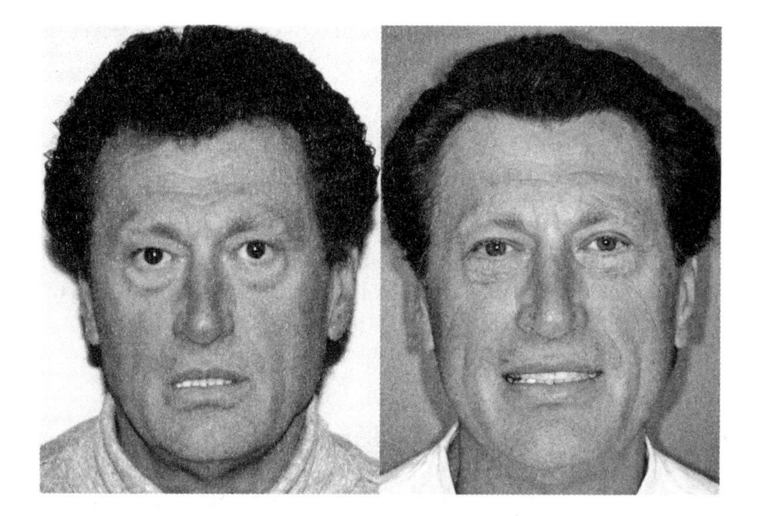

In the photo on the left, this 58-year-old man shows a tremendous amount of extra upper and lower eyelid skin and fat, as well as laxity of the lower lids. The photo on the right was taken six months following eyelid lift and canthopexy. I make sure to not overdo men's lids to avoid feminizing them. See color plates.

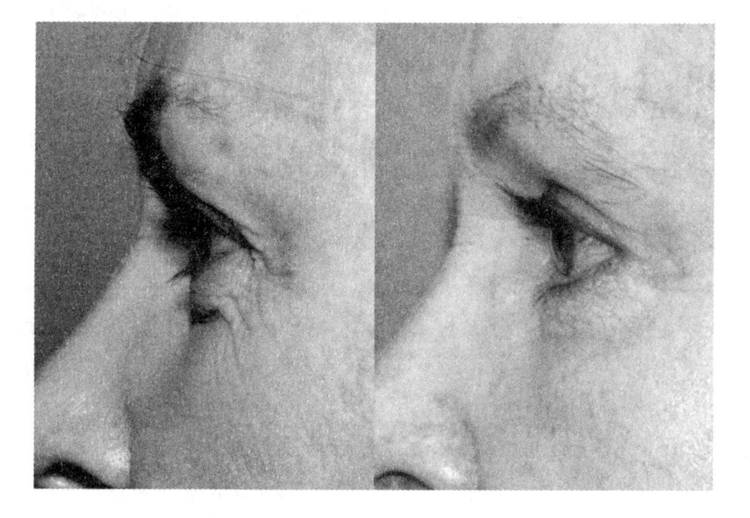

This woman had excess skin and fat of the lids. The photo on the left shows this extra skin. The right side shows the desired clean appearance of a feminine eyelid. See color plates.

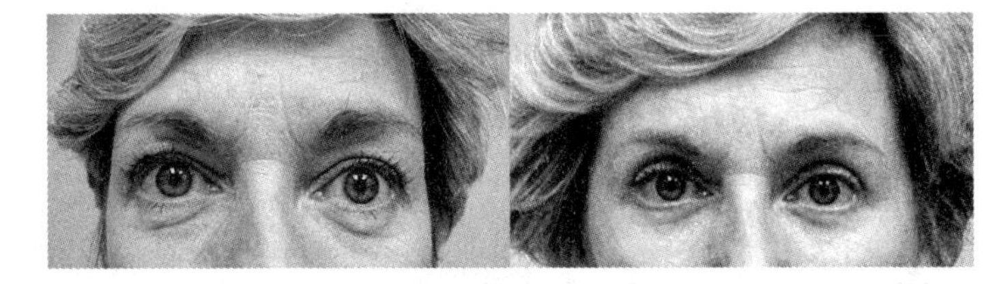

In a woman, the upper eyelid lift should remove all of the drooping skin, to allow for the application of makeup without smearing. This woman also had a lower lid lift with a canthopexy. Preop on the left, postop on the right. See color plates.

that purportedly fix them really work. Surgeons have suctioned, lasered, lifted, and excised festoons.

The Risks

As with all procedures, infection and bleeding are possible. Bleeding around the eye can be severe and can result in disaster. With blepharoplasty, there is a 1 in 10,000 chance of blindness following the surgery, usually after severe bleeding behind the eyeball. If the bleeding is recognized early, rapid action by the surgeon (using both surgery and medications) can salvage vision. This rare but devastating complication underscores that this is real surgery, not just television entertainment.

Another risk is development of a pull-down of the lid. This ectropion can be cosmetically disfiguring and can endanger vision. Most surgeons have had patients who developed this complication temporarily following surgery; most ectropions resolve within a few weeks. Some need help by the surgeon, perhaps a temporary sewing shut of the outer portion of the eyelid. In rare cases, additional, more invasive surgery is necessary. Because not all plastic surgeons can perform this type of surgery, the patient may be sent to a regional specialist, such as a plastic surgeon or ophthalmologist who handles these types of problems.

Symptoms of dry eye can appear after eyelid surgery. This problem is usually present before the surgery, but the removal of eyelid skin and the swelling that accompanies surgery can cause symptoms. Saltwater eyedrops during the day and a thick, protective ointment in the eyes at night can buy a few weeks until the swelling settles down. Dry eye problems are likely when eyelid lifting is performed in the first six months after surgery to correct vision (LASIK). The surgery must be performed carefully, if at all, during this early period.[4]

A rare complication of lower eyelid lifting is injury to the muscles that move the eyeball.

Cosmetic problems are due to the removal of too much or too little skin or fat. It is better to err on the conservative side, because more skin or fat can always be removed later on. It may be nearly impossible to put back skin or fat.

Common problems after eyelid surgery relate to stitches. The lids heal so quickly that little cysts can occur along the incisions, even if the stitches are removed by five days. The cysts are easily treated in the office by numbing the skin and snipping the cyst or overgrown suture tract. White scars are common in brown pigmented skin.

The Cost

Typical fees for a blepharoplasty range from $2,500 to $8,000, with the average being $2,800. Remember that with all surgical procedures, you will be charged additional fees for the facility ($1,000–$2,000 range), for the anesthesiologist ($700–$1,500 range), and for laboratory tests.

13

Brow-Lifts

The first brow-lifts in 1910 removed small amounts of forehead skin. By the 1920s, more aggressive procedures had been introduced. Since then, the popularity of the procedure has waxed and waned. In 2005 in the United States, 72,000 brow-lifts were performed, all but 7,000 in women.

The eyebrows must be evaluated together with the eyelids. If the eyebrow droops and only excess eyelid skin is removed, early recurrence is likely. In a woman, the eyebrow should be positioned *at or just above* the bony rim. In a man, it should be *at* the bony rim. A variety of brow shapes are considered appropriate. Most commonly, the side of the brow near the ear (*lateral*) arches slightly above the side closer to the nose (*medial*).

Be wary of a surgeon who wants to position your brow higher than these guidelines. One of the strangest things I have seen in plastic surgery is the public's willingness to go along with the aesthetic tastes of plastic surgeons. Eyebrows too high up the forehead only look good on Lieutenant Uhura of *Star Trek*. There is great variability in the desired position of the brows, but in nature only a rare human will have eyebrows much higher than the bony rim.

Many different techniques can be used to lift the brows. Older procedures placed incisions in front of the hairline, in the middle of the forehead, or above the eyebrows. None of these locations is now considered acceptable.

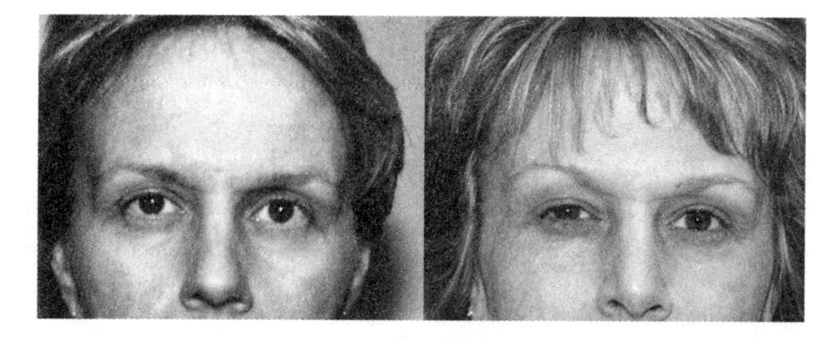

This 38-year-old had low brows and a prominent forehead. On the right, she is six years after a coronal brow-lift and burring down of the bone of the forehead, and four months after a rhinoplasty.

The Techniques

Coronal Brow Lifting

This is a tried and proven technique of brow lifting. A long incision is made inside the hairline, from ear to ear. The skin of the forehead is lifted up, elevating the brows. The muscles that cause frowning are usually trimmed, giving a long-lasting "Botox effect."

This procedure, although reliable, is decreasing in popularity because of the long incision with possible loss of hair and loss of sensation of the scalp.

Endoscopic Brow Lifting

An endoscope is an instrument with a camera on its end. It allows smaller incisions, since the operated area does not have to be seen directly. Instead, the operation is watched on a video screen adjacent to the patient. Hand-eye skills are different in endoscopic surgery, and surgeons who played a lot of video games as teens may have a learning advantage with this procedure. Through forehead and scalp incisions, the brow is lifted and suspended, usually with sutures and sometimes with a new Velcro-like device.

Although this procedure sounds very high tech, after learning how to perform the operation, many surgeons began using the smaller incisions and brow suspension techniques *without actually utilizing the endoscope!*

Interestingly, twenty-one New York plastic surgeons showed that the number of endoscopic brow-lifting procedures declined by 70 percent from 1997 to 2001. Only half of the surgeons were pleased with the results after two years. They reported complications such as brow asymmetry, loss of hair, hairline changes, loss of sensation, and injury to the important nerve that controls the forehead muscles. Backing up my contention that Botox is better than surgery for control over the frown lines between the brows, in this study over 70 percent of surgeons who performed endoscopic brow-lifts still injected Botox into the wrinkles after surgery! In most cases surgery simply doesn't work well enough. Probably the body is too efficient at repairing itself after surgeons injure it. It laces together remnants of muscle, allowing wrinkles to reappear even after surgery.[1]

The decline in popularity of the endoscopic brow-lift is an excellent example of surgeons and patients jumping onto the bandwagon of a new procedure before the risks, benefits, and long-term complications and results are known.

Browpexy

A browpexy lifts the eyebrow through an upper-eyelid incision. The brow is raised just above the level of the bone and physically sewn to a higher position on the forehead. Some surgeons do not believe that browpexies can be used for dramatic lifting and rely on them for more subtle refinements. I am a big believer in this technique; it is simple and it works.

A new variant of the browpexy is the Endotine TransBleph procedure, which uses an innovative dissolving Velcro-like device. After the brow has been lifted, the device is secured to a small drill hole made in the bone above the eye. The brow is physically lifted to a higher position, becoming impaled on the device. This procedure makes a lot of sense and uses hidden eyelid incisions.

Suture Lifting Techniques

This newer type of brow-lifting procedure does not dissect the tissue. Rather, tiny incisions are made in the scalp, and special barbed plastic sutures are embedded into the undersurface of the eyebrow. The suture is then lifted and embedded into the scalp, causing a lift of the brow. This technique shows much promise and has very low risk. Long-term results are unknown, however.

The Surgery

Most brow-lifts are performed under local anesthesia with sedation. The procedure takes thirty minutes to two hours, depending on the technique used. Sutures or staples are placed in the incisions. Some techniques remove skin. Some suspend the brow one way or another.

Aftercare

Swelling in the eyelids can be profound, but is usually short-lived. Most patients can go back to work in about a week.

The Risks

Depending on the technique, the risks include injury to the nerves that control sensation to the forehead and to the muscles that lift the brow and close the eyes. If the nerves are injured, the damage is usually temporary. Scars are inevitable with most techniques, but the majority are hidden in the scalp. Whenever an incision is made in the scalp, hair loss is possible. Asymmetry is also a problem: recurrence of the brow droop is common, and it may not occur on both sides. (Many, if not most, patients have eyebrow asymmetry normally.)

Longevity is the real question with brow-lifting. Recurrence of brow droop can occur as early as six months, or as late as ten years.

The type of brow-lifting procedure chosen depends largely on the preference of the surgeon. Every responsible surgeon learns many procedures and eventually settles on one that is effective and is reproducible. When deciding on a surgeon, the patient needs to determine what procedure is used and whether the risks of that particular surgery are justified.

The Cost

The cost of brow-lifting procedures varies greatly, depending on the technique. Endoscopic brow-lifts or Endotine brow-lifts and barbed sutures use expensive equipment and devices and cost more. The average fee for the procedure is $3,200, with a range from $2,000 to $8,000.

14

Face-Lifts and Lifting with Stitches

The thought of face-lifts conjures up images of Hollywood matrons, stretched so tight that their eyes and mouths are frozen like those of the Joker in *Batman.* Perhaps the most notorious of plastic surgical procedures, the procedure ranks seventh in popularity among cosmetic surgical procedures. There's something secretive about face-lifts, as if aging only happens to *some* people and reversing its changes is something to hide. In 2005 there were 150,000 face-lifts in the United States, all but 13,000 in women.

Face-lifts would be even more popular were it not for their long recovery and the relative high frequency of complications that seriously impair the quality of life.

The First Face-Lift Was Performed in 1901

The subsequent history is filled with intrigue. The first face-lifts were really just removals of a little skin in front of the ears. These procedures were very popular and were performed under local anesthesia in office operating rooms. Early surgeons were secretive about their techniques, to protect their turf from competitors. By 1920, incisions were similar to those we use today. With each passing decade, surgeons became more and more aggressive, lifting more, rearranging more, and removing more skin.

Until the 1980s, face-lifts tightened only skin. But all the soft tissue of the face sags as we age. These drooping fibrous and fatty tissues and the muscles of the neck began to be lifted as a second layer. The most common technique for face lifting, the SMAS lift, began in the 1980s. The SMAS, or superficial musculo-

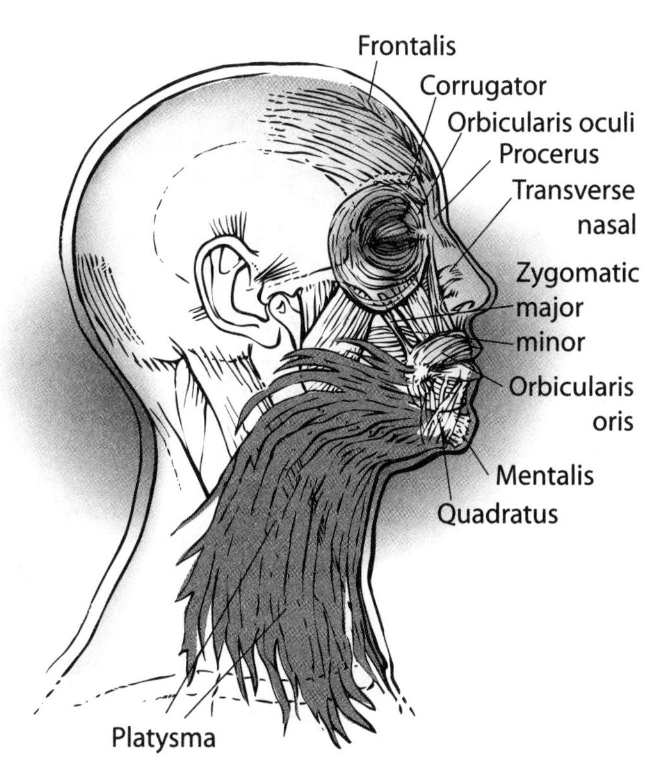

This drawing shows the complex array of muscles of the face.

aponeurotic system, is a collection of muscles and fibrous tissue stretching from the forehead to the neck. By lifting this deeper layer, surgeons can be more aggressive with the jowls and have longer-lasting results.

A traditional face-lift is not actually a face-lift—it is a *neck lift.* A face-lift lifts the sagging turkey gobbler of the neck, improves the "Katharine Hepburn" bands (*platysma*), and lifts the jowls. It does not help the nasolabial folds as much as we would like.

The face-lift does nothing for the fine wrinkles of the face. In fact, when the skin is pulled back toward the ears, existing wrinkles can have their direction changed, leading to an odd appearance.

So Many Face-Lifts, So Many Choices

An explosion of face-lifting procedures occurred in the 1990s. Different plastic surgeons began hawking their favorite procedures. Some favored the SMAS lift, some simply tightened the SMAS, and some removed a section and repaired the remaining tissue. Other plastic surgeons took a more complex approach, lifting the "deep planes" of the face, sometimes tightening muscles individually.

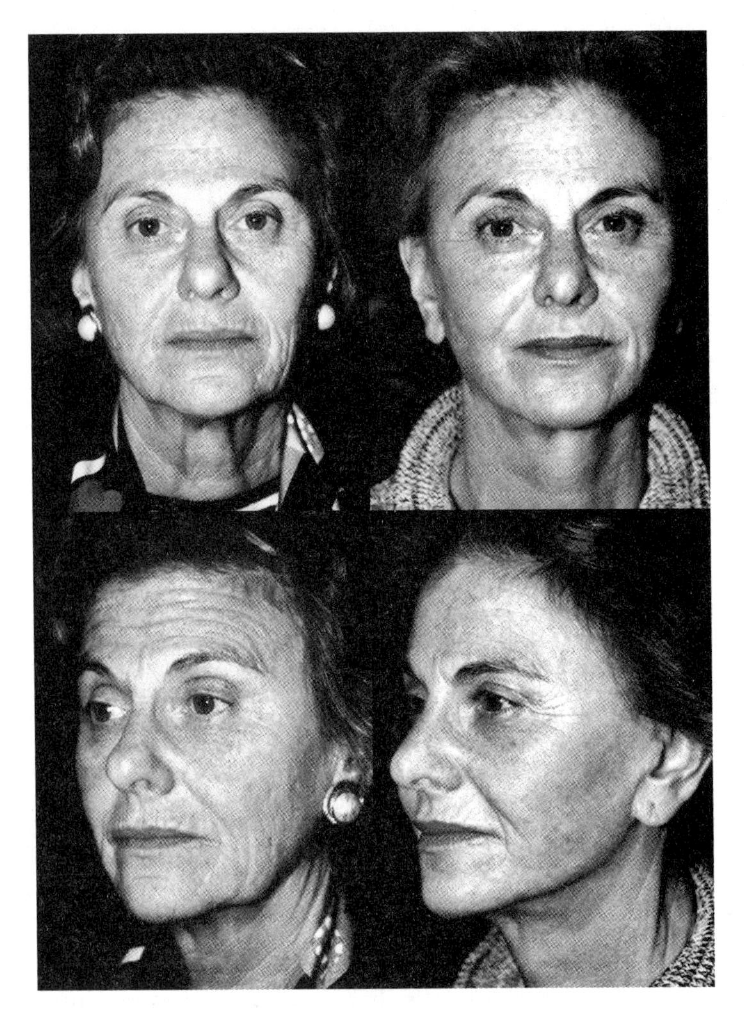

This 61-year-old woman had bands of the neck and jowls and wrinkles around the mouth. On the right, her appearance after face-lift and tip rhinoplasty. See color plates.

More aggressive surgeons advocated the "subperiosteal" face-lift, going all the way down to the bone. Coming full circle, other plastic surgeons felt this was overkill and reverted to the single-layer, skin-only face-lift.

While some plastic surgeons lift the tissue and reposition the fat volume of the face, others remove this tissue. Still others *add* fat to the face. Some cut the muscles of the neck to allow for an uplifting of the mouth. Others warn never to do this.

How can the public determine which procedure to have? For that matter, how are plastic surgeons to figure out which is the best procedure to use?

When Critical Plastic Surgeons Analyze These New Techniques, Many Reach an Interesting Conclusion: All of Them Work!

Whenever a plastic surgeon presents his photos of face-lifts at a national meeting, no matter what the technique, they all look good. But there's no "superbowl" of face-lifts, pitting one technique against another. This type of comparison is difficult, if not impossible to make. Humans have only one face, and we all age differently.

A bold group of surgeons at New York's Manhattan Eye, Ear, and Throat Hospital performed a standard two-layered face-lift on one side of the face, and a more extensive face-lift on the other. They found no difference between the sides and concluded that the more extensive procedures were not worth the increased risks. Paul Lorenc, M.D., F.A.C.S., New York plastic surgeon and one of the researchers, says thirty-two different face-lifting techniques are currently in common practice. He tries to analyze the scientific basis of each before altering his technique.[1]

In truth, many variants of face-lifts probably are equally effective. On the other hand, when so many different operations are proposed for the same problem, there are usually issues with all of them.

It turns out that the *art* portion of plastic surgery allows for enormous variation in techniques. Thoughtful plastic surgeons learn new procedures by attending plastic surgery meetings and courses and by reading journals. They then alter a procedure that they have already performed. They may first operate on a cadaver. Small new steps gradually evolve into a procedure that the surgeon feels comfortable with—one that can be performed without excessive risk.

The Techniques

In a 2004 poll of members by the American Society for Aesthetic Plastic Surgery, over 40 percent of plastic surgeons performed a two-layered face-lift and lifted the jowls by folding the SMAS tissue onto itself. Nearly 30 percent cut and lifted the SMAS, 20 percent lifted the various muscle layers with a deep-plane lift, and the rest lifted only the skin or lifted all the way to the bone (subperiosteal). Confusing? You bet. There are a lot of different ways to do things in plastic surgery. Patients should not choose a surgeon based on the procedure he performs. They should choose a surgeon based on his credentials and his *results*.

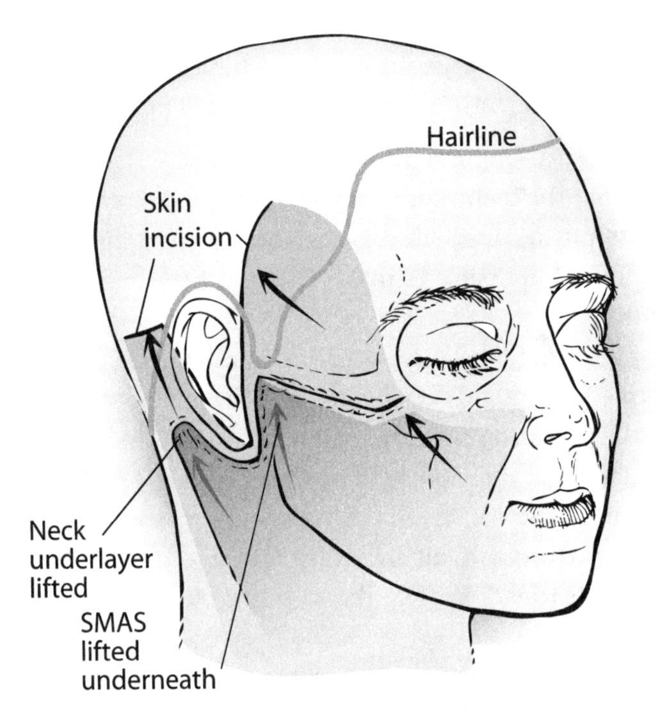

The typical two-layered face-lift. The skin incision goes in front of and behind the ear, and the second layer (SMAS) is lifted and pulled in a separate layer. The arrows show the direction of pull on the tissues.

The SMAS Lift

The incision begins within the hair above the eyebrow. It continues in front of the ear, goes inside the ear, returns to the front of the ear, continues behind the ear, and ends in the hair. The skin is lifted all the way to the nasolabial fold. Under the chin, the skin lifting connects with the lifted skin from the other side. Another incision is sometimes made under the chin. Like a scene out of a horror movie, the skin is lifted under the chin from ear to ear. Through a chin incision, fat can be suctioned and the bands of the neck can be tied together with long-lasting stitches. This portion of the operation is called a platysmaplasty, named for the two platysma muscles of the neck.

The lower layer of tissue, the SMAS, is then lifted. The skin is pulled as the surgeon watches the direction of the skin wrinkles and the color of the skin. Too tight, and the skin will die. Too loose, and the patient will complain. Strong stitches above and behind the ears secure the new face. A little tailoring is done, as extra skin is cut and discarded. A dazzling array of stitches and staples is

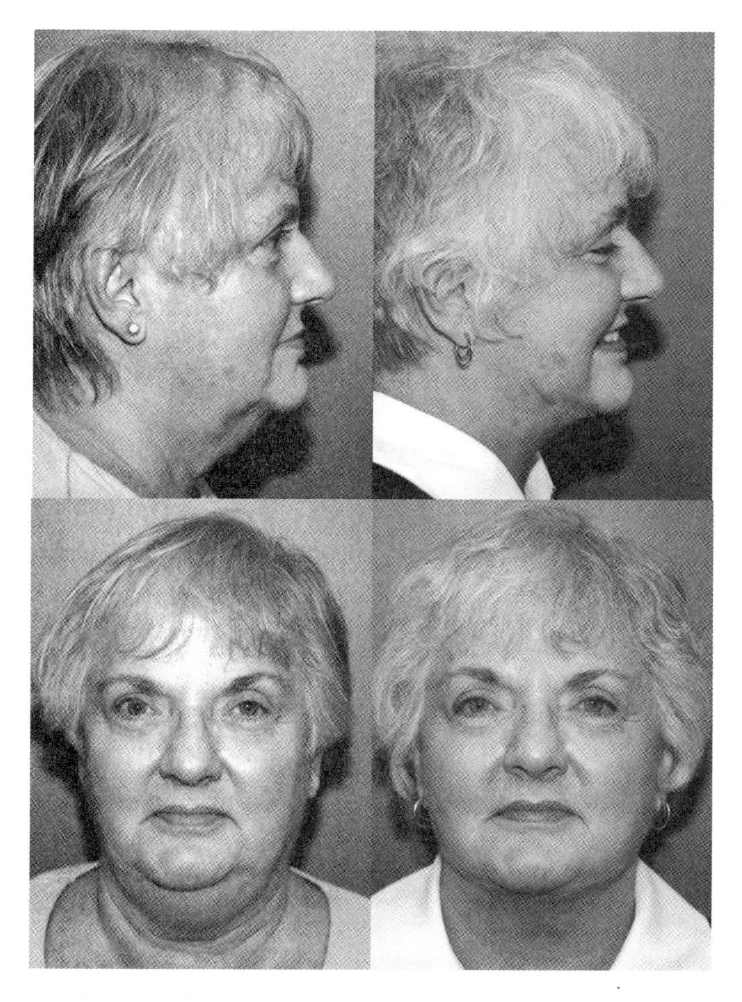

This 68-year-old woman had a large amount of neck fat and excess skin. A standard SMAS lift was performed, with removal of fat of the neck and repair of the neck muscles. Preop on the left, six months postop on the right. See color plates.

placed. In front of the ear, two layers of fine stitches are used. Dissolving stitches are used behind the ear, because the ear should not be pulled forward to remove stitches after surgery. The stretch of skin behind the ear connecting to the scalp has either Dermabond glue or buried stitches. This telltale area of the face-lift always shows cross-marks if stitches are placed there. Staples are used in the scalp, as they are gentler to the hair than stitches. After the second side is completed, the chin incision is closed and a dressing is placed.

The Art of the Lift

Plastic surgeons are really sculptors, using the human body as their medium. No one has defined this art better than the Washington, D.C., plastic surgeon William J. Little, M.D., F.A.C.S. Dr. Little has brilliantly advanced the face-lift into the next century. He says the lips sag as we age, causing less visible upper teeth and more visible lower teeth. His "stomapexy" fixes that. He also fills the dark circles under the eyelid with fat and builds up the cheekbones with lifted SMAS. His techniques are teaching the current generation of plastic surgeons the artistry of the face-lift.

Mini Face-Lifts

Whenever I hear the words "mini face-lift," I am reminded of a woman who is just a little pregnant. Sometimes a mini face-lift (S-lift) is a marketing ploy. It can be an attempt to attract patients who do not want a full face-lift but would settle for a smaller procedure. And certain patients truly are better served with a smaller operation.

Mini face-lifts use short incisions. The skin is not aggressively lifted. The incision loops around the ear. A small amount of skin is removed, limiting the amount of tightening. The underlying tissue usually is lifted. The procedure generally is used when there isn't a lot of extra skin. Some surgeons feel that liposuction and a tightening of the bands of the neck may accomplish the same goal.

New York plastic surgeon Daniel Baker, M.D., now performs his short scar face-lift in *most* patients, with compelling results. He has eliminated the most visible portion of the scar, which travels from the back of the ear into the hairline. He calls his new face-lift ponytail friendly. His signature face-lift removes a long section of the underlying SMAS and lifts the deep tissues of the face up and back. In very thin women, he just lifts and stitches the SMAS. While less skin is removed, he is able to redrape the excess skin.

The Endoscopic Face-Lift

Certainly if very expensive, high-tech equipment is used during a surgical procedure, it must be a better procedure. That's what some patients and even some plastic surgeons count on. The endoscope, a tube with a camera on its end, allows surgery through small incisions. The whole concept of "minimal incision surgery" depends on the endoscope. In abdominal and gynecologic surgery, scope procedures have become the preferred method in many cases. In plastic surgery, though, the endoscope is a machine searching for a reason for existence.

Consider the face-lift. The very reason for the operation is to remove the extra tissue in the neck. The only way to remove skin is to make big incisions and snip it out. Any seamstress knows that the only way to make a dress smaller is to cut out material and create a seam. It's the same with a face-lift. The use of tiny incisions in this operation is therefore oxymoronic, or mutually exclusive. And proving that endoscopes are not needed in plastic surgery, procedures with small incisions that were originally developed for the scope are being retooled into short incision procedures *without* the endoscope!

The Midface-Lift

Since the face-lift really is a neck-lift, what do we do with sagging cheeks and plummeting nasolabial folds? Plastic surgeons have focused for the last decade on these problems, creating the midface-lift or cheek-lift. Again, there are a huge number of variants of this procedure. By now the reader should realize that when there are many different procedures that treat a particular condition, none is usually fully successful. The most common midface-lift uses an incision under the eyelid. The substance of the cheek (*malar fat pad*) is then separated from the skin above and the bone below. This fibrous fat pad is lifted to a higher level. In theory, it sounds great. In practice, there are many complications.

The midface-lift hangs a lot of weight on the lower eyelid, threatening to pull it down. Because of this, the lid must be stabilized with an aggressive canthopexy. When extra skin gets pulled up to the area outside the eyelids near the scalp, an "alien" look can result. To fix this, extra skin is removed from the scalp.

The Endotine Midface Suspension

This Endotine device, approved by the FDA in 2003, is like a piece of Velcro placed on top of the cheekbone. Through incisions in the mouth or in the hairline, the cheek is lifted at the level of the bone and placed on the device. Its tines impale the cheek tissue, boosting it higher. The device dissolves after about a year, relying on scar tissue to hold the cheek in its new position.

The system seems brilliant. Of course, many new procedures and devices have flamed out over the long term. Only time will tell how popular the Endotine device will become.

Lifting with Barbed Sutures

The current buzz in plastic surgery circles is the Threadlift. Also called the Featherlift, or the one-stitch face-lift, it is an intriguing concept. Barbed stitches, like fishhooks, are embedded into the cheek fat. The stitches are physically lifted, ele-

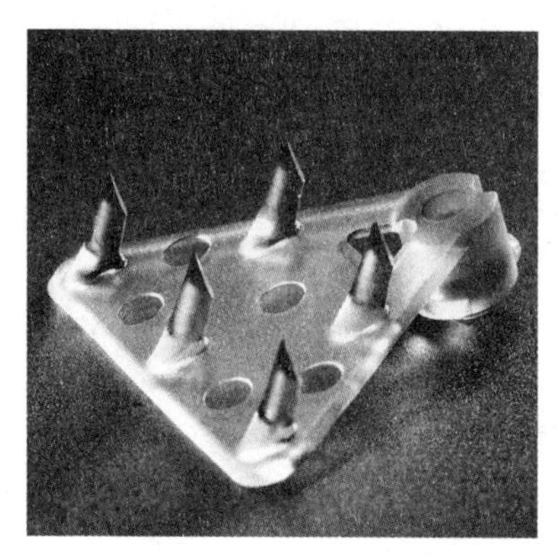

The Endotine unit. Photo courtesy of Coapt Systems.

vating the cheek. No tissue is cut, so the risks are low. Some surgeons use these barbed stitches to assist them in traditional face-lifts. The brow, the midcheek, and the jowl can all be lifted by means of several threads in each area.

While the concept is innovative, there are predictable problems. The fewer stitches that are placed, the more the skin will hang like a curtain between the stitches. To avoid this problem, a series of stitches has to be placed, lifting over a broad area. Other problems can be slippage or stretching of sutures. Eventually, a technique may or may not emerge that propels cosmetic surgery to another level.

An interesting lesson for skeptics is that proponents of these suture-lifting techniques are already forecasting how long the procedure will last. An article in *USA Today* said, "A threadlift typically lasts two to five years, but—depending on a patient's skin tone . . . it can hold for as long as a traditional face lift, up to 10 years." I want to know how anyone can make these predictions when the procedure itself wasn't even available until a few years ago. In fact, the longest follow-up that I know of is a year and a half.[2]

Barbed sutures can be placed with the patient under local anesthesia. In less than an hour, the threads can lift the jowls, the midcheeks and nasolabial folds, and the brows. Right after the procedure, the skin may bunch around the stitches, but it settles in about two weeks. The patient plays a large role in making this procedure work. If the skin is moved around before the sutures have scarred in, the stitches can pull through. This takes about three weeks. Any rolling onto the face or pulling on the cheek will destroy the results of this operation.

Complications are now starting to be reported, although the procedure is too new to know what surprises are in store. There has already been a report of the stitches cutting through the salivary gland duct. The stitches may work their way to the surface and become visible or infected. They may fail early and cause asymmetry. An asymmetry of the face is worse than a symmetrical drooping, and will certainly be a problem that many patients will see. The midface sutures must be placed perilously close to the important facial nerve. If this nerve is speared, twitching of the face or even paralysis will result.

The suture lifting procedure is sure to become popular, because there is an industry poised to create the special sutures and the special equipment that will be needed to place them. Competing designs are already lining up, even before scientific publications have determined which one works best, or at all. The Threadlift, Contour Threads, Aptos ("antiptosis") sutures, and Johnson & Johnson's Nuvance system have either just been approved by the FDA or are on the near horizon. Already a second-generation suture that can be hiked up a second time, years later, is being developed.

I believe that these barbed sutures will be extremely popular in men, in that they address a pent-up need. And they will be popular in the huge portion of the population that would have surgery, if it just didn't require surgery . . .

Like all new procedures, expect a learning curve by plastic surgeons and product developers. Also expect some early "pioneers" to have their faces plastered on television programs and advertisements. Because this technique is deceptively simple, expect to see family doctors, dentists, and even nurse practitioners taking weekend courses and trying to perform it. And then expect to hear about the inevitable complications and lawsuits.

The Platysmaplasty (Band-Lift™)

When bands in the neck are the only problem, a procedure called a platysmaplasty may be appropriate. No informal name yet exists for this operation, so I'll coin the term: band-lift.

In this operation, an incision is made under the chin and the skin is lifted all the way to the Adam's apple (*larynx*). It is lifted several inches wide of the midline. Fat is cleaned off the paired muscles of the neck. These muscles are fairly useless, allowing the subtle tightening of the skin when a man shaves, but just annoying women. The muscles are sewn together with long-term dissolving stitches, bringing them back toward the midline. The neck emerges with a more youthful appearance.

The band-lift will not help sagging skin or jowls, although it may be combined with liposuction of the jowls. Sooner or later the rest of the face will sag, and there may be no option other than a traditional face-lift.

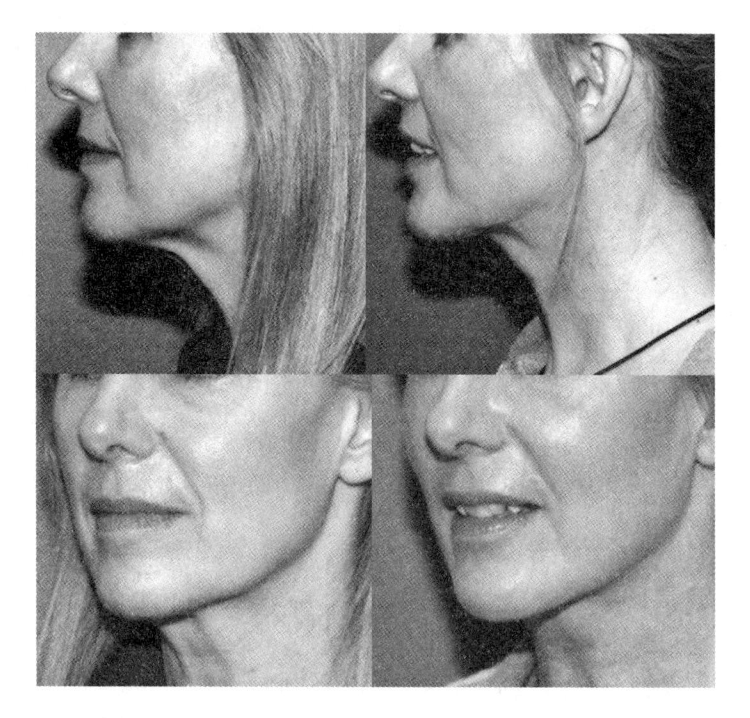

This 51-year-old woman underwent a band-lift to correct her neck banding. See color plates.

Direct Excision of the Nasolabial Folds

This is a powerful way to reduce deep nasolabial folds—better than face-lifting and more long-lasting than filling. Under local anesthesia, the folds are cut out as if they were moles.

The cheek fat can be slid under the fold and then the skin is brought together, using dissolving deep stitches and Dermabond glue on the surface. The technique relies on creating an exceptional scar that looks like a wrinkle. Certainly if a red, raised scar results, the patient will be troubled for months or years. Men with significant sun damage are the prime candidates for this procedure, because women's scars will be more noticeable.

Deep nasolabial folds trouble many people. Some patients ask for complete eradication. I point out that humans, unlike apes, normally have folds—check out any child.

Direct Excision of the Neck Skin

This seldom-used technique may be making a comeback. It can eliminate hanging neck skin, but it leaves a zigzag scar in the neck. Through this incision, fat can

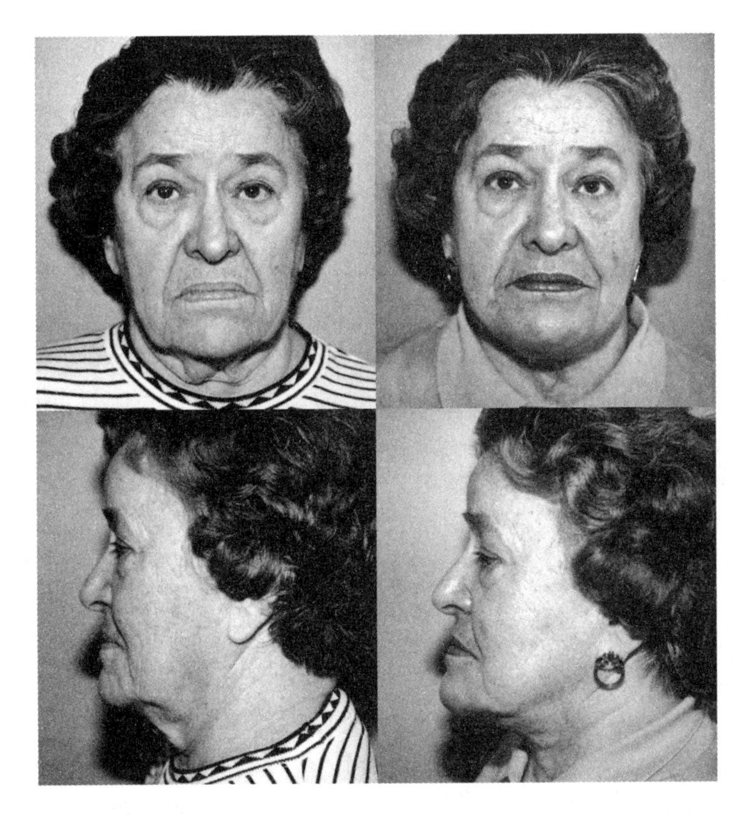

On the left, preop. On the right, following direct excision of neck skin, a band-lift, liposuction of the neck and jowls, and a deep chemical peel around the mouth. See color plates.

be removed and the neck bands eliminated. The procedure takes less time than a face- or neck-lift and can be performed under local anesthesia. Patients who are at high risk for face-lifting complications and are willing to accept noticeable neck scars are candidates. They must be willing to wear opaque makeup over the neck scar forever.[3]

Suture Suspension of the Neck

How compelling it would be to place two sutures strung from ear to ear under the chin, tightening until it looked just right. That is basically what suture suspension is. Only a few plastic surgeons use this technique. I have removed several of these stitches from other doctors' patients. If they work, they feel like a noose around the neck; if they aren't tight, they are useless.

The Procedure

Preparation for a face-lift is no different from that for any other surgical procedure. If your skin is very thin, your surgeon may try and plump it up with several months of Retin-A and glycolic acid prior to surgery.

On the day of the surgery, the surgeon will draw on your face with you in a sitting position. Once in the operating room, general anesthesia or intravenous sedation with local anesthesia will be administered, depending on the preference of the surgeon and the skills of the anesthesiologist. Some plastic surgeons prefer sedation, in that there is a lower chance of nausea. Nausea and vomiting are big problems after a face-lift and increase the chance of bleeding.

A solution of local anesthetic and epinephrine, to stop bleeding, is injected into the face. Incisions are made and the tissues are lifted, trimmed of excess skin, and sutured into place. Drains may be placed, in the hope of decreasing fluid collection after surgery. A bulky dressing is then applied, protecting the face and placing gentle pressure to decrease the collection of blood and serum.

The procedure takes between two and four hours. You'll spend a few hours in the recovery room before you are discharged to your room or home.

It is critically important to keep your head elevated above the level of your heart for three weeks following this surgery. If you bend to pick up something only *once,* or turn over onto your face in your sleep, you could have bleeding. It might range from excess bruising to major collection (*hematoma*) that requires reoperation as an emergency. You'll know that the bleeding is serious if you have new severe pain, rapid swelling, and perhaps nausea. The skin may blister and bleed.

The next day, the surgeon will change your dressings and clean your wounds. A lighter dressing will be placed. Sometimes no dressing is needed by this time. You may shower on the following day.

After the face-lift you will look like a victim of assault and battery. You will be swollen and bruised beyond recognition. The swelling peaks at thirty-six to forty-eight hours and then begins to subside. The surgeon will remove some sutures five to seven days after surgery and the rest by two weeks. By ten days, you will be able to shop for groceries in the next county. By two weeks, your swelling and bruising will have settled down to the point where you can go about your business, with some camouflaging makeup.

Your swelling will continue to settle for months after surgery. While the swelling resolves, scarring starts up. It occurs not only in the incisions around the ears and chin, but also underneath the skin, in the interface between the

Dr. Perry's Tips for Not Turning Over When Sleeping

- Use a U-shaped travel pillow or a pillow with arms—a reminder not to turn.
- Sleep in a reclining chair—it makes it hard for you to turn over.
- Try self-hypnosis: five minutes in a dark, quiet room, intensely focused on not turning over.

lifted skin and the underlying tissue. Scarring increases and peaks at about three months, then takes the next nine months to settle. During this time the nerves will grow back, causing funny electric sensations in your cheeks. Small lumps are common and require massage.

Your surgeon will see you in three weeks and then again three months after surgery. Additional visits may be scheduled at six months or a year.

During the months following surgery, you should wear cover-up on the incisions. A visit to an aesthetician for a makeup lesson is well worth the expense. Sunblock is crucial during the first year after your surgery.

The Risks

A multitude of potential complications can occur with face-lifts. Infection and bleeding can take place. Because the open area under the lifted skin is so large, a great deal of blood can collect. In fact, this hematoma can get big enough to destroy the skin.

Hematomas usually occur in the first twenty-four hours following surgery, but have been reported for up to three weeks. When they are diagnosed, the patient returns to the operating room as rapidly as possible. Some plastic surgeons think that drains, or tubes under the skin, can decrease hematomas, but others do not believe in them. A new technique sprays the tissue with human glue, fibrin. It doesn't decrease the chance of a hematoma, however, and the increased cost and the theoretical chance of contracting an infectious disease from a human product are the downsides. Hematomas are more likely in men than in women. Hypertension is a major cause of this problem and must be carefully controlled before, during, and after surgery.[4]

Scars will, of course, result from this procedure. The quality of the scar depends on the skill of the surgeon and the genes of the patient. Certain spots are notorious for bad scarring. One is the span behind the ear, extending into the hairline. We try to not use sutures in the skin in this area, relying on deep stitches and Dermabond instead. Sutures leave unattractive, usually white, cross-marks.

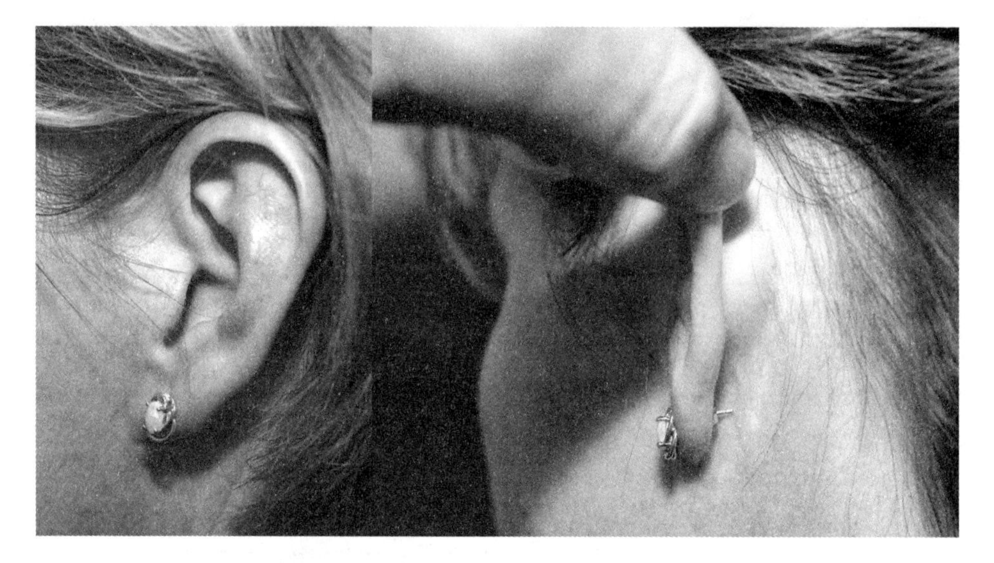

This woman had a prior face-lift by another surgeon. She was left with an unattractive scar. These photos show her appearance after a revision face-lift. Note that the scars are hard to find, even without makeup. See color plates.

It's a "Battle Between Beauty and Blood Supply"

Dr. John Converse, former chairman of plastic surgery at New York University and one of the great innovators in plastic surgery, taught this to a generation of plastic surgeons. Many people wonder why their skin can't be tighter after a lift. They often contort their faces and necks in an effort to show me their extra skin. I spend much of my life explaining that we can only lift a certain amount. The harder we pull on the skin, the better the neck and face can look. But the more we pull, the more we damage the blood supply to the skin. First, the skin turns white, then it blisters, then it turns black. Then it is replaced by scar. *Not pretty.* The seasoned surgeon knows exactly how tight he can pull without destroying the skin.

Several important nerves of the face can be injured during a face-lift. Those that control the muscles of the lower mouth, and another nerve that controls the forehead and the eyelids, are at risk. Plastic surgery residents spend a great deal of time learning anatomy, to be able to protect these nerves. They learn where these nerves are located in 95 percent of people. Unfortunately, the other 5 percent of people don't follow the rules. Because of this, nerve injuries can occur during face-lifts with even the best surgeons. The deep-plane and subperiosteal face-lifts are more likely to cause nerve injuries, however. While most such injuries are temporary, permanent paralysis can occur.

The nerve that supplies sensation to the earlobe also can be injured in a face-lift. If it is cut, the bottom third of the ear will be numb. This is not such a big deal, but it can make the placement of earrings difficult. When this nerve is cut, it often grows back.

How Long the Face-Lift Lasts

Some women believe that if they have a face-lift, they are doomed to repeat the operation in a few years. In fact, a lift lasts between five and ten years, with seven years being the average. By then some of the changes of aging will have returned. Surgery for facial rejuvenation is an ongoing process, with periodic maintenance. And ten years later a woman who has had a face-lift will be better off than if she had not. As New York plastic surgeon Darrick E. Antell, M.D., says, "This means the face-lift really lasts a lifetime."

Face-Lifts in Men

Thirteen thousand men had face-lifts in 2005, a small fraction of the number in women. In all modern societies men do not undergo as much cosmetic surgery as women. Among the very practical reasons why men don't have face-lifts is that they simply do not like to wear makeup. Cosmetics are required for weeks and even months after face-lifts. Scars in front of and behind the ears are easily covered by women with makeup and hair. Men are reluctant to use makeup and usually do not have hair long enough to cover the scars.

In addition, face-lifts remove the hairless skin between the sideburn and the ear giving men a bizarre appearance. With the advent of laser hair removal, this may be less of an issue in the future, however.

The Acupuncture Face-Lift

How could sticking dozens of needles into a person's face improve its appearance? As reported in the "definitive medical journal," *USA Today,* cosmetic acupuncture is the latest evidence that people will do just about anything to look younger. Some acupuncturists are charging $1,200 to $1,800 for twelve treatments over five weeks. Each session takes about an hour. After the course of therapy, monthly maintenance sessions are necessary.[5]

Acupuncture seems to work for certain pain syndromes. Apart from relieving pain and stress, there is no evidence that it can make you look better. Acupuncture is yet another example of wishful thinking that gets in the way of science and common sense.

The Cost

Face-lifts vary in cost, depending on the type of lift and the reputation of the surgeon. While these procedures average $6,300, they range from $4,000 to nearly $20,000. One might think short-scar face-lifts would be less expensive. In actuality, they are more difficult to perform and the costs are about the same as for traditional lifts.

There is great variability in the cost of face-lifts. In fact, it is hard to compare surgeons. There are many different types of lifts, and some surgeons add other procedures. They may dermabrade or laser the wrinkles around the mouth, light peel the skin to improve color, and fill the nasolabial folds with fat. The bands of the neck may be treated or fat may be removed. When you are comparing surgeons, be sure you understand precisely what they plan to do in the operating room to address *your* concerns. Then compare apples to apples.

15

Fat Grafting

By the time a patient hits her late 30s, aging begins to show. By her 40s, the eyelids wrinkle and the jowls make their first appearance, breaking up the smooth jaw line. The nasolabial folds (the creases between the upper lip and the cheeks) deepen. These are actually present from birth; however, they noticeably deepen during the 40s, bringing attention to the aging process.

In many people, the nasolabial folds also have actual wrinkles at their bases. Their shadows accentuate the folds. The visibility of wrinkles increases with sun damage.

When nasolabial folds extend downward below the lips, they are called marionette lines, aptly named for the wooden puppets with movable jaws. These are particularly distressing since, unlike the nasolabial folds, they are not present in children.

"Pulling procedures" such as face-lifts don't have any dramatic effects on the folds, since the point of pull, in front of the ear, is nearly four inches from the fold. Because the skin stretches, pulling only moves the fold a fraction of an inch. As the skin stretches out, the fold reappears within six months. More aggressive procedures, such as the midface-lift, pull from the lower eyelids, closer to the fold. But they have a high complication rate and require a three-week recovery from the huge swelling.

The closer the incision is to the fold, the more effectively the fold is lessened. The extreme is direct removal, in which the fold along with some skin is excised and the skin is closed. Some plastic surgeons slide in some cheek fat to fill the fold. Dissolving stitches are placed under the skin and the upper layer of skin

This 13-year-old has nasolabial folds, but they are not associated with aging until age 30.

is repaired with Dermabond or very fine stitches. The results are dramatic, but patients must be prepared for the scar that replaces the folds. In most people the scar will heal and will not be visible, particularly with makeup. This procedure is more common in men, because the scar blends with the rough skin texture.

Peeling procedures such as deep chemical peels, lasering, or dermabrasion have minimal effect on the folds. However, when the folds are filled from below *and* shrunk from above with peeling, the results may be dramatic.

Temporary fixes include filling the folds with chemicals such as collagen, Restylane, or Radiesse. When a more permanent result is desired, fat grafting should be considered. Fat is an ideal filling material for the nasolabial fold. Fat grafting has been around since 1893, with chunks of fat cut out of the hips and inserted into skin depressions. To get more fat to survive, in 1944 the fat was moved together with the attached skin. By 2005, fat grafting procedures were performed on 91,000 patients.

The Technique

Modern grafting uses fat from any convenient location. Most people have plenty to give up. I prefer the abdomen or the hip. I make a quarter-inch incision inside the belly button or under the bathing suit line. I numb the fat and use tiny tubes to extract it in a manner similar to liposuction. Several teaspoons of fat are all that is needed. The blood and local anesthetic are then removed by placing the fat on an absorbent Telfa gauze pad. The Jello-like fat is scooped up and placed

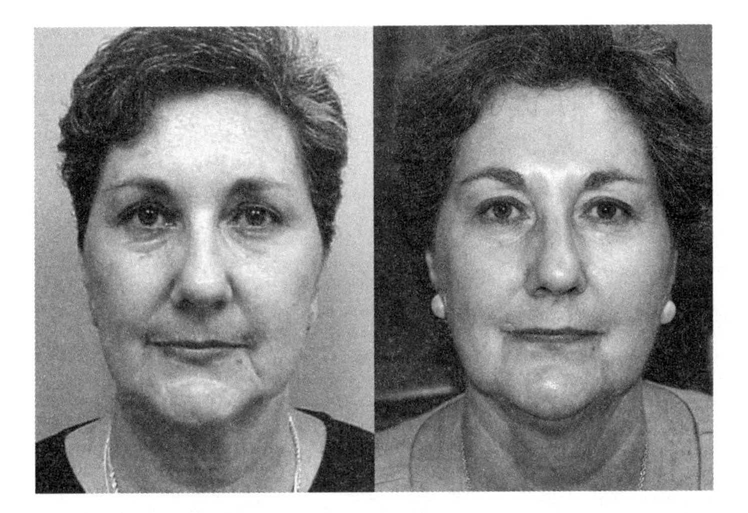

This woman underwent a laser peel of the face and two fat grafting procedures to lessen her marionette lines. On the right, her appearance after three and a half years. (Note that there is a slight lighting difference.) See color plates.

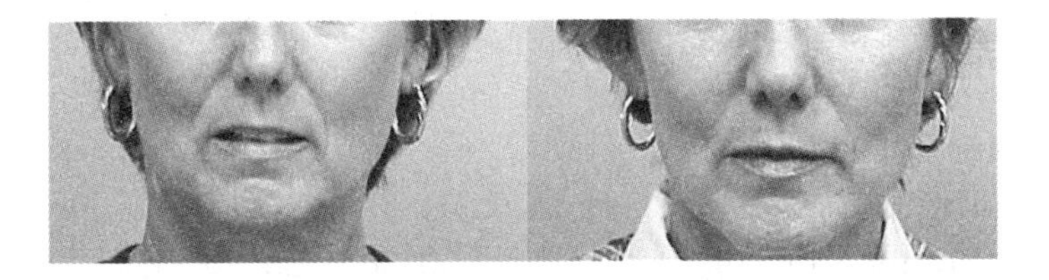

This woman underwent fat grafting to the nasolabial fold.

into small syringes. A blunt-tipped needle on the syringe breaks up the fat in and around the nasolabial fold and marionette lines. This step is crucial, for it creates a space for the fat to live. The technique not only adds fat to the fold, but also redistributes the adjacent "shoulders" of fat into the fold. The harvested fat is then distributed throughout the tissues under the fold. Part of the benefit probably comes from scarring underneath the skin, which may shrink the fold.

The incisions for fat grafting can be made in the fold, at the corner of the nose, or inside the lips. The cuts are usually only about an eighth of an inch in length and are repaired with one or two stitches.

The Results of Fat Grafting Are Highly Variable

Some plastic surgeons do not believe it works; others swear by it. I am a strong believer in fat grafting.

Very few studies are available to guide the clinician. The variability in this technique comes from many factors. From the surgeon's perspective, dozens of steps in the harvesting and injecting of the fat can affect its survival. The force of suction, the method of cleansing the cells, and the amount of time the fat sits outside the body before being reinjected all make a difference.

And then there's the patient. If the procedure were performed on rats, the animals would all be bandaged, restrained, and controlled in cages. As soon as human patients leave the operating room, though, they are out of my control. Some patients simply cannot keep their fingers off the newly operated area. They need to feel the fat that was injected. This constant touching of the fat probably destroys it.

Fat is very delicate. Like any graft, it takes about three days to get its blood supply from the surrounding tissue. Once blood vessels grow into it, it will be exceedingly fragile for about three weeks. After that time, scar tissue grows into and stabilizes the fat. Think of fat like a seed. When planted, the seed can be handled roughly. Once it germinates, the new seedling can be killed with even a light touch. Eventually it will become durable.

If the fat is present at six months, it should be there forever. The fat behaves as if it still existed in the area from which it was taken. If you lose weight in the abdomen, you'll lose volume in the nasolabial fold.

Dr. Louis Bucky, plastic surgeon at the University of Pennsylvania, has shown that the harvested fat really does survive in both experiments and patients. It is not inert like injected collagen or Restylane. About half of the fat survives—more if it is cut out of the body in pieces and transplanted than if it is suctioned out and reinjected.[1]

Fat can be used to fill in nearly any depression. It can fill the space between the chin and the jowls, smoothing out the jaw line. It can enlarge small earlobes, fill in the depressions in the lower eyelid and sunken cheeks, and plump crumpled chins. Fat can also be used to fix depressions created after liposuction and can make buttocks larger.

The Surgiwire

Get ready for this one. The Georgia plastic surgeon Miles Graivier, M.D., has come up with the Surgiwire incisionless dissector, a brilliant device that lifts the skin through two tiny puncture wounds. The Surgiwire is a very thin wire that saws through tissue, separating the skin from the underlying fat. Lifted nasolabial folds or other scarred areas are usually then plumped up with fat.

This technique wins my award for the simplest new technique that really works. Expect it to boom in popularity.

Fat in Aging Hands

The hands are the only area of the body other than the face that cannot be hidden beneath clothing. The skin thins and gets splotchy as aging hands lose fat, exposing veins and tendons.

Fat grafting can fill in the layer between the skin and the tendons of the hand. While not a common procedure, if done properly it can rejuvenate the hands. I would be hesitant to undergo this procedure myself because if infection were to occur, or skin were to be lost, a devastating loss of hand function would result. Besides fat, doctors are starting to inject Restylane and Sculptra into hands. These injectables are probably safer.[2]

Like the face, sun-damaged hands can be helped with the use of tretinoin or alpha hydroxy acids. Light chemical peels with the chemical trichloroacetic acid can lighten brown spots. Resistant brown spots can be reduced with laser or intense pulsed light (IPL) devices.

Some people have gone to the extreme of removing veins from their hands to improve their appearance. In the latest twist, a laser is actually placed inside the veins, frying them. The stakes are high; if anything goes wrong, the hands could be permanently crippled.[3]

The entire field of hand cosmetic surgery is very new and to date few papers have appeared in the literature.

Banking on Fat—A Word of Caution

When removing fat for a grafting procedure, the surgeon may be tempted to take a little extra and store it for future use. In fact, many plastic surgeons and dermatologists are doing exactly that. This scares the heck out of me. There are two good reasons *not* to do so.

First, a prime advantage of fat over synthetic fillers is the possibility of permanence. When fat is frozen for later use, most of it is killed, as has been proven in mice. Ice forms inside the cells and explodes them, in much the same way a pipe bursts when it freezes. Storing and freezing fat cancels the advantage of fat over synthetic fillers. And bacteria may grow in the fat, creating a dangerous situation when it is reinjected.

Second, *only licensed tissue banks should store tissue.* There is always a chance of mixing your fat with someone else's. Consider blood banking: When you donate blood, the blood is labeled, bar coded and tracked while it is typed and tested for disease. An intensive identification process goes on from donation to transfusion, all with the goal of preventing mixups of blood. Yet each year, some blood does get mixed up and is transfused into the wrong patient. In a two-year period in the mid-1990s, for instance, the United Kingdom saw 191 cases of incorrect transfusion of blood to patients.

When fat is stored in an average plastic surgeon's office, a technician, secretary, or nurse scribbles the fat donor's name onto a container. It is placed on the shelf in a refrigerator without stringent temperature controls, emergency power backup, antibiotics, or computerized identification. The chance of a mishap—the wrong fat being injected into the wrong patient—is real. If your surgeon saves your fat for injection at a later date, ask if he follows the regulations of his country's Association of Tissue Banks and is accredited by them.

The issue goes even further. The U.S. Food and Drug Administration (FDA) requires tissue-banking facilities to perform medical screening and infectious disease testing. Important rules must be followed by such facilities. I know of no *office* facility where fat is banked that follows FDA regulations or is accredited by the country's Association of Tissue Banks.[4]

My final word is, don't let your doctor store your fat for later use. In fact, I'll go one step further and recommend steering clear of any doctor who even suggests banking your fat.

The Cost

The average cost for fat grafting is $1,400, although this amount is extremely variable and depends on how much fat is used and the number of areas into which it is injected.

16

Liposuction of the Neck and Jowls

Of all the cosmetic surgical procedures, perhaps the biggest "bang for the buck" is liposuction of the neck and jowls. Any weight gain causes the definition between the face and neck to become obscured. With a short procedure under local anesthesia, fat can be suctioned from under the chin, the jowls, and the neck. The result can be striking and can simulate massive weight loss. Because fat commonly collects in the jowls and under the chin as we age, liposuction can produce a more youthful definition of the neck and jaw line.

In the 1980s neck liposuction was billed as a procedure for people under age 35. It was felt that skin would hang in older people after fat removal. Indeed, loose skin was a reasonable expectation. But because of a tenth-grade geometry principle, when the fat is removed, the skin must redrape over a longer distance. (The sum of two sides of a right triangle is longer than the hypotenuse!) So when fat is removed, by stretching over that longer distance the skin appears to tighten. Add to this the scarring that always occurs following surgery, and the skin really looks tighter. In only a small number of people does the skin hang.

And so, because of living geometry, in the early 1990s liposuction of the neck and jowls began to be performed in older people. I have successfully performed this procedure on women as old as 68.

Sometimes the neck bands are hidden by overlying fat. When the fat is removed, the bands may become visible—usually in a few years. If the bands are visible before the surgery despite the fat, they should be tightened at the time of the liposuction. This procedure, called a platysmaplasty or "band-lift™," is dealt with in Chapter 14.

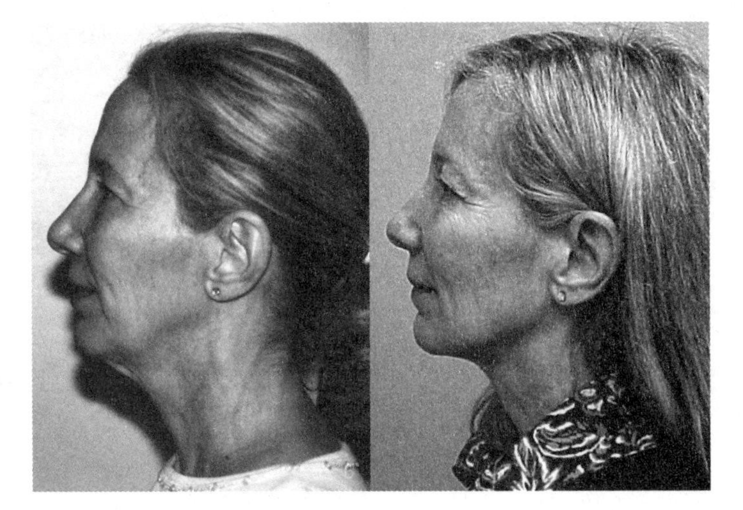

This 56-year-old woman underwent liposuction of the neck and jowls. It looks as though she had a face-lift (right). (Note lighting difference.)

Your plastic surgeon needs to examine you thoroughly prior to surgery. Sometimes the fat of the jowls really comes from inside the mouth (*buccal fat*). You'll feel it when you suck in your cheeks. It cannot be removed with liposuction. And sometimes the fat is underneath the muscles of the neck, requiring more than just liposuction.

Liposuction of the neck and jowls can be done on overweight people, and on people with early jowls from aging.

The Procedure

A topographic map of the fat of the neck and jowls, and even the cheeks, will be drawn with marker in the preop area. Liposuction of this area can be performed either under local anesthesia with intravenous sedation or under general anesthesia. I have performed this procedure under local anesthesia without sedation on patients whom I have already operated on, who I know will not move or talk during the procedure.

Once in the operating room, your face will be painted with an antibacterial Betadine solution and sterile drapes will be used. The surgeon numbs up the 4-millimeter locations of the incisions, in the crease just under the chin and under each earlobe. Through these incisions, a solution of saline and lidocaine anesthetic with epinephrine is infused. The epinephrine is crucial, as it significantly reduces bleeding during and after the procedure. Of course, if your procedure is performed under general anesthesia, lidocaine is unnecessary.

ıbes, called cannulae, to break up and then
ıally 2–3 millimeters in diameter and have
:ked. A liposuction machine, really a fancy
ıves the fat. More than one incision is used
t. Incisions under the ears lower the chance
: lives at the angle of the jaw, which would

to 150 millimeters (0.5 to 6 ounces) of fat
enough to allow the skin to slide over the
:oncept is aggressive removal of the fat, al-
: the neck looks more natural with some fat

Usually, one stitch under the chin is all that is needed. The ear wounds heal better without any stitches.

After the Surgery

Liposuction of the neck and jowls is a surprisingly pain-free procedure. My patients generally require a few Percocet tablets the day of surgery. By the next day, Tylenol is usually all that is necessary. On the other hand, you'll look absolutely horrible after the surgery. Your neck will swell and bruise tremendously, and you'll need a tight-fitting garment that reminds me of a bunny costume. Patients tolerate this garment for about a week. There really isn't much science to the garments, although it seems to make sense. I believe it decreases swelling and blood collection after the surgery.

It is critically important to keep your head up following this procedure. Blunt procedures such as liposuction disrupt hundreds of tiny blood vessels. These generally stop oozing within a few hours. If your head drops below the level of your heart anytime in the first three weeks, the pressure in the blood vessels can blow off newly formed clots and restart the bleeding. At the least, you'll have new areas of black and blue. In a worst case, you'll have lumps that can persist for months.

The bruising is likely to be gone in two weeks, rarely persisting for three weeks. Because of gravity, however, over the days and weeks the blood percolates into the lower neck and even the upper chest. After about a week, creative placement of scarves and turtleneck shirts can hide residual bruising and allow the patient to return to the public eye.

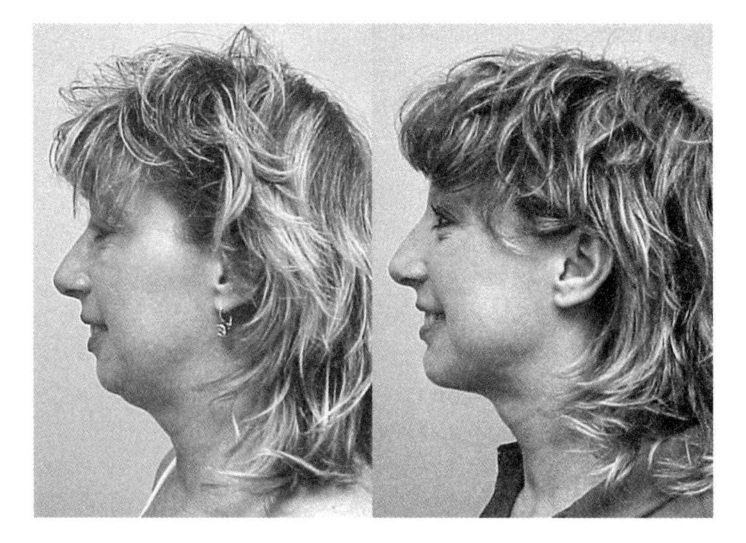

This 38-year-old woman underwent liposuction of the neck and jowls and the placement of a silicone chin implant to balance the face.

The Risks

As with any surgery, there is a risk of infection after this surgery. The risk is very low, however. Serious bleeding is rare with liposuction and would require an injury to one of the major blood vessels of the neck. Such damage is, however, possible. Any surgery can cause blood clots in the legs and clots that travel to the lungs.

Obviously, critically important structures lie in the neck, such as nerves, blood vessels, and the trachea. Any surgeon can spear these, although the risk is less with a fully trained, board-certified plastic surgeon.

Aesthetic or cosmetic problems are possible. Rippling, waviness, dimpling, fat left behind, too much fat removed, and asymmetry can occur.

The Cost

Liposuction will average about $3,000 in surgeon's fees; the range is from $2,000 to $5,000. The operating room and anesthesiologist add $1,000 to $2,000 to the bill. If the platysma muscle is operated on, or if fat is retrieved from deep areas, the cost will rise by several thousand dollars.

17

Rhinoplasty: Nasal Reshaping

Of all the procedures plastic surgeons perform, rhinoplasty is the most difficult. And it is with this procedure that society judges the skills of the surgeon.

Even though the first rhinoplasty was performed in 1887, the operation didn't become popular until the 1950s. In those days, radical operations whittled down the nose. With the use of gross instruments the bones were chiseled down, the cartilage was shaved, and the tip was bobbed. Initially those noses looked acceptable. But over the years and decades, plastic surgeons noticed that the noses continued to evolve. As the swelling and scarring decreased and as the skin thinned, the noses began to look overdone—in some cases, horrific.

In the 1960s, more refined and conservative rhinoplasties were developed. The mistakes of prior decades were noted by the 1980s, and a new rhinoplasty technique evolved. Rasps were used instead of chisels, and a new focus emerged on the proportions, rather than the size, of the nose. Incisions were moved from inside to outside the nose to allow more accuracy.

By 2005, there were 201,000 rhinoplasties performed in the United States, making it the fourth most common cosmetic surgical operation.

The Nose Is the First Thing You See When You Look at Someone

Like it or not, people are judged by the size and shape of their noses. A 1-millimeter hump changes a straight nose to a rounded nose and is visible from across the room.

The Los Angeles rhinoplasty surgeon Rollin Daniel, M.D., claims that it takes

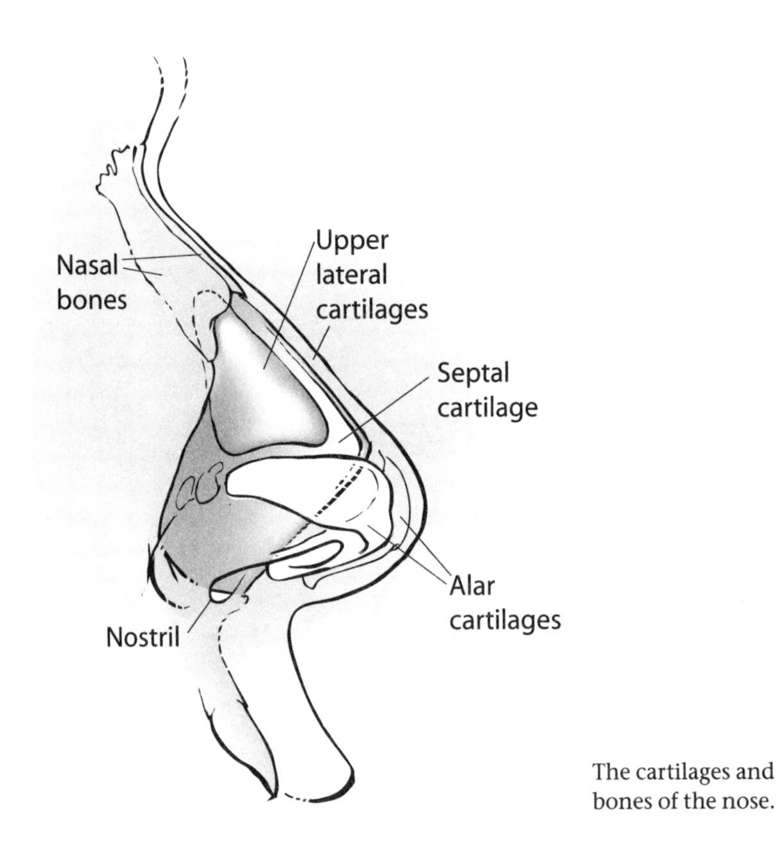

The cartilages and bones of the nose.

a surgeon a hundred noses to begin to have some facility with the operation. Very few surgeons have ever done more than a thousand rhinoplasties. One of the reasons why rhinoplasty is so challenging is that every nose is different and every result should be different, fitting the patient's face. Bad rhinoplasties are everywhere, many of them on people in the public eye. There is no way to hide the nose, before or after the procedure. People speculate that Michael Jackson has had many rhinoplasties, ultimately giving his nose a "plastic" appearance. It is very possible that Michael lost the tip of his nose as the result of a procedure and wears a prosthesis to hide the defect.

Noses are so variable in size and shape that it is hard to define what is normal. Different ethnic groups have different nasal characteristics. Within a given group are variations in what is aesthetically pleasing. Our celebrities echo what our society feels is beautiful. In the 1960s, the scooped nose was desirable. By the 1980s, the ideal nose was straight. Neither of these variations occurs commonly in nature, so the current "nose du jour" is one with a slight hump. Think Tom Cruise, the Olsen twins, and Jennifer Garner.

The Consultation

In a rhinoplasty consultation, the surgeon examines the inside and outside of the nose. The bones, cartilage, and breathing are evaluated. Digital photographs are then viewed on the computer.

The next half hour is a critically important time. The surgeon will ask the patient what she does not like about her nose. The images on the computer will be manipulated, in an attempt to create a better-looking nose. If the surgeon and patient cannot agree on what the nose should look like through computer simulations, it is unlikely that she will be pleased with the ultimate outcome. It is better to not undergo surgery than to wake up with an unexpected result.

If the surgeon and patient do agree on what the nose should look like, an operative plan is designed. I calibrate the computer, creating actual measurements on the computer image. I make decisions with the patient in front of me, then check my plan on the computer.

A surgeon often spends hours preparing for a rhinoplasty, using his photographs in much the same way an architect uses blueprints to design a house. The computerized digital imaging system has made rhinoplasty planning much easier, accurate to the half millimeter.

In a second consultation prior to surgery, we review the plan. I expect a fair amount of discussion with the patient during this visit. If she is too immature to air her feelings about the surgery, then it should not be performed. I canceled one 15-year-old on the morning of surgery because I did not feel that she had a real grasp of what was going to happen.

The Procedure

In the preop area, the cartilage and bones are marked on the patient, with the proposed new lines. Surgery is usually performed under general anesthesia. Most anesthesiologists believe this is safer for the patient, since the airway is well protected by the breathing tube. Even though gauze packing is placed during surgery to prevent any blood from dripping down the throat, some does get through. If local anesthesia and heavy sedation are used, the gag reflex is decreased and there is a chance that the patient could breathe in (aspirate) blood. With general anesthesia, this is less likely.

Open and Closed Rhinoplasties

The closed rhinoplasty is the traditional operation. All incisions are made inside the nose and the operation is performed, upside down and inside out. The

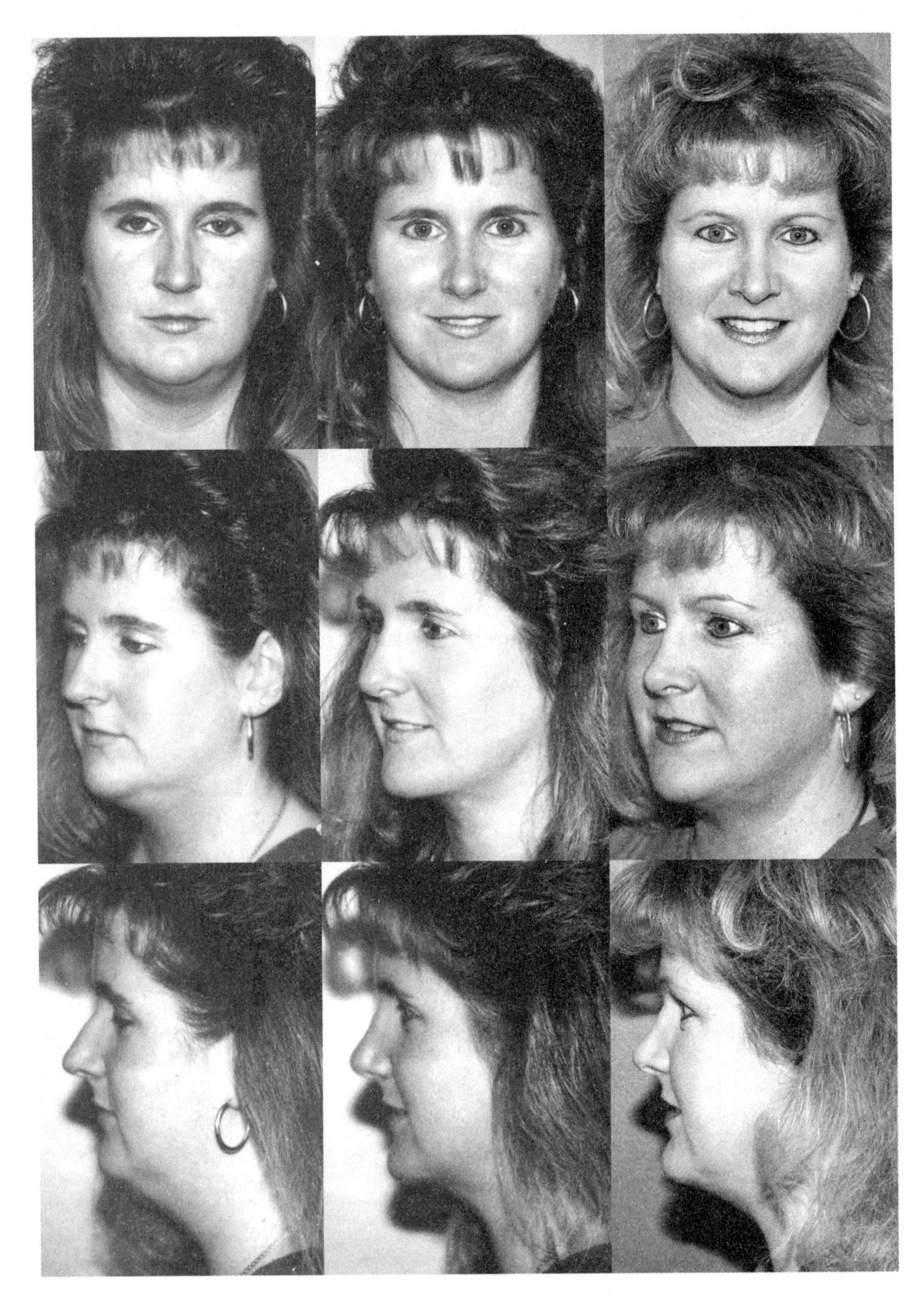

Preop rhinoplasty and neck liposuction on the left, three months postop in the middle, eight years postop on the right.

rhinoplasty masters of the 1960s through the 1980s performed closed rhinoplasty. Some still do. This old technique is one of the reasons for the difficulty of this operation and for the plethora of bad rhinoplasties. In the 1980s, with the push of the well-known Dallas rhinoplasty surgeons Drs. Jack Gunter and John Tebbets, and Drs. Rollin Daniel and Ronald Gruber of California, the open rhinoplasty was taught to a new generation of plastic surgeons.[1]

The open rhinoplasty cuts across the columella. The columella is the skin between the nostrils. Most other incisions are made inside the nose, as in the closed technique. The extra 4-millimeter external incision allows the skin to be lifted up off of the nose, exposing the cartilage and bones much the same way as raising the hood of a car exposes the engine. Bleeding can be directly stopped with this technique, unlike the closed operation that requires old-fashioned pressure to stop bleeding. Sutures can be placed between cartilages. Grafts can be accurately sewn in, and techniques that could only be performed by the great masters of rhinoplasty can be implemented by the average plastic surgeon.

In a rhinoplasty, typical steps include rasping down the bony hump, shaving the cartilaginous hump, sculpting the tip cartilage, and making cuts in the bones and pushing them inward. Cuts can be made from inside or outside of the nose. Some noses require removal of fat from the tip, and some require removal of skin from the nostrils. Various steps may be performed to improve the breathing, such as operating on the septum between the nostrils or the turbinates (the air baffles of the nose).

With hundreds of possible steps, rhinoplasty is the most individualized operation in plastic surgery. No two noses are alike, and no two will need exactly the same procedure. The rhinoplasty demands creativity, since the actual structures of the nose often turn out to be different than expected. The plastic surgeon has no x-ray to give him a look at the cartilage prior to surgery, so he must be able to think on his feet. Surprises such as cracked or buckled cartilage and scar tissue make this operation challenging to even the most experienced surgeon.

Sometimes a long nose will make the upper lip look short. The opposite is also true: a short nose makes the upper lip look long. Techniques are available to either lengthen or shorten the nose. The upper lip can be dropped or shortened.

Cartilage Grafts

Increasingly common in rhinoplasty, grafts are pieces of tissue taken out of the body, altered, and then replaced. Cartilage grafts can be used to build up or straighten sections of the nose or to replace missing cartilage (which usually results from a prior, overzealous rhinoplasty). Sometimes so much cartilage or bone is missing that rib cartilage or bone from the skull must be used to help make a properly shaped nose.

More commonly, grafts are taken from other cartilage in the nose. If not enough is present, cartilage can be taken from the ears. The surgeon shapes the cartilage and sews pieces together to create a perfectly shaped graft—a perilous and tedious process. Pieces of cartilage smaller than matchsticks must be sewn together in a precise order. The surgeon gets no more than two chances to make it perfect; after that, the cartilage cracks and becomes useless. This is one of the reasons many plastic surgeons don't perform rhinoplasties.

When a Nose Has Already Been Operated on, the Second Operation Is Called a Secondary Rhinoplasty

Fifteen percent of rhinoplasties require revisions. All sorts of problems can occur after the surgery. Healing can result in bumps in the bones, or the bones may not have been moved symmetrically. Too much or too little cartilage may have been removed. Stitches that were placed to hold cartilage together or in a different position may tear through. Tiny remnants of bone, cartilage, or fat can scar, causing visible bumps in the nose. Cartilage grafts can move, warp, or disappear. And the skin can thicken to the point where it is impossible to see just what was changed surgically. The unpredictability of this operation is one of its challenges.

Plastic surgeons say that only two types of surgeons don't ever have to redo a rhinoplasty—liars and neophytes. Honest surgeons will tell their patients that "redo" rhinoplasties are common.

Because of the huge array of deformities that can occur after rhinoplasty, only the most expert plastic surgeon is not challenged by a secondary rhinoplasty. Tissues can be absent or fused together with scar. Noses can be twisted, overshortened, and fragile. New plastic surgeons should perform only *easy* secondary rhinoplasties, if there are such things. The mark of a mature, honest surgeon is knowing when to refer difficult cases to more experienced surgeons. In plastic surgery, we all know the surgeons to whom to send the really difficult cases. I keep their phone numbers on my Rolodex. But the more experienced the plastic surgeon becomes, the more difficult cases he can handle. There is no operation in all of plastic surgery more demanding than a secondary rhinoplasty.

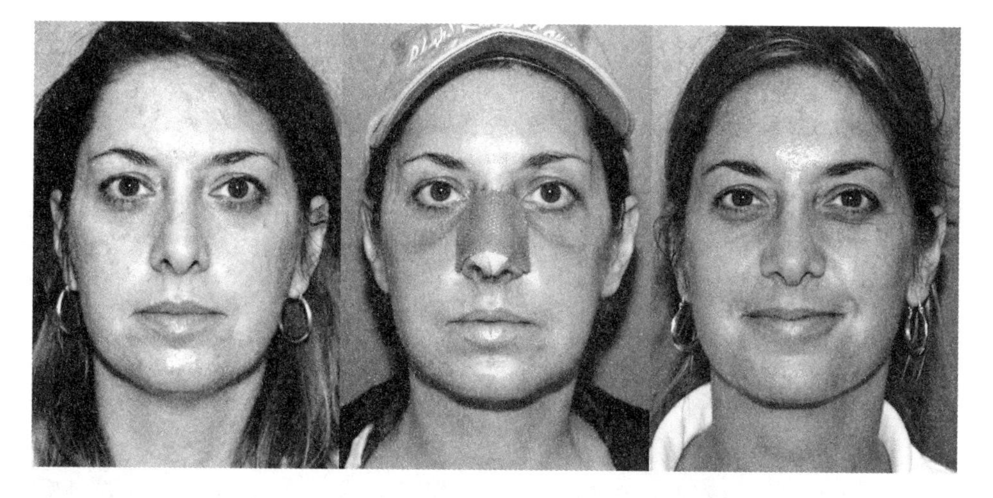

On the left, preop. In the middle, appearance with splint one week after rhinoplasty, chin implant, neck liposuction, and TCA peel. Postop on the right. See color plates.

After the Surgery

You will wake up from surgery with a splint on your nose and possibly with packing inside your nose. The packing is eliminated when possible, since it is the most uncomfortable aspect of the procedure. You will have stitches in the various incisions in your nose. A drip pad, a piece of gauze held in place with a dressing that drapes around your ears, catches the oozing blood after surgery. At home you will change the drip pad often the first night; by the next day, the oozing stops.

You'll see your surgeon a day or two after surgery and have your nose cleaned. The bruising and swelling around your eyes, forehead, and lips will remind you of the 1988 *Newsweek* photo of Geraldo Rivera after his television-show guest smashed a chair over his nose. Your swelling will subside in a few days.

The splint and sutures are removed a week after surgery. At that time, your nose will still be quite swollen. Brave people go back to work after the splint is off; others wait two weeks, until the swelling settles down and any residual bruising can be covered by makeup.

Now comes the hard part—waiting. Unlike a breast augmentation where you wake up from surgery with C cups, your final nose will not make its appearance for many months, if not years. While the swelling decreases over the months following surgery, the internal scarring *increases* over the first three months, and then takes the next nine months to settle down. Even though we say that the

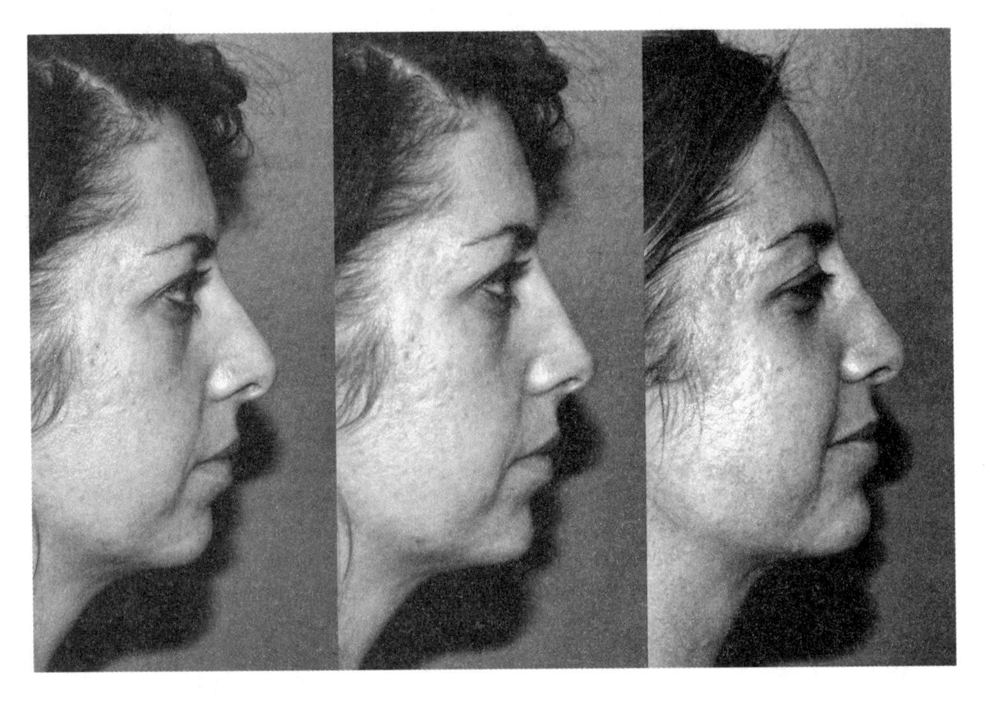

Side view of the same patient shown on the previous page. The preop appearance is on the left, the computer-simulated "goal" in the middle, and the actual postop image on the right. In this case, the actual result looks better than the computer simulation.

nose is done a year following surgery, in actuality subtle changes occur for years and even decades afterward.

The overdone rhinoplasties of the 1960s didn't all look bad in the 1960s. In fact, they didn't look bad until ten or twenty years later. While the scar tissue continues to settle, the skin thins, resulting in progressively more detail revealed—good or bad.

The Risks

As in all cosmetic surgical procedures, the risks divide into medical and aesthetic. Medical risks include the rare rhinoplasty infection and the common (1 in 50) nosebleed. Blood clots in the legs, or clots that travel to the lungs, and death are rare complications. Compression boots on the legs lessen these risks. Extremely rare complications include bone fractures and eye and brain injuries. Blindness has occurred. Skin can be lost in rhinoplasty, usually in smokers or those who have undergone prior nasal procedures. The result is horrible deformity that requires complicated reconstructive procedures.

Cosmetic risks are the highest of all cosmetic surgical procedures, with 15

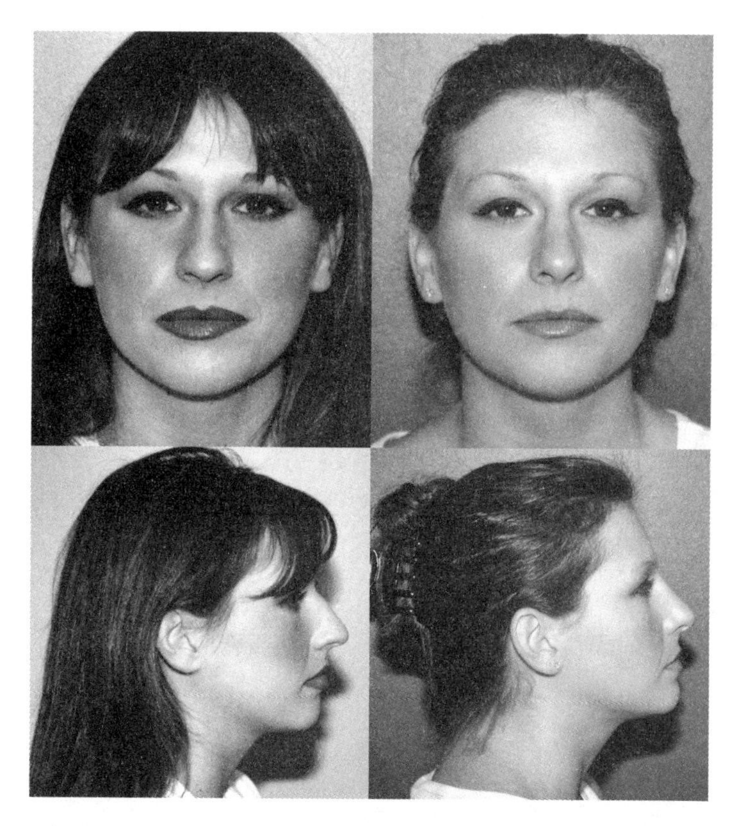

This 25-year-old underwent a rhinoplasty to correct an over-projecting large nose. Preop on the left, six months postop on the right.

percent of all patients requiring a revision. Reasons for revisions might be bumps that return following surgery, warping cartilage, moving grafts, stitches that rip out, and incisions that separate. Some of the problems are random. Some are caused by the patient. Probing fingers and nose blowing can destroy the delicate array of surgically altered cartilage. And some poor results are simply due to poor surgical techniques.

Plastic Surgeons Know That There Are Two Types of Rhinoplasties—Difficult Ones and Tremendously Difficult Ones

Imagine having to work in a space smaller than one square inch. And that space keeps filling up with blood, so that your view is constantly blocked. You have one, maybe two, chances at altering the cartilage before it cracks and becomes useless. Then imagine the pressure of knowing that whatever you create in the next few hours, with a degree of precision measured in quarter millimeters, will

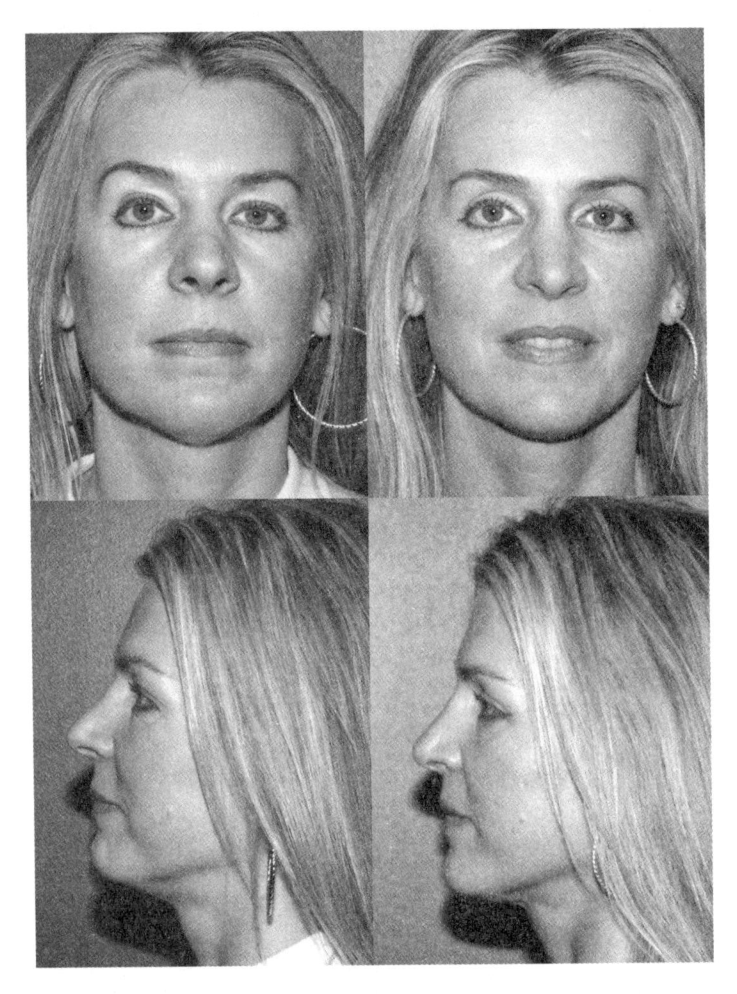

This woman had a prior rhinoplasty by another surgeon and was unhappy with the result. Preop on the left; three months after rhinoplasty, liposuction of the neck, and upper blepharoplasty on the right.

be visible to the patient and her friends and family for a lifetime. Welcome to the world of the rhinoplasty surgeon. While this scenario is present for most procedures in cosmetic surgery, it is most acute in rhinoplasty because the margin of error is almost nonexistent.

The Nonsurgical Rhinoplasty

The filler Radiesse has been touted as a method for improving the shape of the nose without surgery. In actuality, Radiesse can only fill in depressions. It can't decrease the size of the nose or alter most angles. Its use in the nose is unfortu-

This 21-year-old woman desired improvement in the hump in her nose and definition of the tip. Preop on the left, postop on the right.

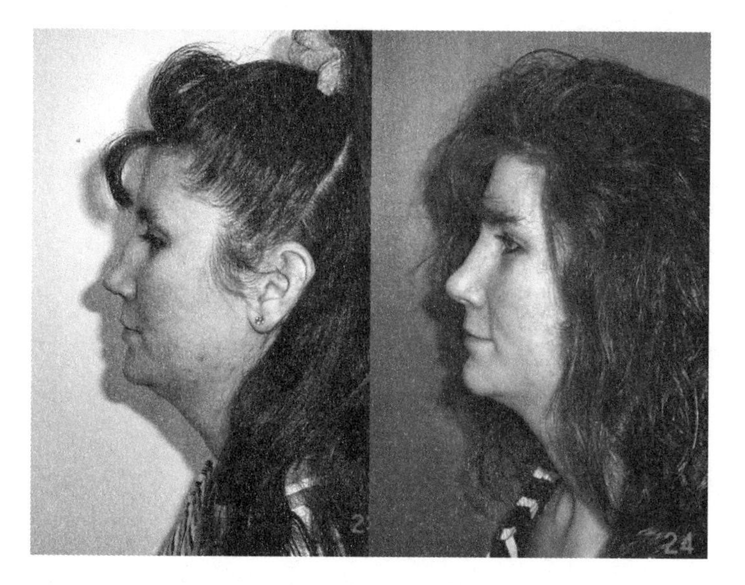

This woman's nose collapsed after a rhinoplasty and septal operation by another surgeon. On the left, preop. On the right, after reconstruction with cartilage graft from the ear to several places in the nose. See color plates.

nately quite limited. Yet when the problem is a small depression, this filler can save the patient a large operation, acting like a sort of modeling clay. It is one of the best applications of Radiesse, which will be used in the nose more and more.

The Cost

While the national average fee for a rhinoplasty is about $4,200, the actual cost ranges from $3,500 to $10,000, depending on what steps need to be performed. Secondary (redo) rhinoplasties are even more expensive. And because rhinoplasty is one of the signature procedures of skilled plastic surgeons, the experts in the field will have significantly higher fees.

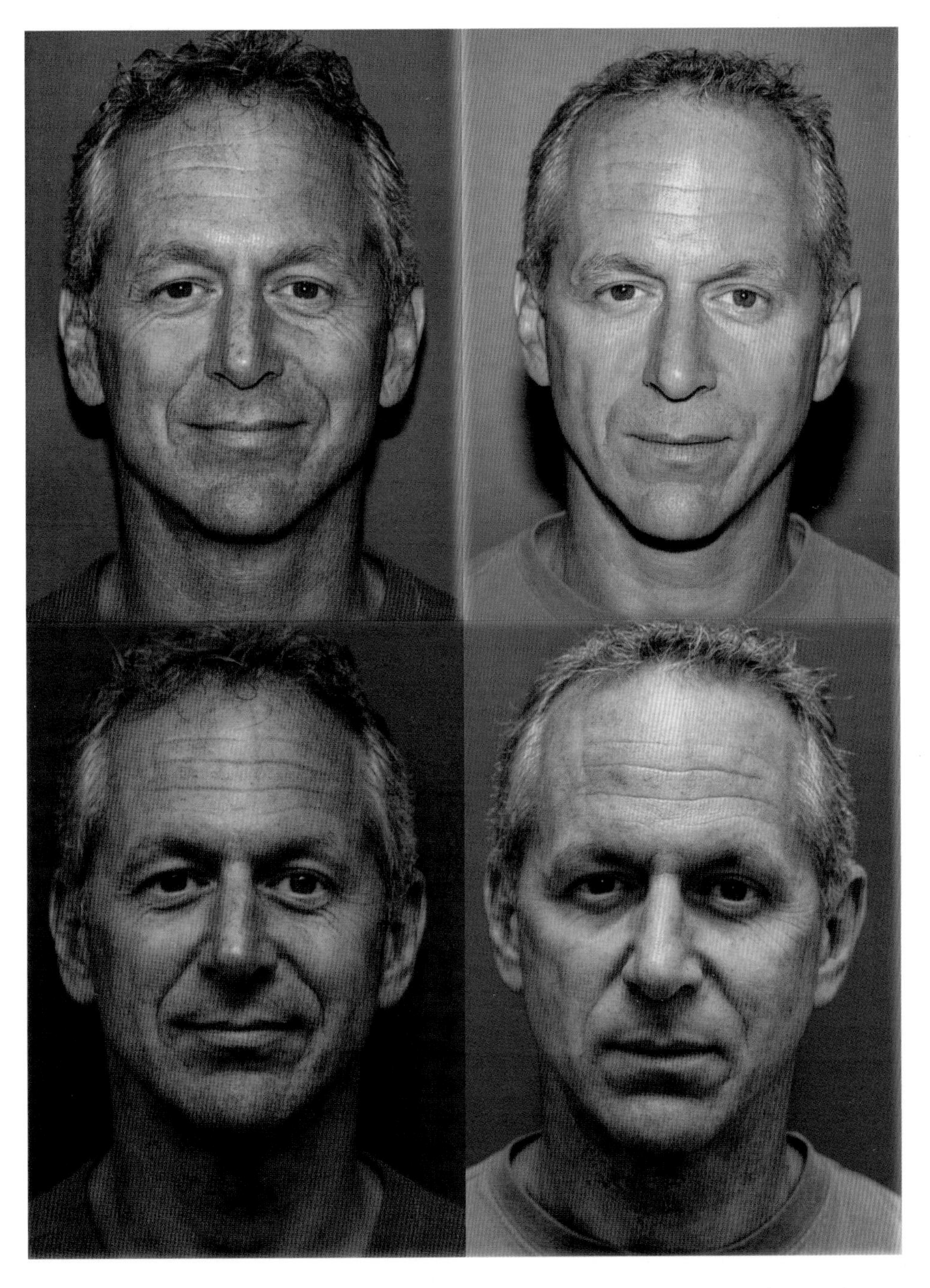

Beware: the flash fills in sags and wrinkles. The top and bottom photos were taken on the same days! The top photos were taken with full flash; on the bottom, the flash was bounced off the ceiling. On the left, the patient's face before upper eyelid lift, browpexy, and fat grafting to the nasolabial fold. The photos on the right were taken a few months after the procedures.

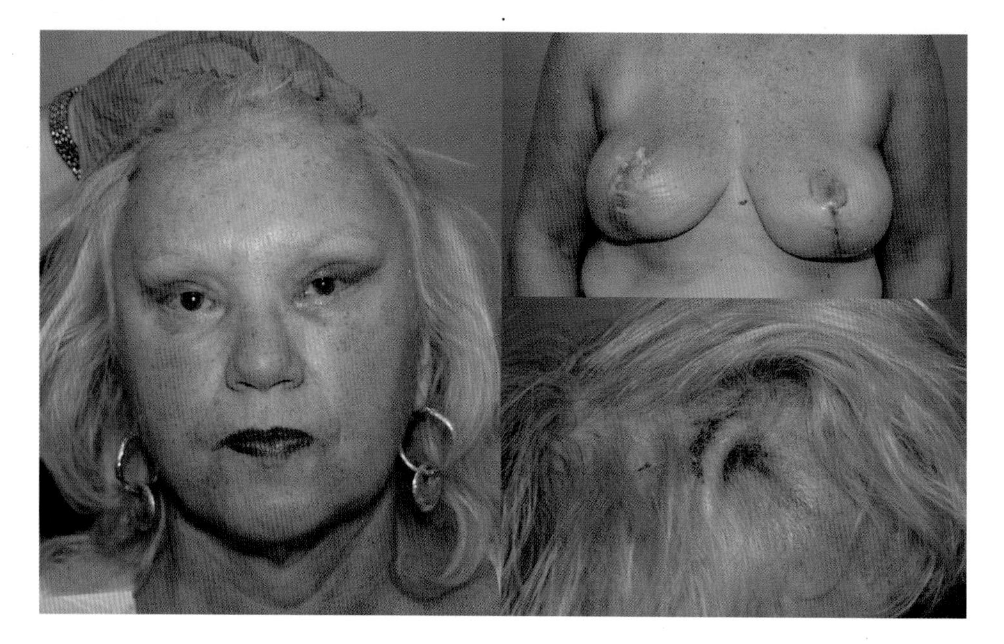

This American woman underwent a breast-lift, face-lift, brow-lift, and eyelid lift in Peru two months prior to these photos, with disastrous results. Her forehead is paralyzed, she has a bizarre clumping of the scalp, arching of the eyelids, and open wounds on her breasts and scalp.

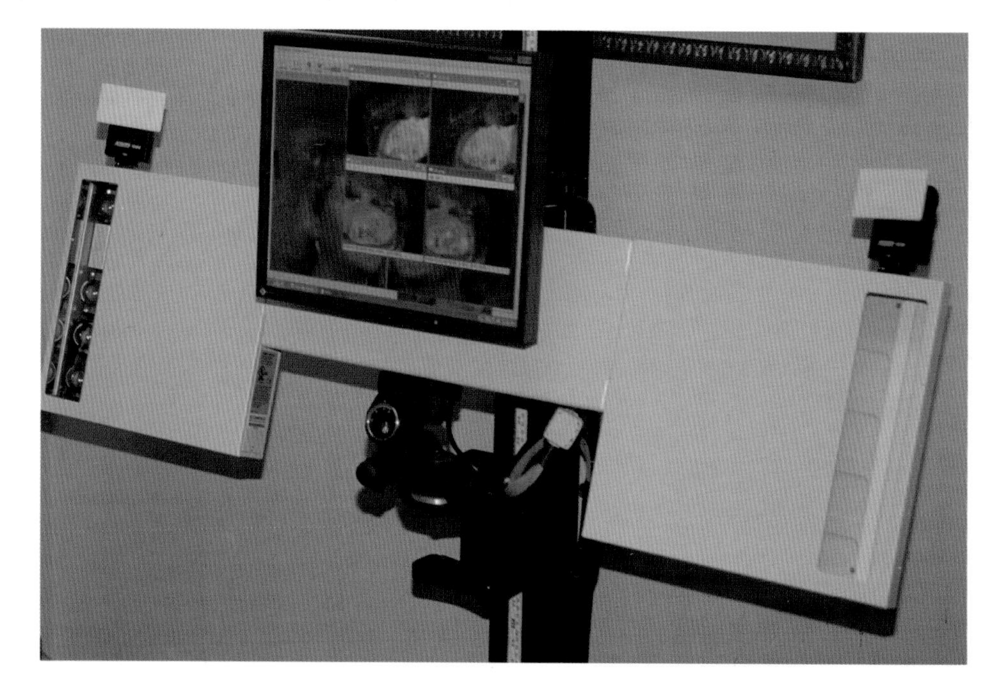

The Canfield Imaging 3D Vectra system.

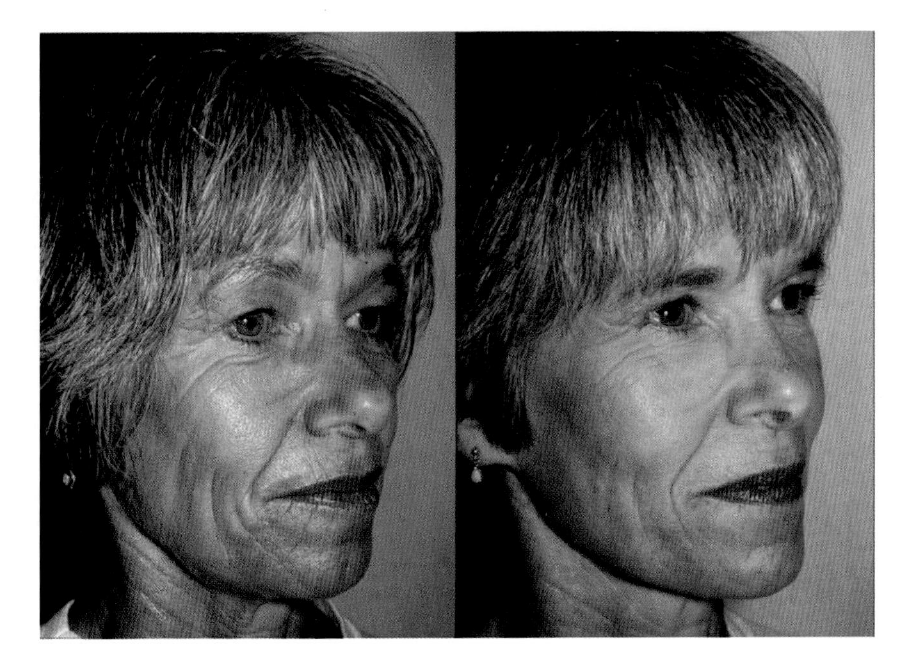

The twin on the left has had significant sun exposure; her sister, on the right, has had much less exposure to the sun. Neither has had any surgery. Photos courtesy of New York City plastic surgeon Dr. Darrick E. Antell.

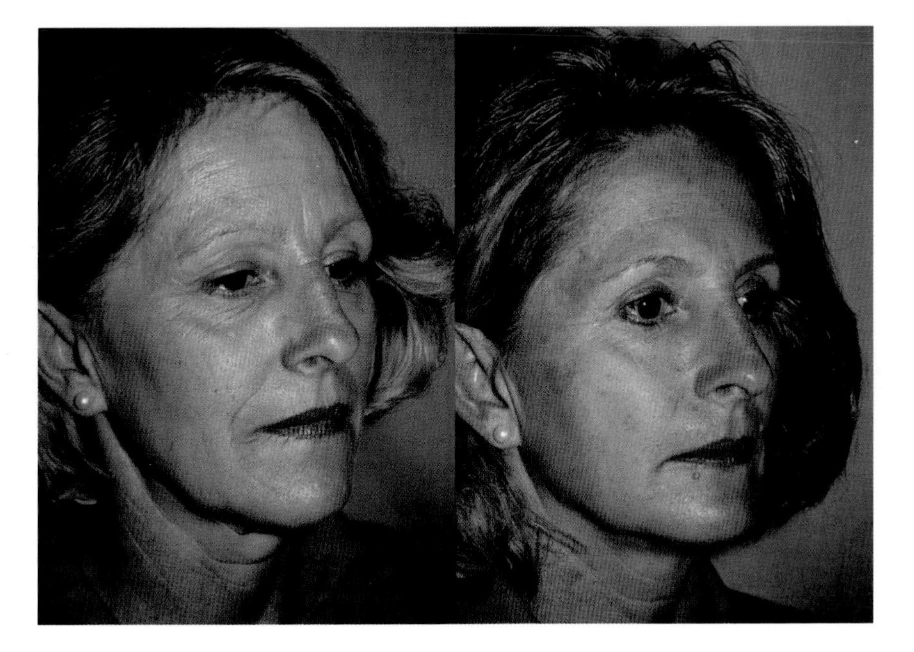

The twin on the left has smoked for thirty years; the twin on the right does not smoke. This environmental influence on aging is reason enough to "kick the habit." Photos courtesy of New York City plastic surgeon Dr. Darrick E. Antell.

Three generations of the same family: the 74-year-old grandmother is on the left, the 48-year-old mother in the middle, and the 15-year-old daughter on the right. Note the progressive descent of the eyebrow, the upper eyelid skin, and the development of the nasolabial folds, jowls, and excess skin of the neck over the decades.

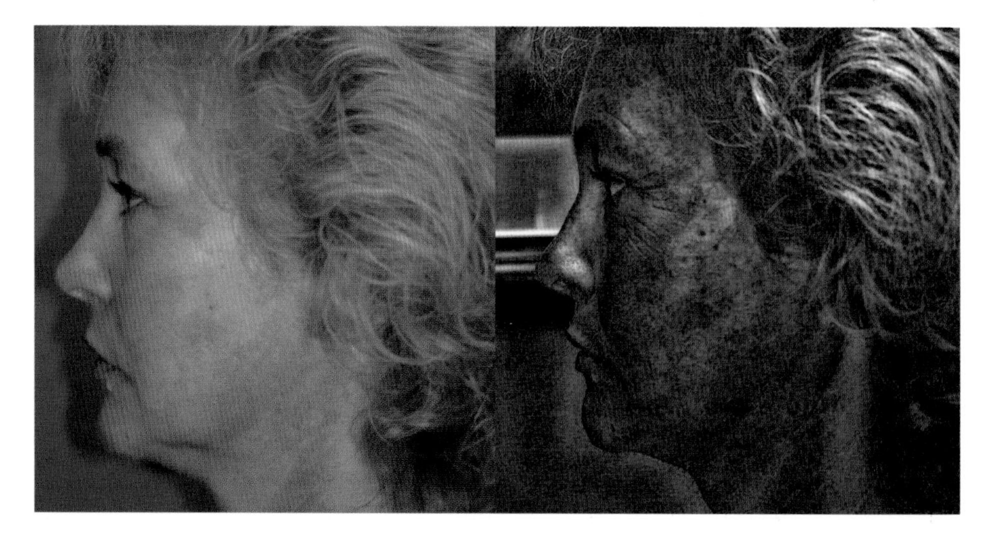

The use of special ultraviolet filters allows the camera to see sun damage years before it is visible to the naked eye. On the left, a 50-year-old woman; on the right, her appearance using ultraviolet filters. Photos such as these are taken prior to beginning a skin care program and at regular intervals thereafter to chart progress.

This man had a mole on his cheek that was distracting. The middle photo demonstrates the method of removal, by placing a scar along the natural fold of the face. On the right, his appearance a year later.

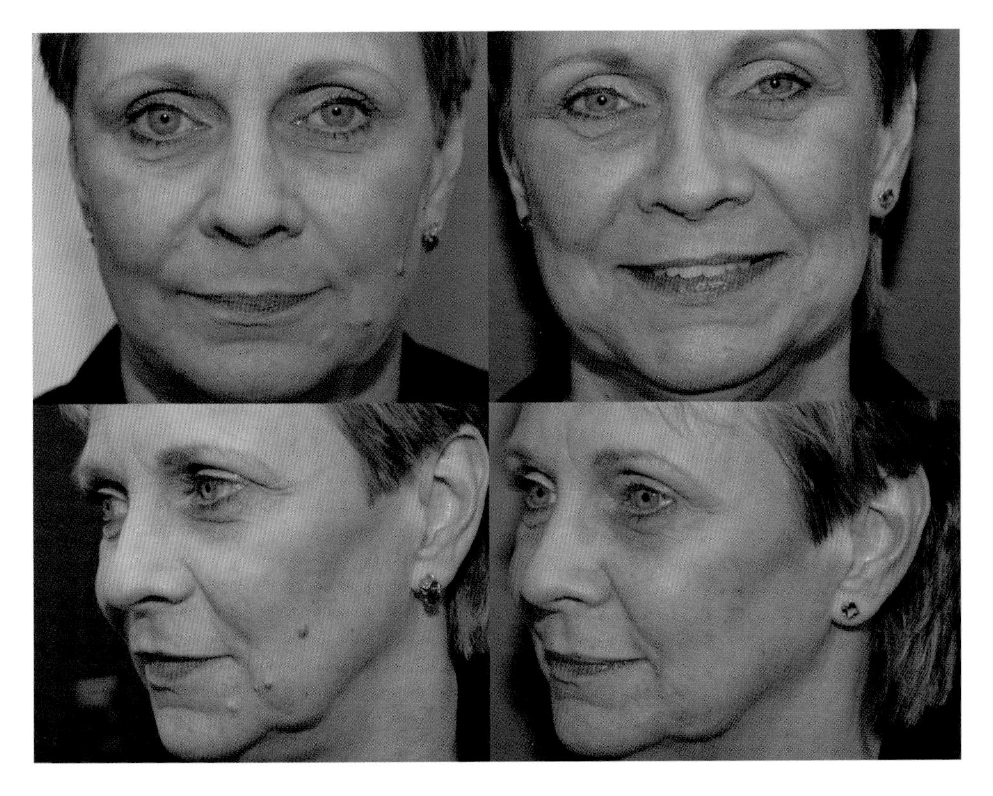

This woman had four facial moles removed, significantly improving her appearance.

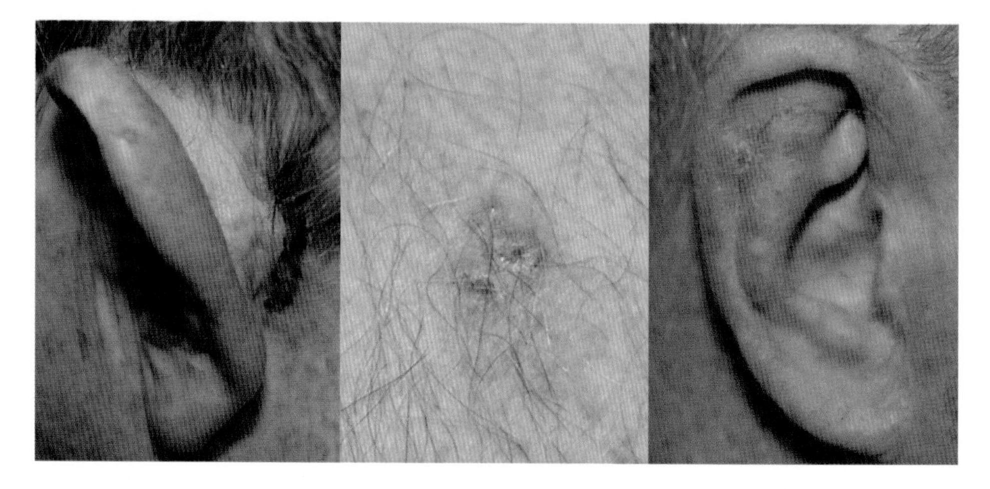

Three appearances of basal cell carcinomas—they can be ulcerations (left), red marks (center), or nodules (right).

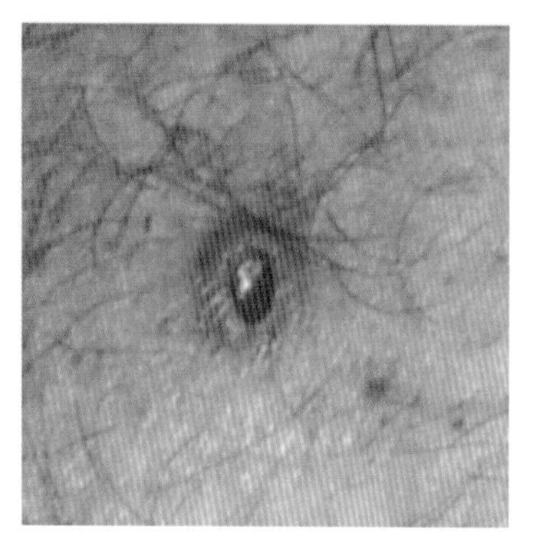

Squamous cell carcinoma.

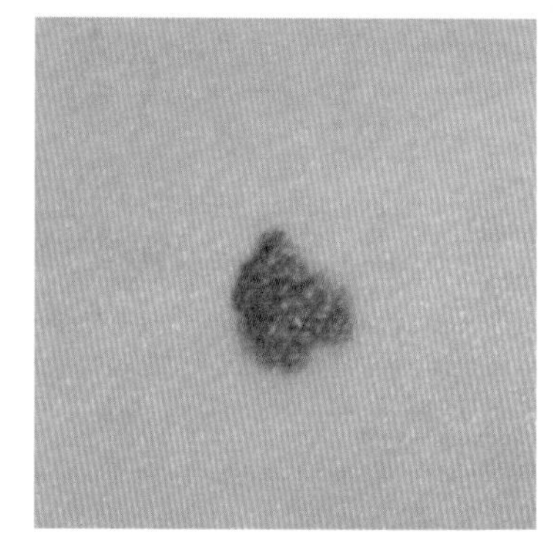

Melanoma.

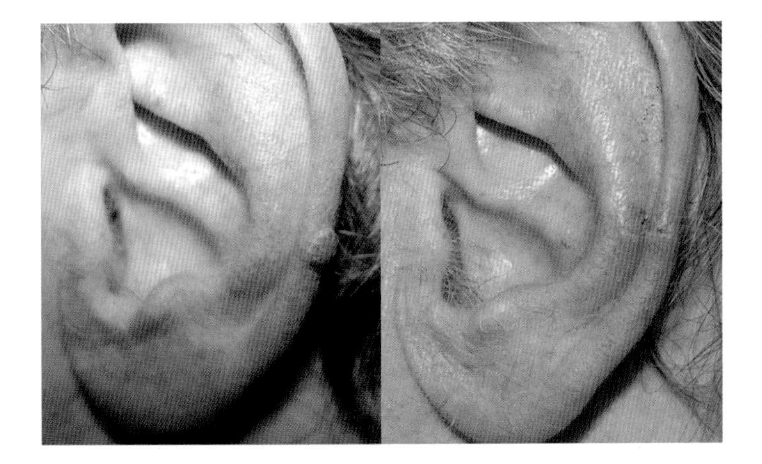

This man had a flesh-colored mole on his ear for three years. It turned out to be a melanoma. On the right is the ear after removal. More skin had to be removed later.

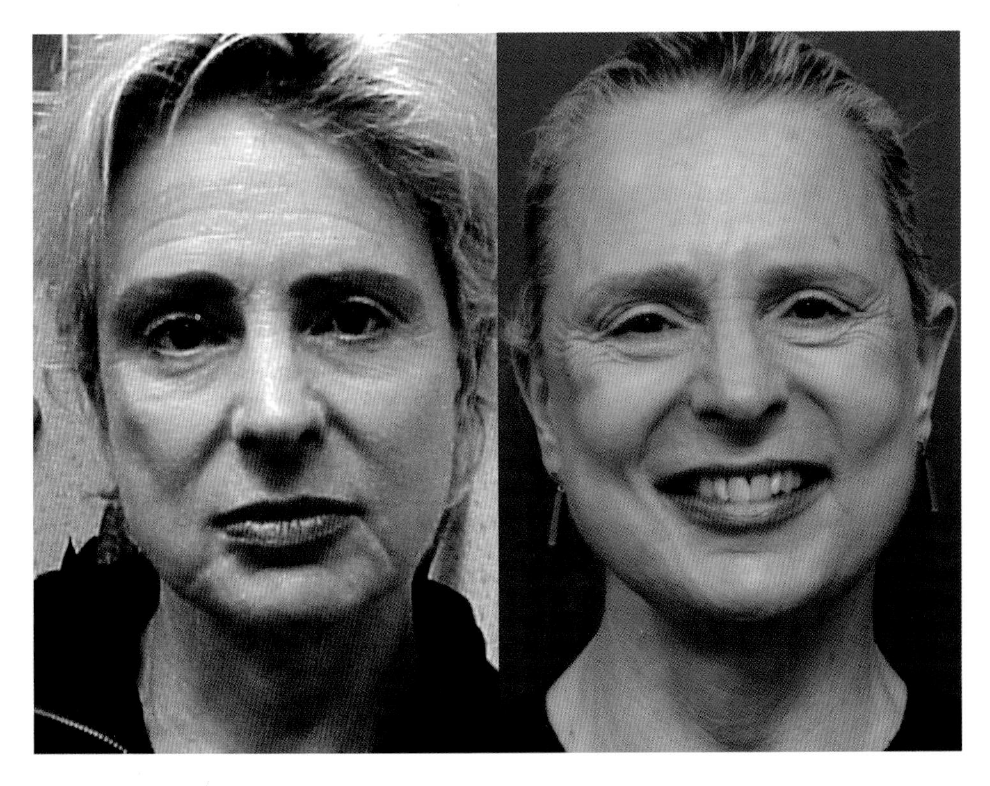

Another surgeon excised cysts below this woman's mouth and filled in the scars with too much fat. I re-excised her depressed scars and filled them with Restylane (on right).

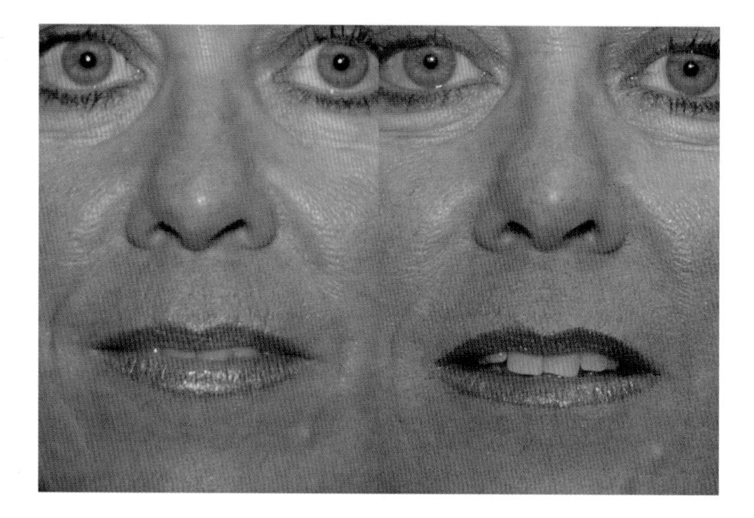

This woman had Restylane injected into the wrinkles around the mouth. On the right, two months after injection.

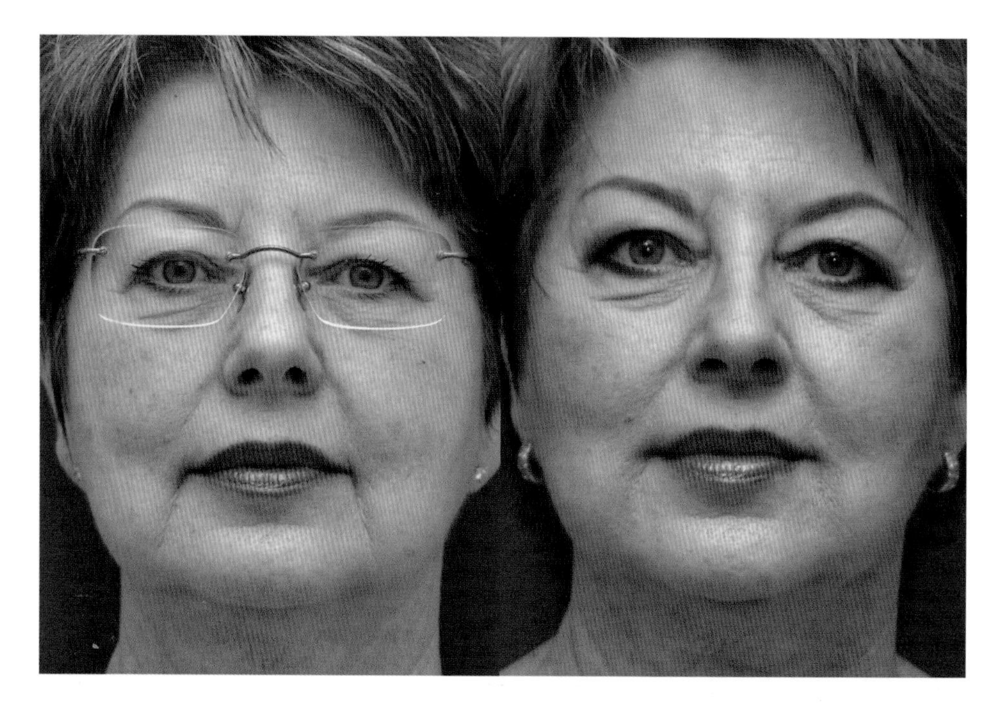

This woman had a combination of Radiesse injected deep into the nasolabial folds and Restylane injected more superficially. Left, before; right, following her second Restylane syringe a week later.

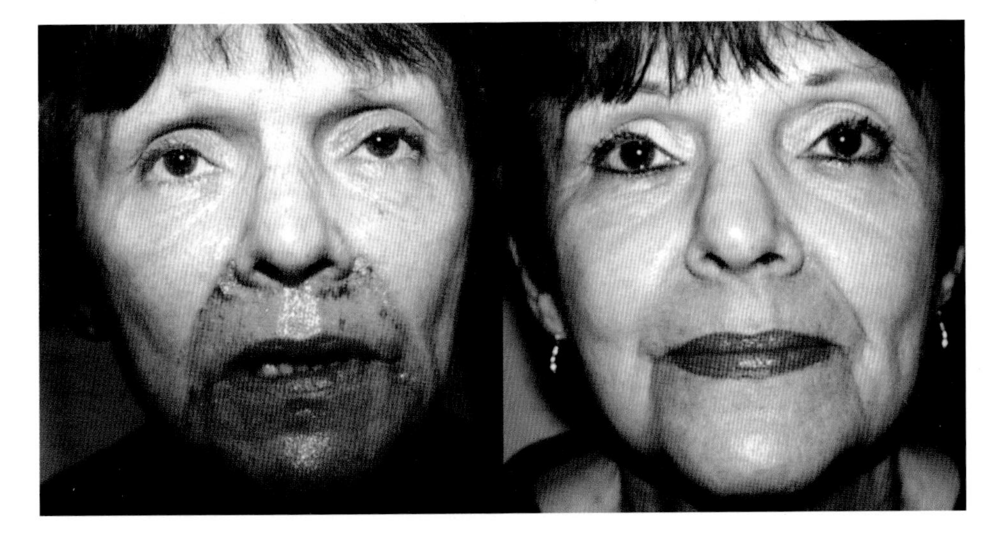

On the left, lasered skin at one week postop; on the right, at one month.

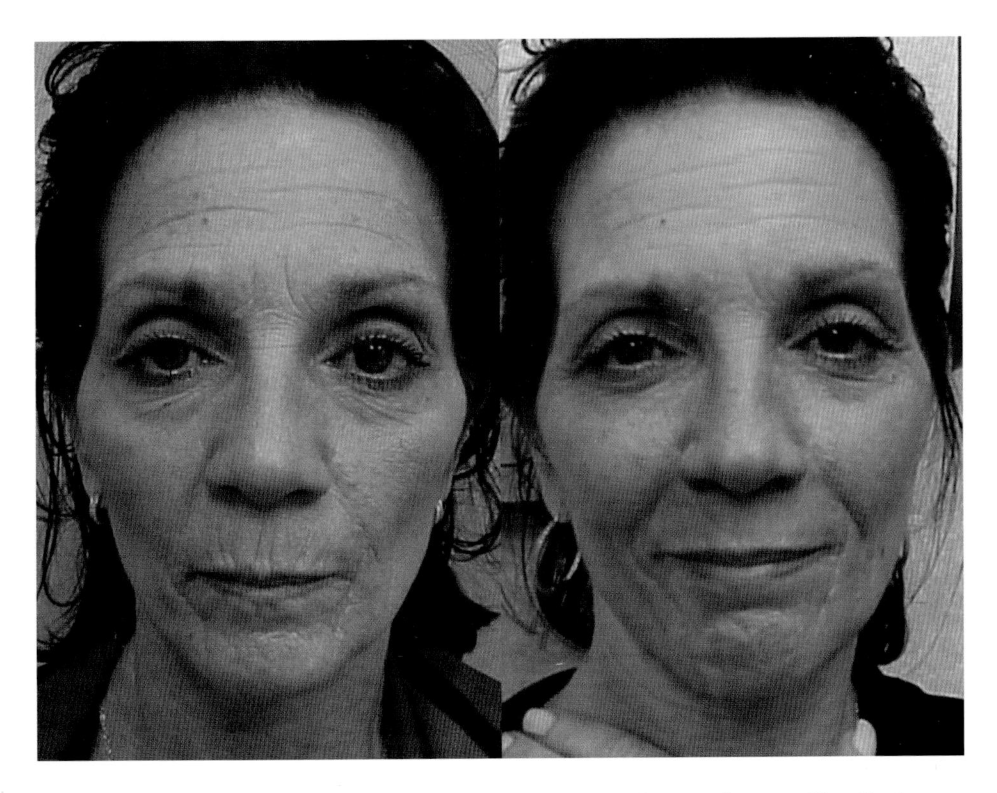

This 49-year-old woman spends a lot of time in the sun, working on her wrinkles. She has undergone a full-face CO_2 laser resurfacing.

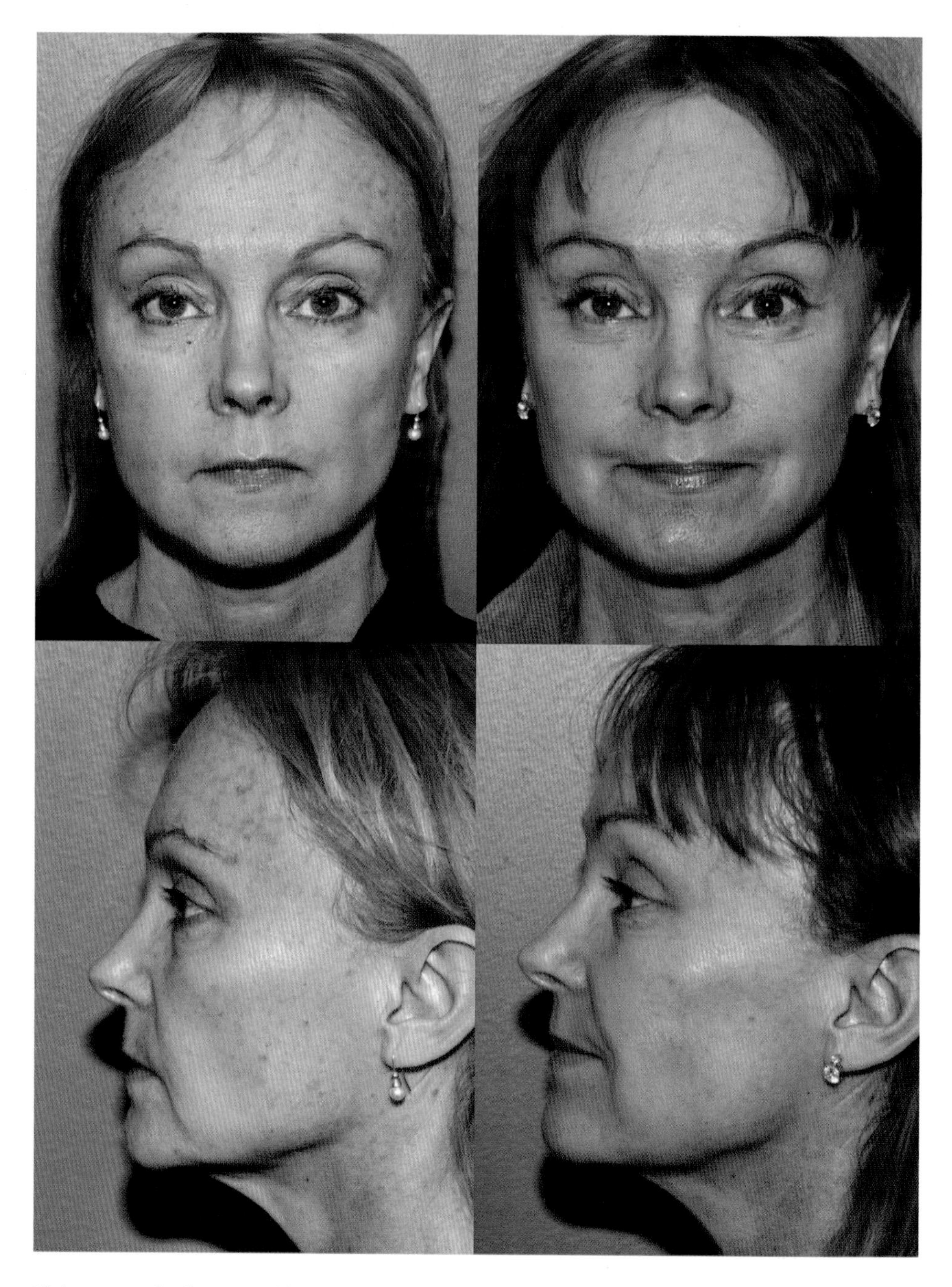

This woman had scattered brown pigmentation of the face and fine wrinkling. Note the improvement in pigmentation and wrinkles on the right, after the TCA peel.

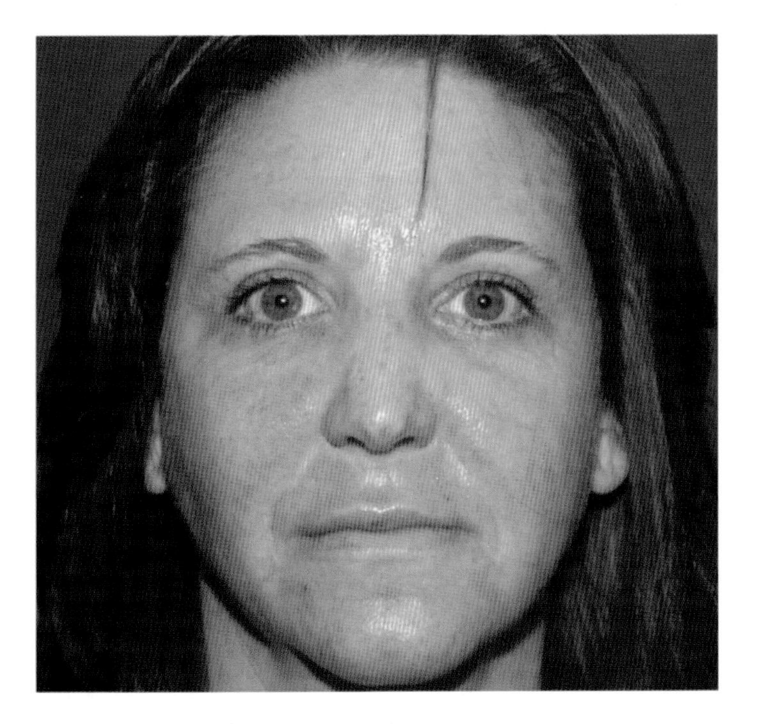

This degree of redness of the face is expected one week after a TCA peel.

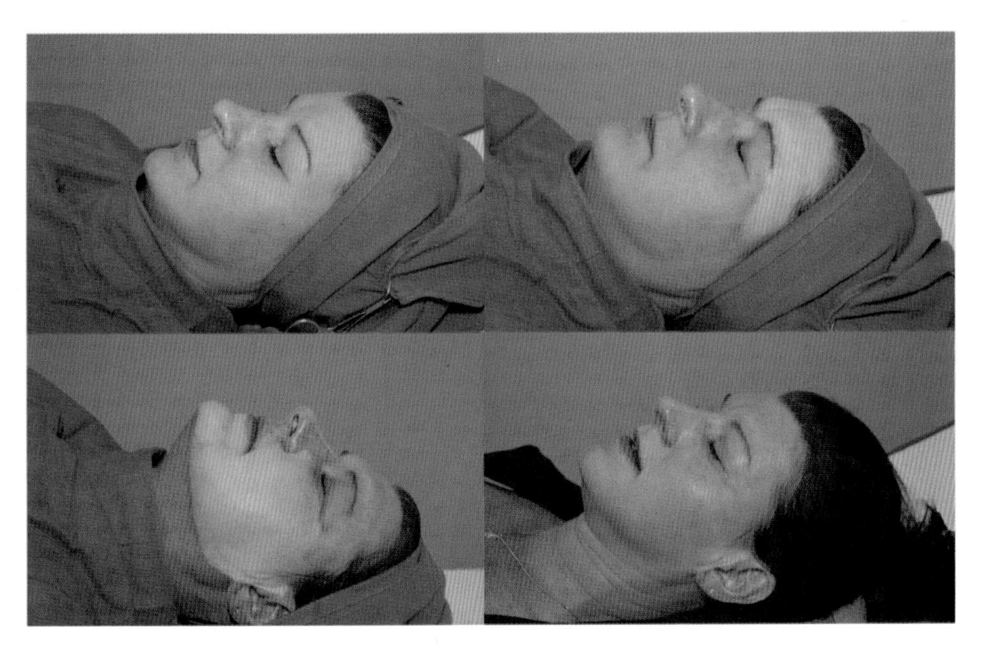

A TCA peel in progress. Upper left, before the peel. Upper right, after peeling the forehead; frosting is visible. Lower left, after peeling the mid and lower face, with the forehead already recovering. Lower right, face recovering.

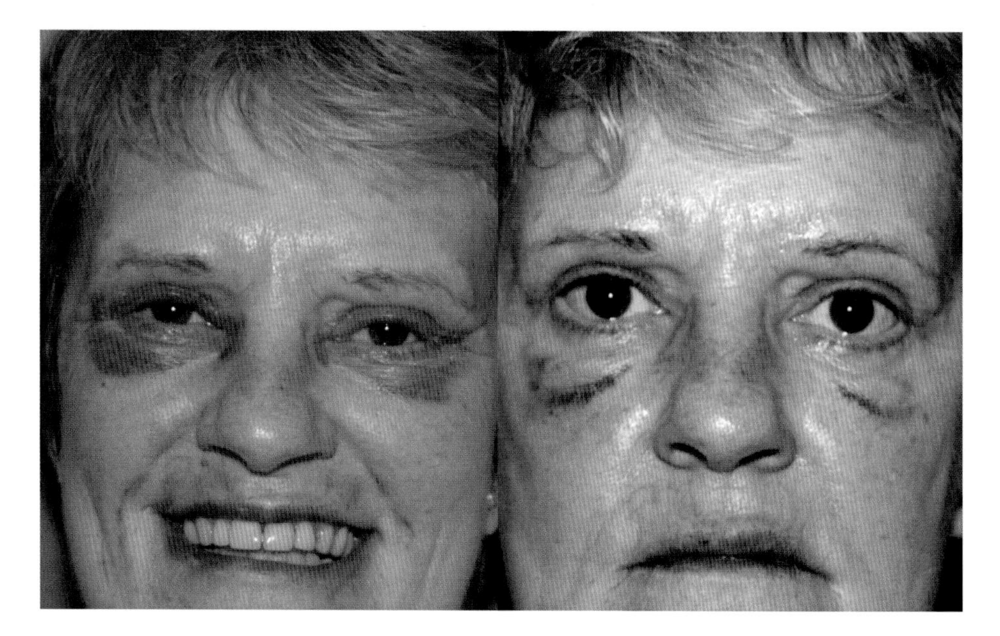

On the left, one week following an upper and lower eyelid lift, canthopexy, browpexy, and Restylane injections to lips and wrinkles around the mouth. On the right, two weeks postop.

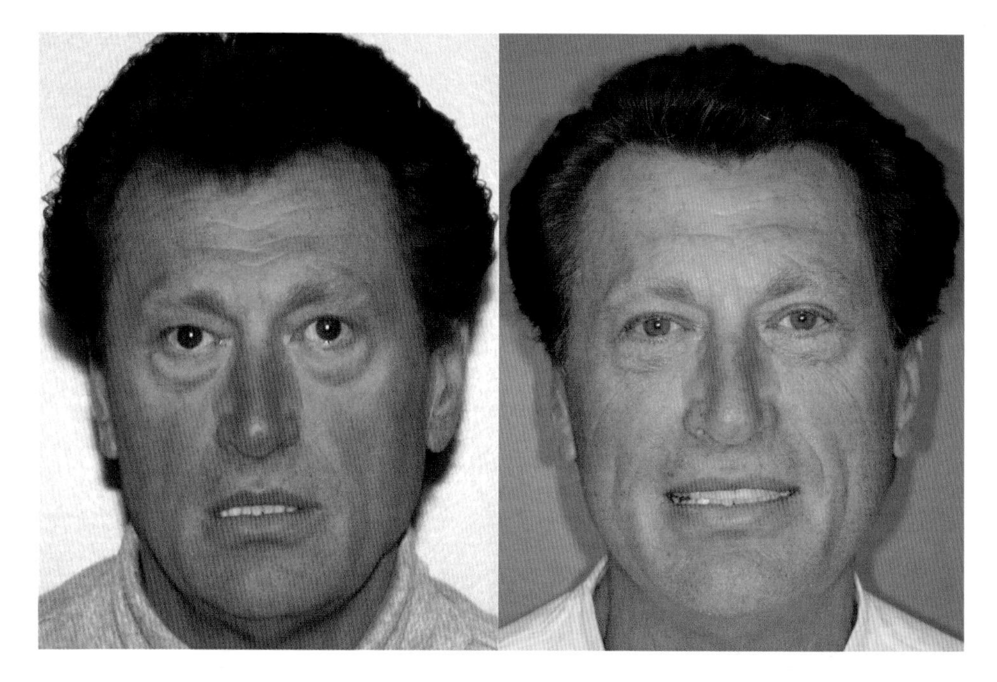

In the photo on the left, this 58-year-old man shows a tremendous amount of extra upper and lower eyelid skin and fat, as well as laxity of the lower lids. The photo on the right was taken six months following eyelid lift and canthopexy. I make sure to not overdo men's lids to avoid feminizing them.

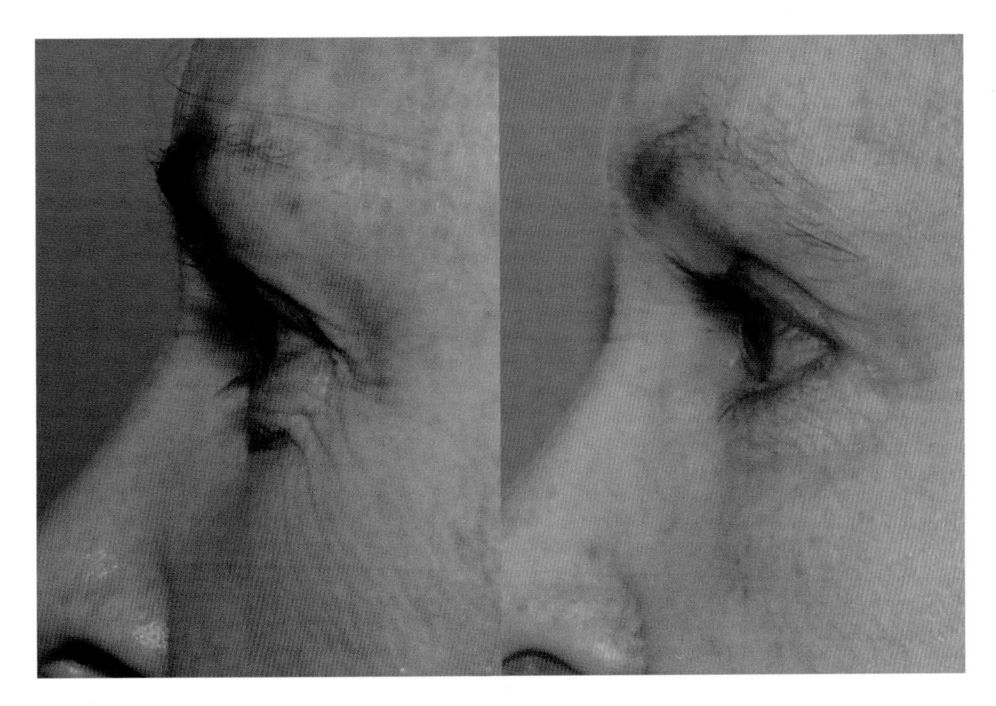

This woman had excess skin and fat of the lids. The photo on the left shows this extra skin. The right side shows the desired clean appearance of a feminine eyelid.

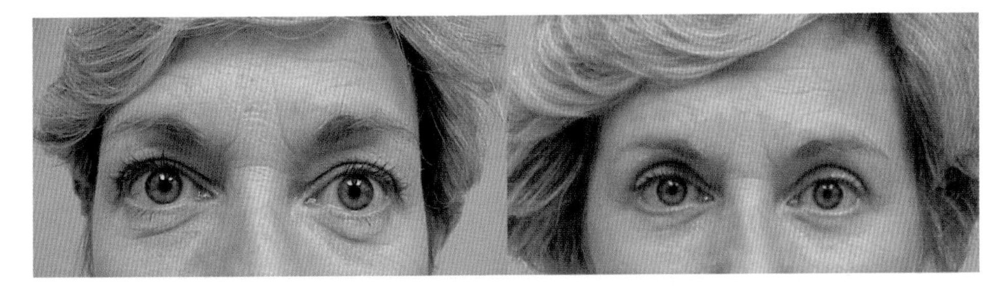

In a woman, the upper eyelid lift should remove all of the drooping skin, to allow for the application of makeup without smearing. This woman also had a lower lid lift with a canthopexy. Preop on the left, postop on the right.

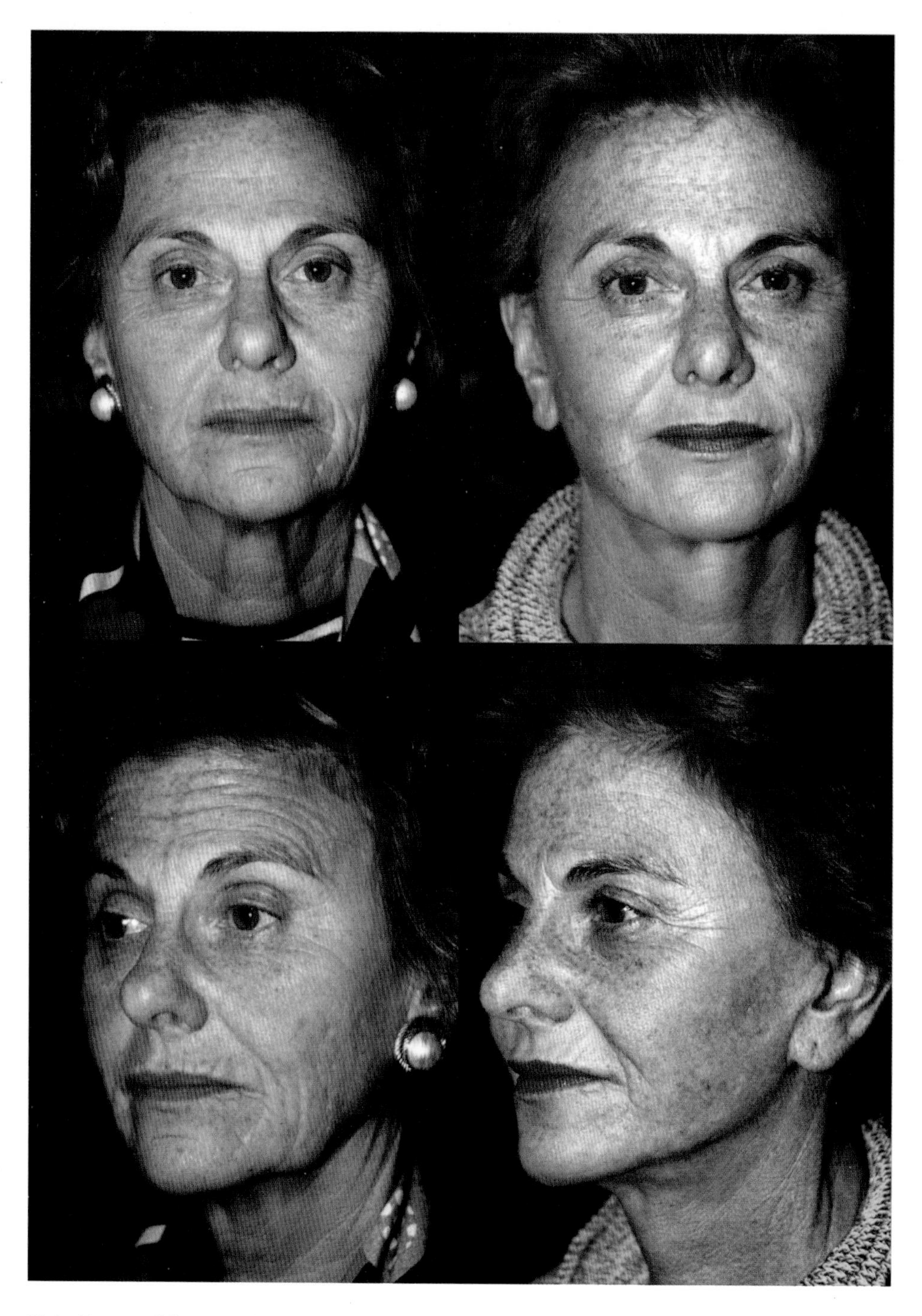

This 61-year-old woman had bands of the neck and jowls and wrinkles around the mouth. On the right, her appearance after face-lift and tip rhinoplasty.

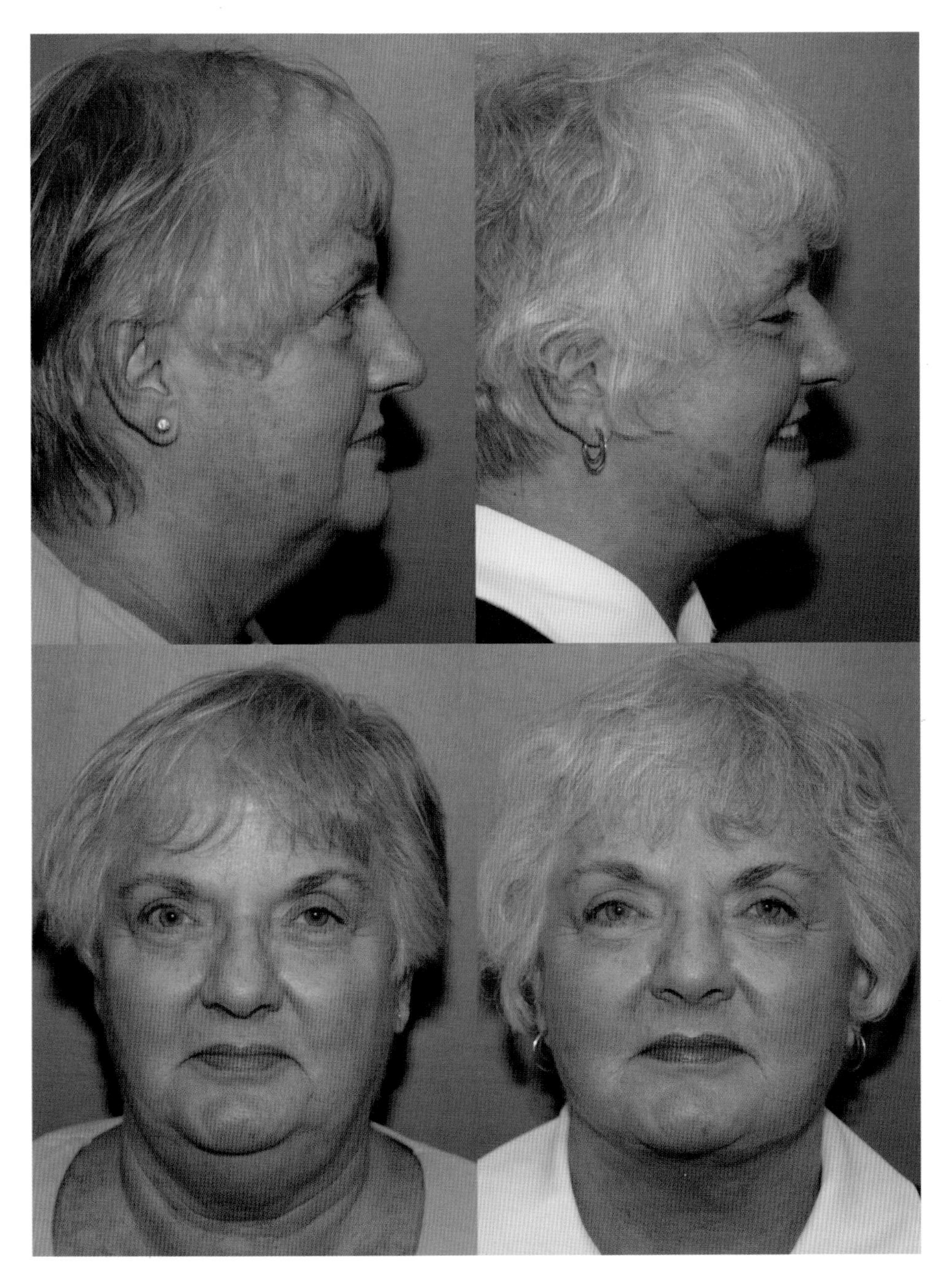

This 68-year-old woman had a large amount of neck fat and excess skin. A standard SMAS lift was performed, with removal of fat of the neck and repair of the neck muscles. Preop on the left, six months postop on the right.

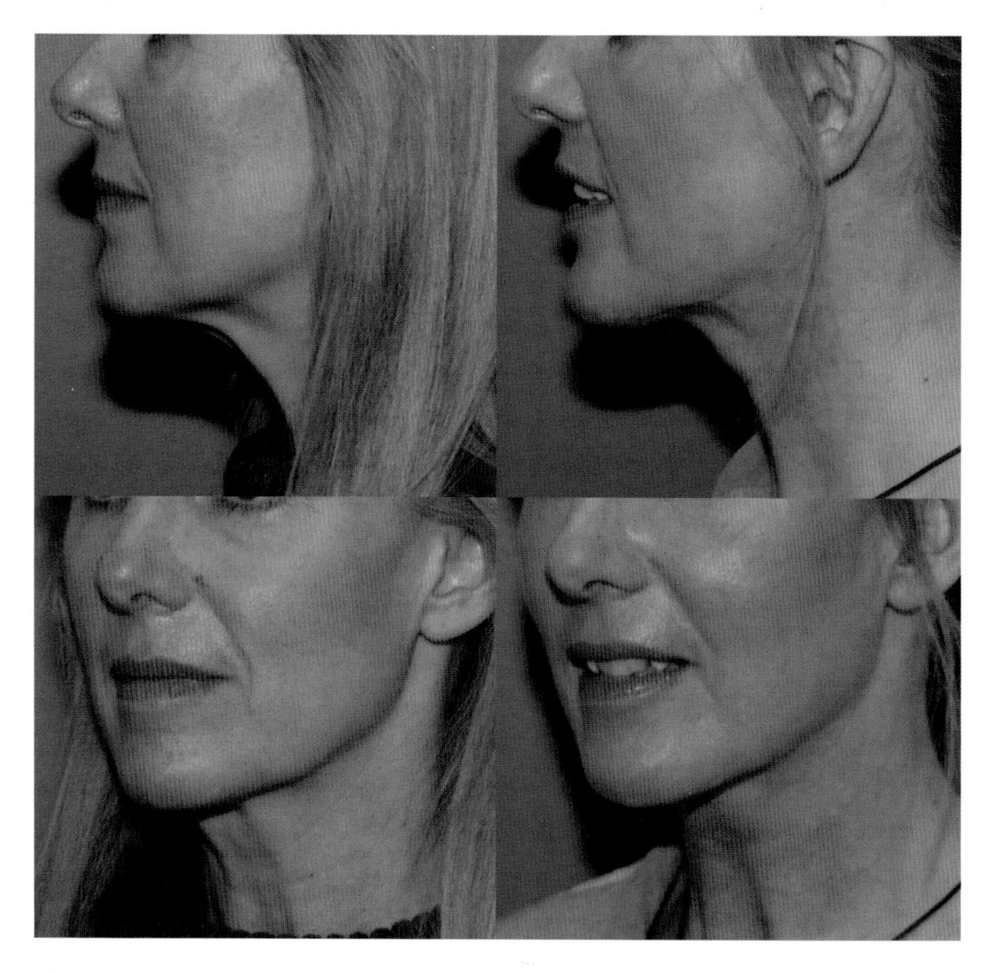

This 51-year-old woman underwent a band-lift to correct her neck banding.

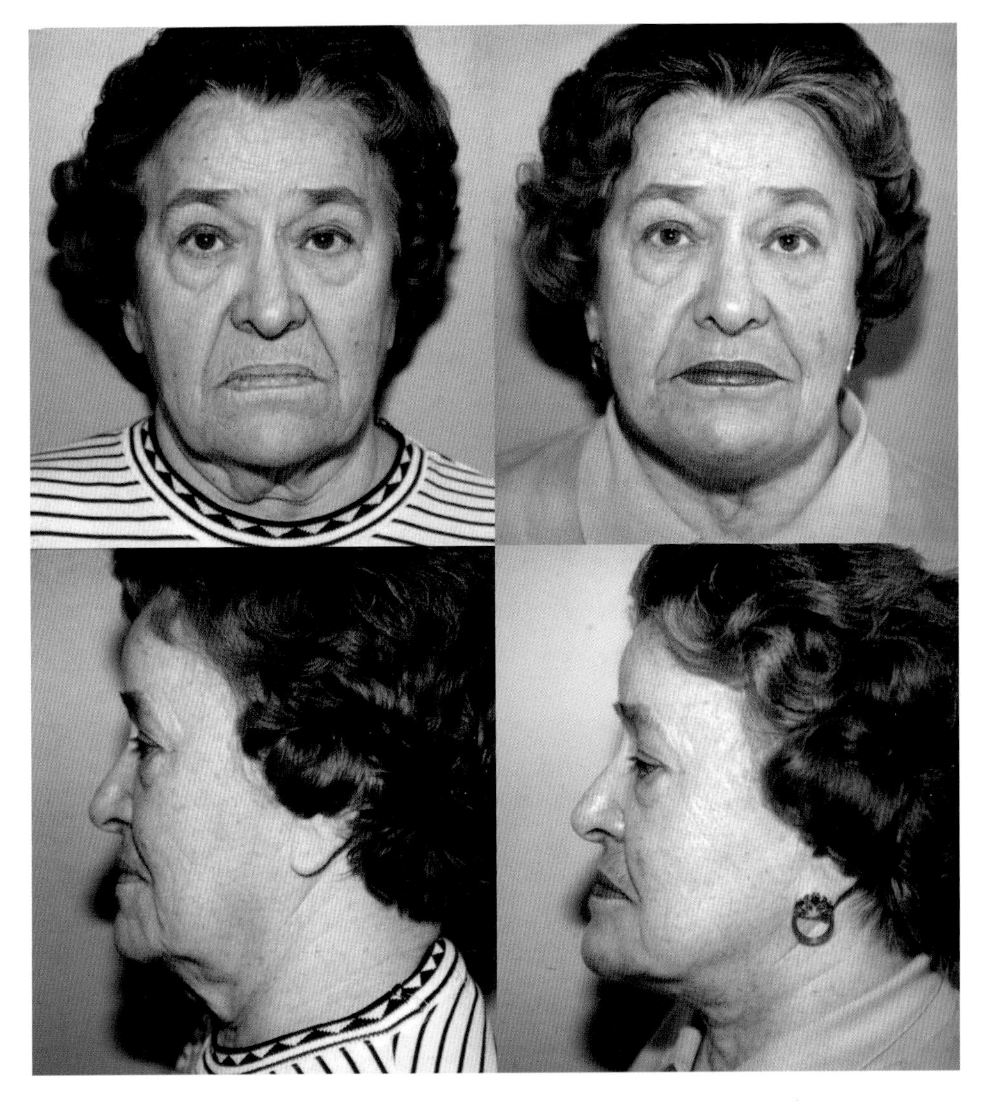

On the left, preop. On the right, following direct excision of neck skin, a band-lift, liposuction of the neck and jowls, and a deep chemical peel around the mouth.

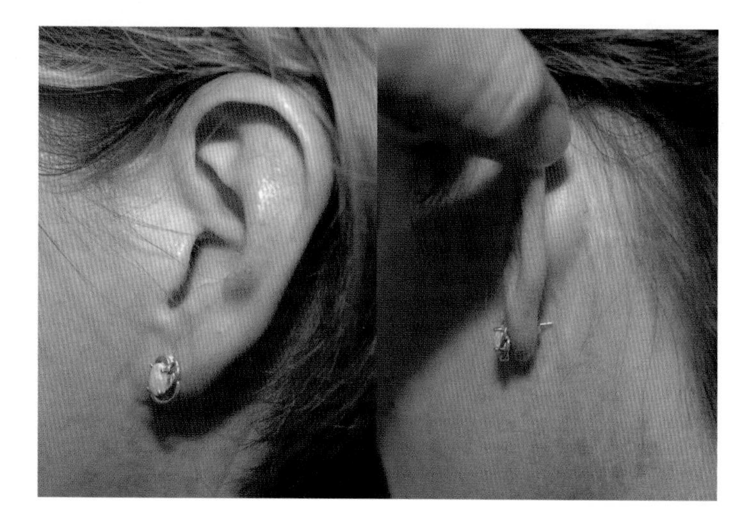

This woman had a prior face-lift by another surgeon. She was left with an unattractive scar. These photos show her appearance after a revision face-lift. Note that the scars are hard to find, even without makeup.

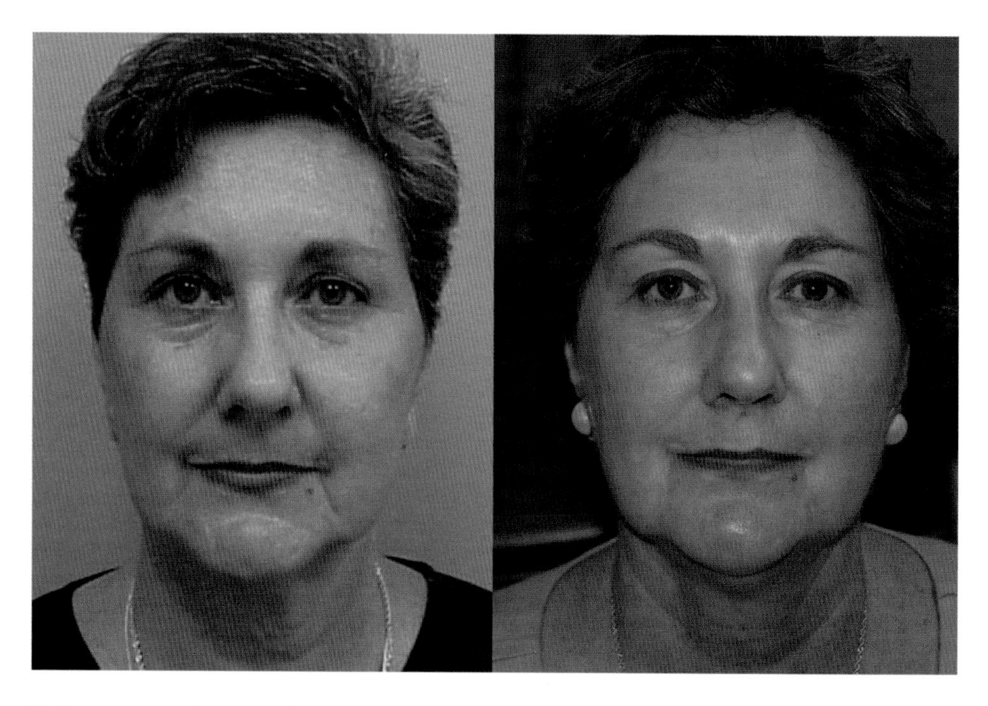

This woman underwent a laser peel of the face and two fat grafting procedures to lessen her marionette lines. On the right, her appearance after three and a half years. (Note that there is a slight lighting difference.)

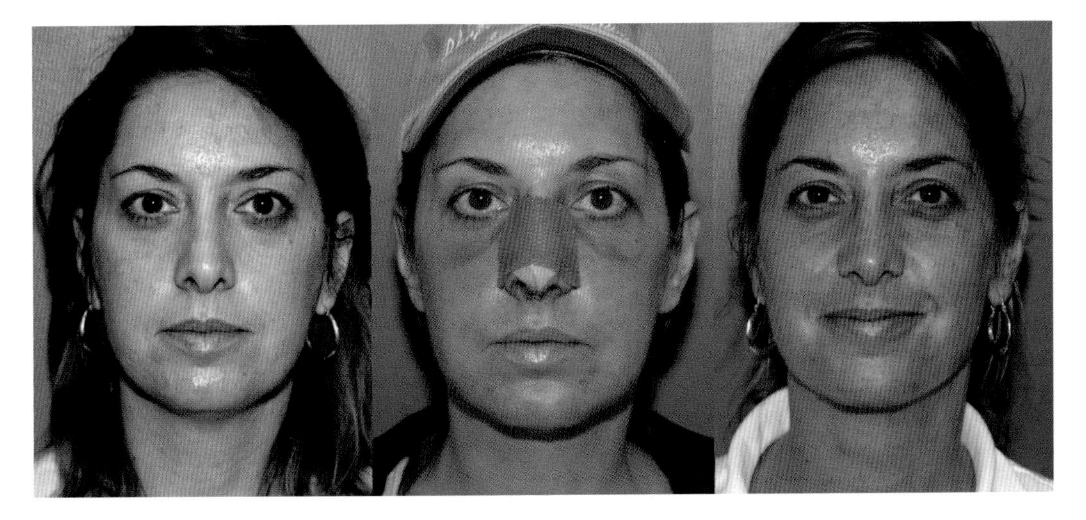

On the left, preop. In the middle, appearance with splint one week after rhinoplasty, chin implant, neck liposuction, and TCA peel. Postop on the right.

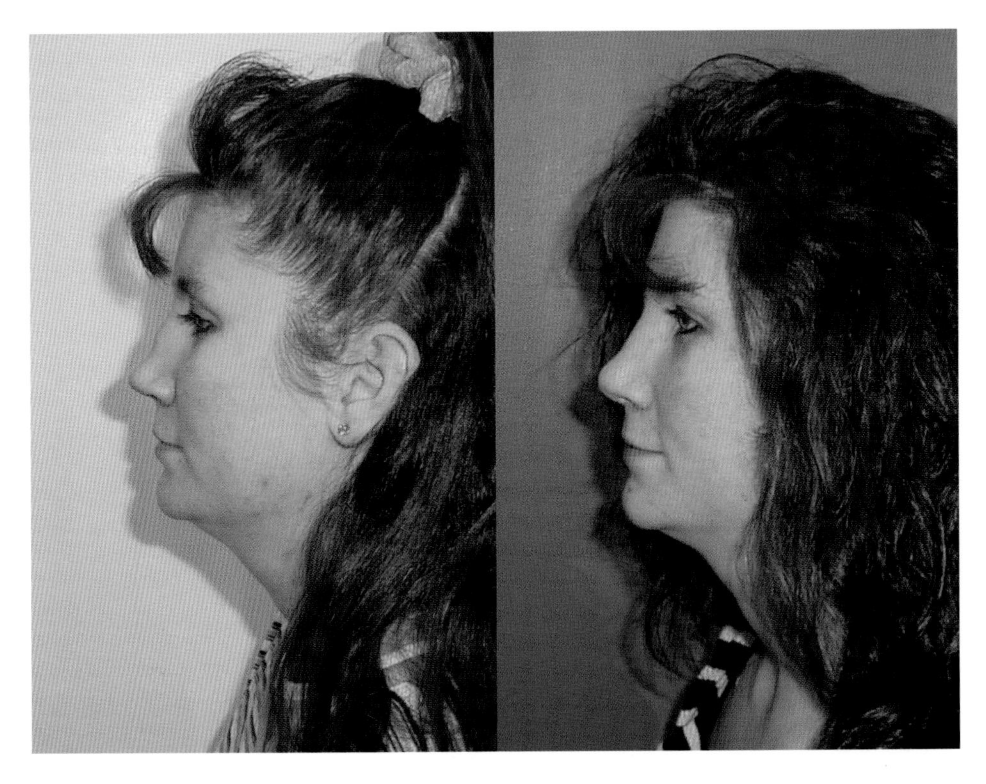

This woman's nose collapsed after a rhinoplasty and septal operation by another surgeon. On the left, preop. On the right, after reconstruction with cartilage graft from the ear to several places in the nose.

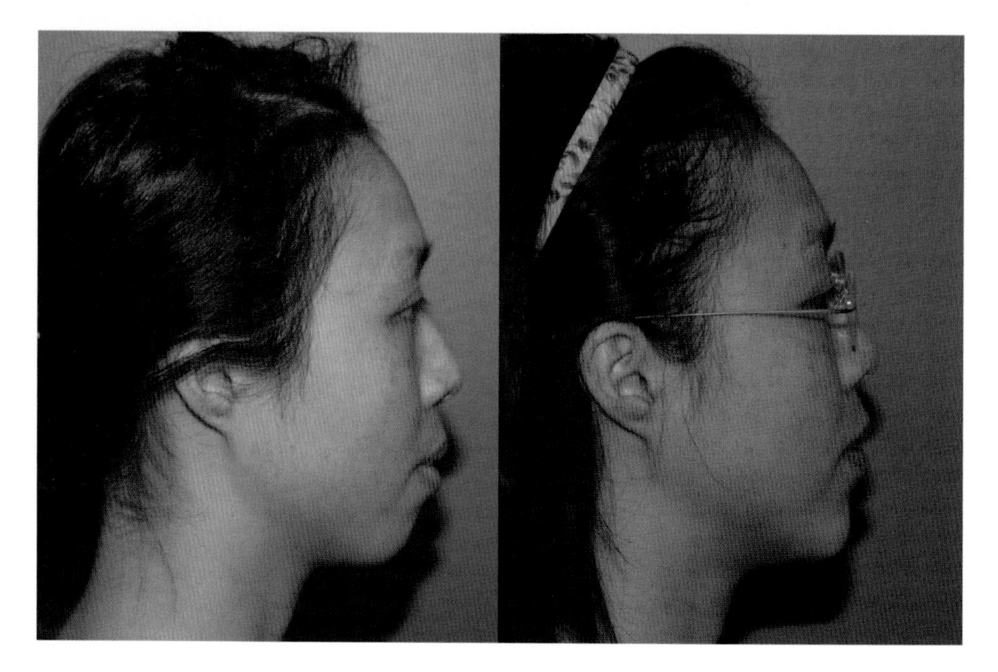

This 28-year-old woman had a very small chin (left). On the right, after chin augmentation with a silicone implant. The operation elevated the lower lip muscles and allowed better closure of her lips.

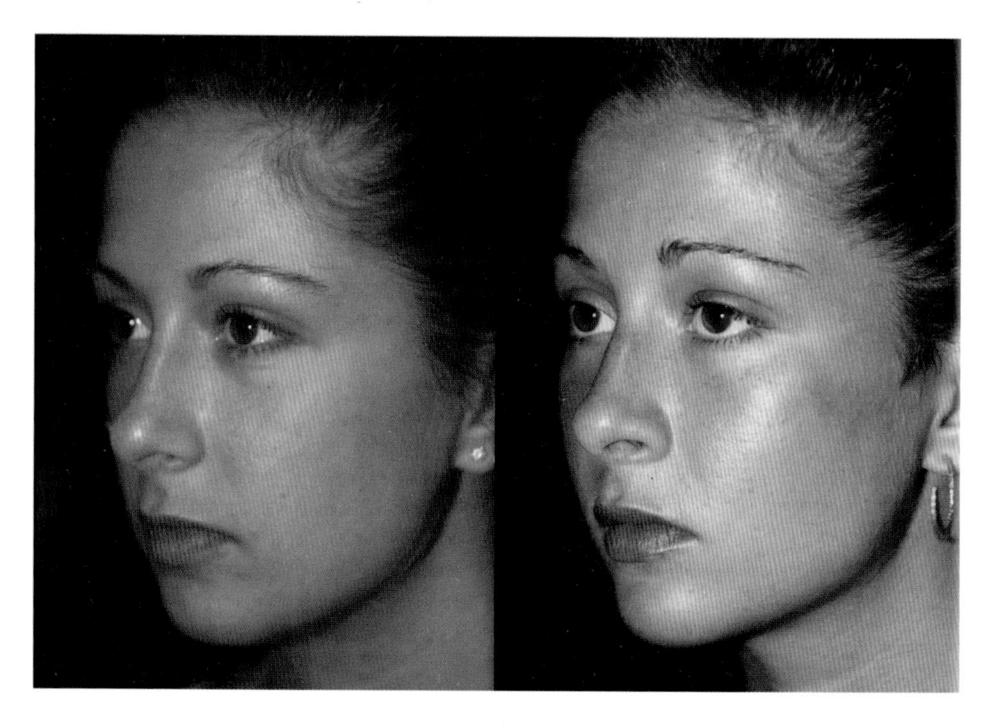

Preop on the left; postop rhinoplasty, chin and cheek augmentation on the right. Facial balance is achieved.

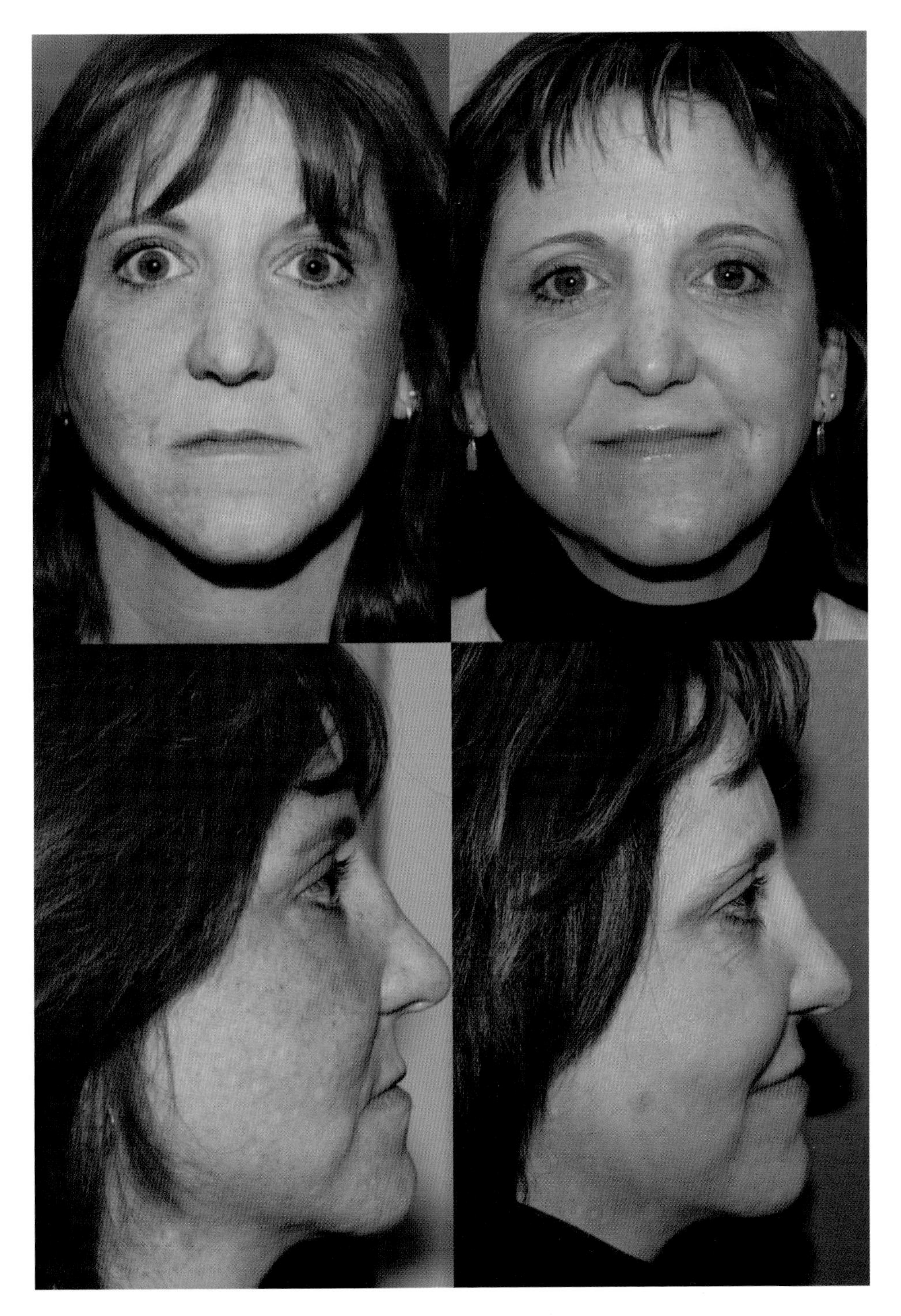

This woman underwent a dermal-fat graft to the upper lip, balancing the lower lip. A TCA peel also improved her skin tone and reduced the number of freckles.

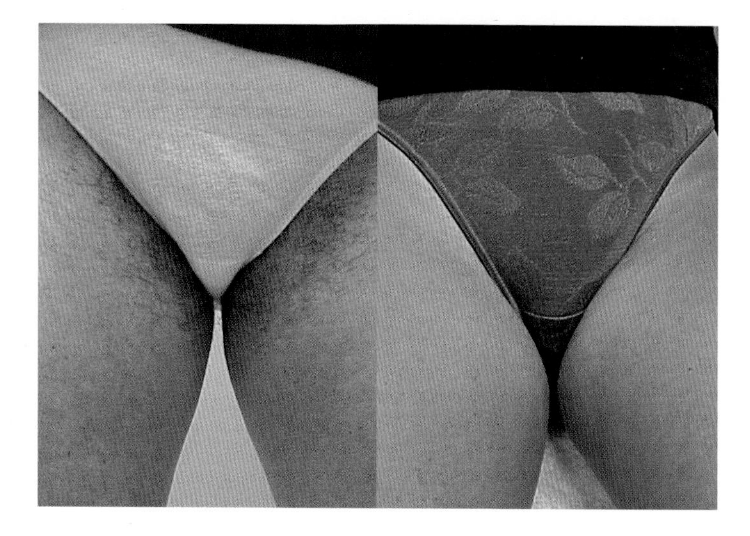

The bikini area is one of the most common areas for laser hair removal. On the left, prior to treatment; on the right, after three treatment sessions.

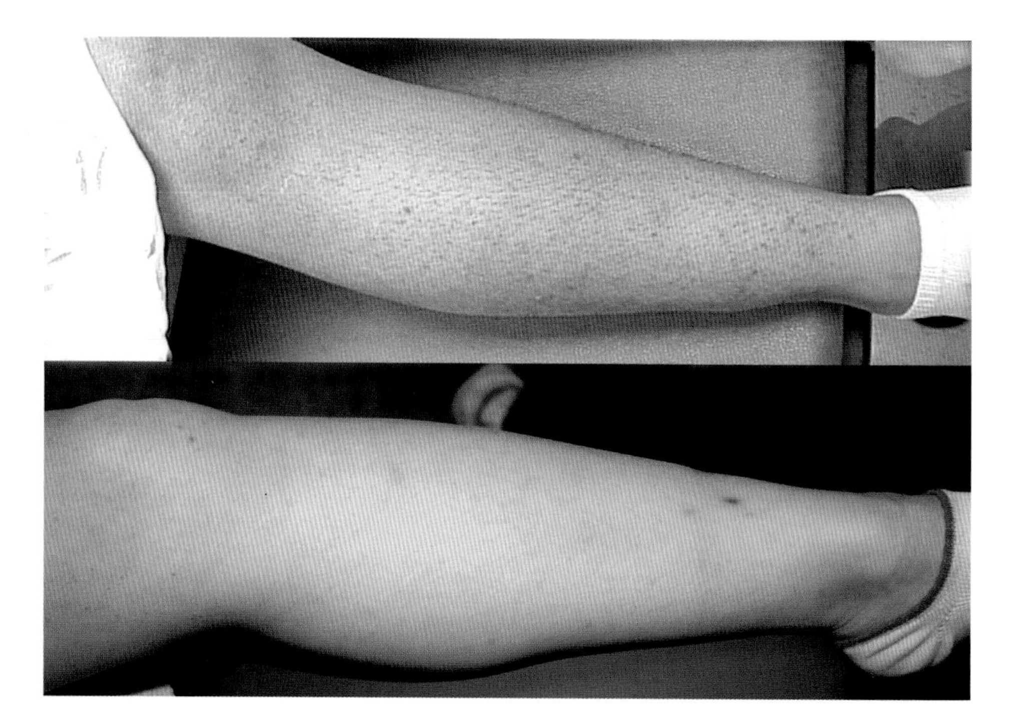

Top photo shows this woman's leg prior to treatment. Bottom photo is after four treatments.

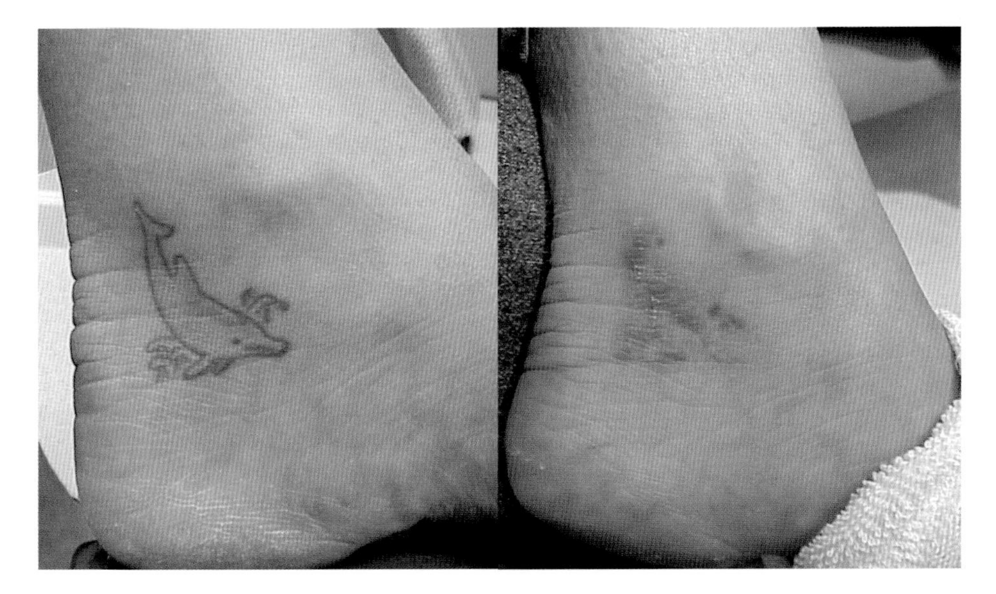

This woman underwent seven laser treatments to get rid of the pesky dolphin. Even so, some pigment remains. Left, before treatment; right, postop.

This tattoo had the most common color, blue-black. The photo on the right is after just three laser treatments. Notice the scarring left behind by the laser, or perhaps by the tattoo itself.

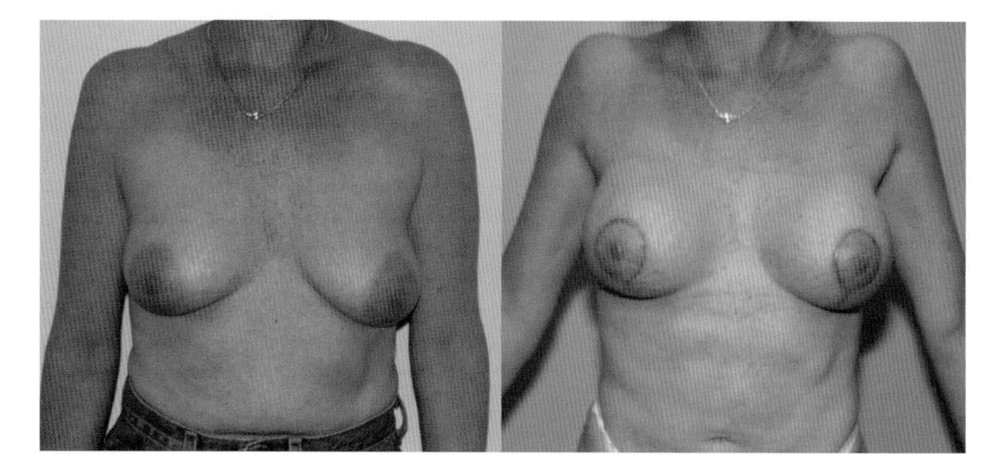

Bad scars occur in 3–5 percent of patients. Assuming your plastic surgeon has closed the wound properly, this hypertrophic scarring is genetically predetermined and largely unavoidable. This woman had small, droopy breasts. On the right, her appearance after a combined lift/augmentation procedure. She developed red, raised scars, although she had soft-feeling breasts.

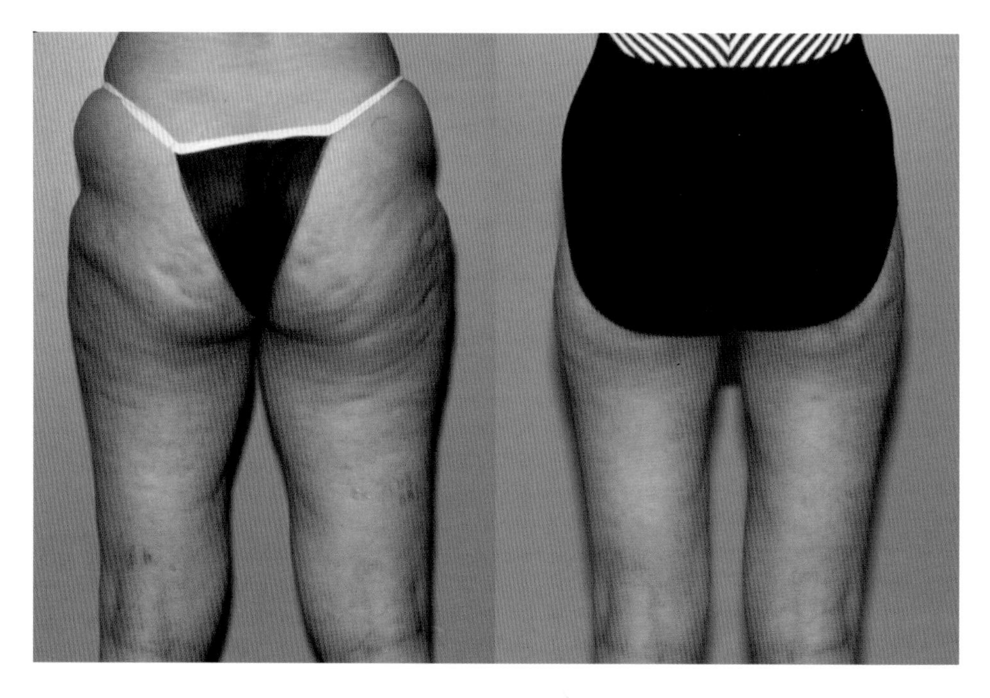

This woman underwent two rounds of sclerotherapy after having liposuction of the hips, thighs, and knees. Postop on the right.

18

Chin and Cheek Implants

Chin Augmentations

The first cosmetic procedures on the chin were actually performed during World War II. Cuts were made in the bone, and the bone was pushed outward and held in place with steel wires. The first chin implants were used in the 1960s and presented a much simpler method of making the chin larger.

A small chin is commonly called a weak chin—and that says it all. The impression is of a weak personality. A strong chin, particularly in a man, is associated with strength and leadership. Chin implants can be placed onto the bone, changing a weak chin into a more powerful, domineering chin. In men, the change increases the image of masculinity. In women, chin implants balance the face. In both sexes, however, a weak chin makes a large nose seem even larger.

Chin implants can be placed on the front of the chin, improving the profile by as much as a centimeter. Extended versions of implants are available that allow the width of the chin to be increased. Depressions in the jaw line can even be corrected with the creative use of these implants. Clefts can be created or removed. Most implants today are made of silicone rubber, although some surgeons use other materials, such as Gore-Tex (polytetrafluoroethylene), Medpor (polyethylene), and a bone-like hydroxyapatite.

Chin augmentations with silicone rubber implants are examples of a low risk, high-benefit procedure. Thirty-two thousand procedures were performed in 2005, making it the eleventh most common cosmetic surgery operation.

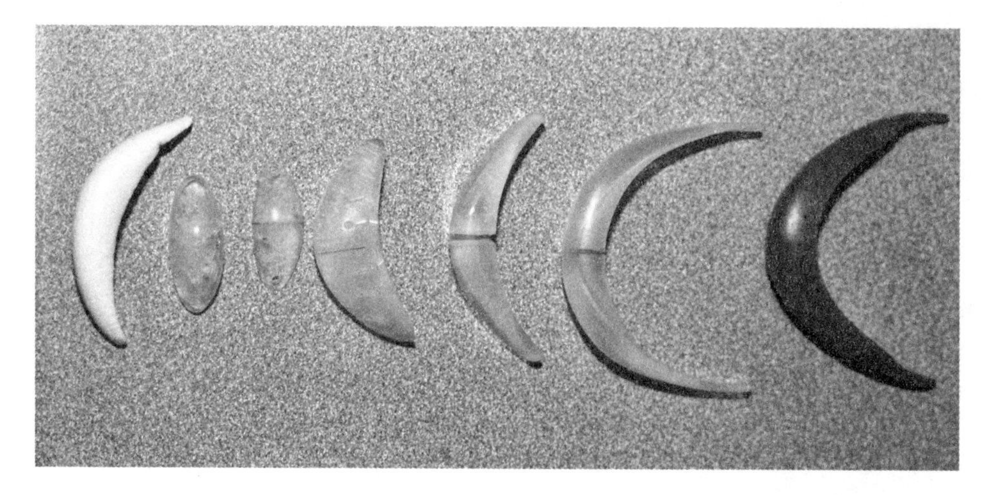

Different types and styles of chin implants. The white implant on the left is made of Teflon, the next one is a silicone gel implant. The other implants are various types of silicone rubber implants. The implant on the far right is used to determine the proper size during the procedure.

When the Chin Is Very Small, It May Be Better to Have the Actual Bone Changed with a "Sliding Genioplasty"

As with all foreign materials, the larger the implant, the more frequent the problems. When a large augmentation is needed, a horizontal cut in the chin bone can be made. The bottom portion is advanced and held in place with tiny titanium plates and screws. Not every plastic surgeon performs this procedure, but those who do claim it is no more risky than an implant and has numerous advantages. No foreign implant is present to move or get infected, although the plates themselves are foreign. And the chin can be vertically shortened or lengthened and advanced forward or backward. The downside, however, is that most genioplasties require general anesthesia.

Once successfully placed, a chin implant should be permanent. After a few weeks, the brain incorporates the implant into its body image. As with a dental filling, the implant no longer feels foreign.

The Procedure

Digital imaging is useful in determining the size of the proposed chin implant.

A chin implant can be performed under local or general anesthesia. A three-centimeter-long incision is made in the crease under the chin and is deepened to the bone. A pocket is made directly over the bone, fitted precisely to the size of the implant. The implant is washed with antibiotic solution and placed in the

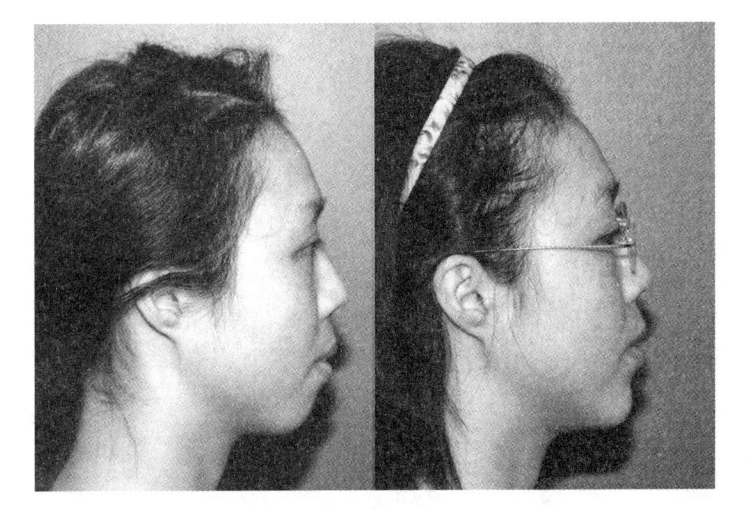

This 28-year-old woman had a very small chin (left). On the right, after chin augmentation with a silicone implant. The operation elevated the lower lip muscles and allowed better closure of her lips. See color plates.

pocket. The muscle and skin are then closed over the implant. Some surgeons prefer to make the incision inside the mouth, below the lower teeth. Although an incision made in this location is invisible, the chance of infection or slippage of the implant is higher.

The Risks

A chin implant can extrude through the skin. If it becomes exposed, it must be removed. Some surgeons recommend using antibiotics before dental procedures if the patient has a chin implant; others disagree. If the implant becomes infected, it must be removed.

An implant can shift in position, requiring a second procedure for repositioning. It needs to be placed carefully, so that the nerve that supplies sensation to the lower lip is not injured. All implants gradually erode the very bone that they were intended to augment. Sometimes the result is a loss of the improved projection of the chin. In extreme cases, the implant can erode into the roots of the teeth. Pain and sensitivity of the teeth result and sometimes a root canal is required.

Cheek Implants

The cheekbones are seen most easily with the head turned 45 degrees, the oblique angle. High cheekbones are considered beautiful by many cultures.

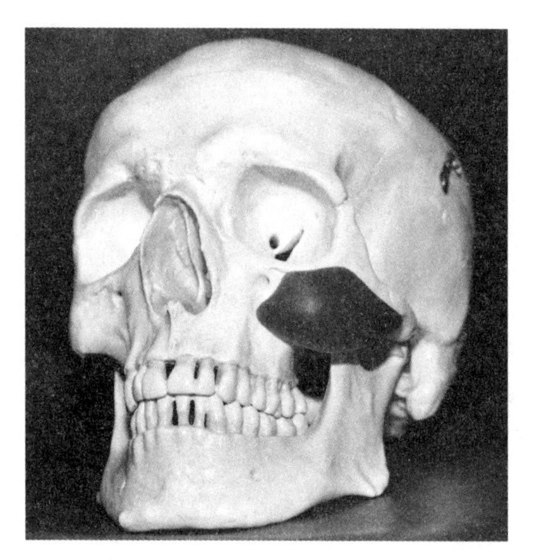

This silicone rubber implant sits on the cheekbone in this skull model.

Cheek implants are one method of restoring the cheek height that is lost with advancing age. Also called malar implants, the implants are usually made of silicone rubber. Midface or thread-lifts are other methods.

Cheek implants receive more mentions in the press than in the operating room. In fact, only 12,000 procedures were performed in 2005, making it the eighteenth most common surgical procedure.

Some have proposed making cuts in the cheekbones and advancing them to increase their prominence. Bone grafts and plates with screws are then necessary to hold the bone in place. This technique can also be used to decrease the size of cheekbones that are too big. Unless there is a major deformity, I would be reluctant to have my facial bones cut and moved if I were a patient. Implants are simple and straightforward.

Current face-lift techniques and minimally invasive suture lifting techniques raise the sagging cheek pads to a more youthful position, eliminating the need for foreign material in the face.

The Procedure

Cheek implants can be inserted either through the mouth, with incisions made above the upper molars, or through a lower-eyelid incision. They can also be placed through the skin incision during a face-lift. Usually general anesthesia is used, although local anesthesia is adequate for a motivated patient. A pocket is made directly on the cheekbone, and an appropriately sized silicone rubber implant is washed in antibiotics and positioned. The skin is then closed.

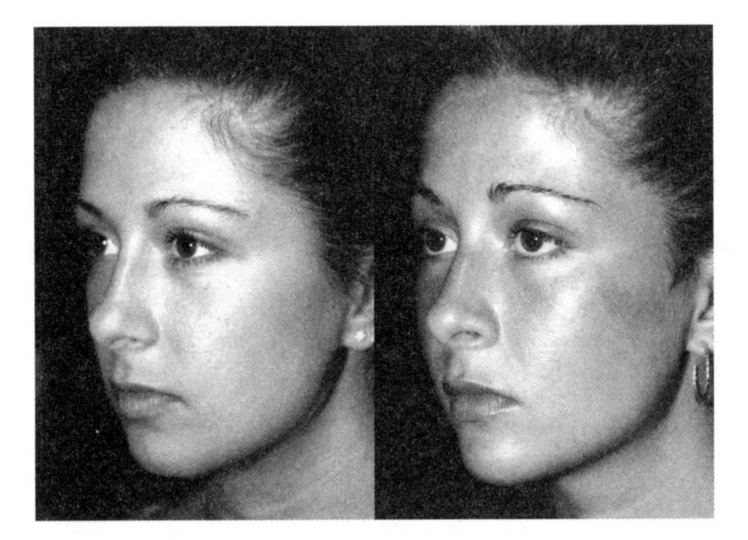

Preop on the left; postop rhinoplasty, chin and cheek augmentation on the right. Facial balance is achieved. See color plates.

The Risks

Like other implants, cheek implants can become infected. If this happens, they will need to be removed. Cheek implants are harder to position than chin implants and are more likely to be placed too high, too low, or off to the side. Since these implants are paired, they must be perfectly symmetrical in order to look good. The cheek implant is placed very close to the nerve that supplies sensation to the cheek, upper lip, and nose. If this nerve is injured, an annoying sensation is experienced until it grows back. Worse, the nerves that supply the muscles of the upper lip and cheek may be injured during the operation, resulting in either weakness or paralysis. Needless to say, this is a potentially devastating complication. Fortunately, when it does occur, it is usually temporary.

The Cost

Cheek implants cost $2,700 on average and chin implants cost $2,100. This does not include the cost of the implants themselves, which averages about $150 apiece.

19

Lip Augmentations

In the mid-1980s, lip augmentations were virtually never performed. Then, sometime around the early 1990s, large lips became desirable. Goldie Hawn's hilarious lip augmentation in the 1996 movie *The First Wives Club* actually increased my lip enlargement business. It is hard to know just how many of these operations are performed annually, since organized plastic surgery counted only the 50,000 *surgical* lip augmentations in 2005, exclusive of augmentations from the injection of fillers. The popularity of lip augmentation certainly has increased over the last decade.

Foreign materials do not belong in the lip. Silicone or Gore-Tex do not do well there. The newer filling materials Artecoll and Radiesse can cause hard nodules in the lips.

Surgical procedures to alter the shape of the lips are not popular. They leave scars and are really not totally successful.

Dermal-Fat Grafts

This type of graft is more durable than other techniques. It has been around for over half a century, and the lip augmentation is simply a new use for an old technique. With dermal-fat grafts, an area of skin somewhere on the body is excised in the shape of an ellipse. The upper skin (epidermis) is removed, leaving the lower level (dermis). The skin is then cut out, keeping a layer of fat attached. The location where the skin was removed is closed as a straight line. Often I remove the skin in a scarred area, in hopes of improving the existing scar, giving a secondary benefit.

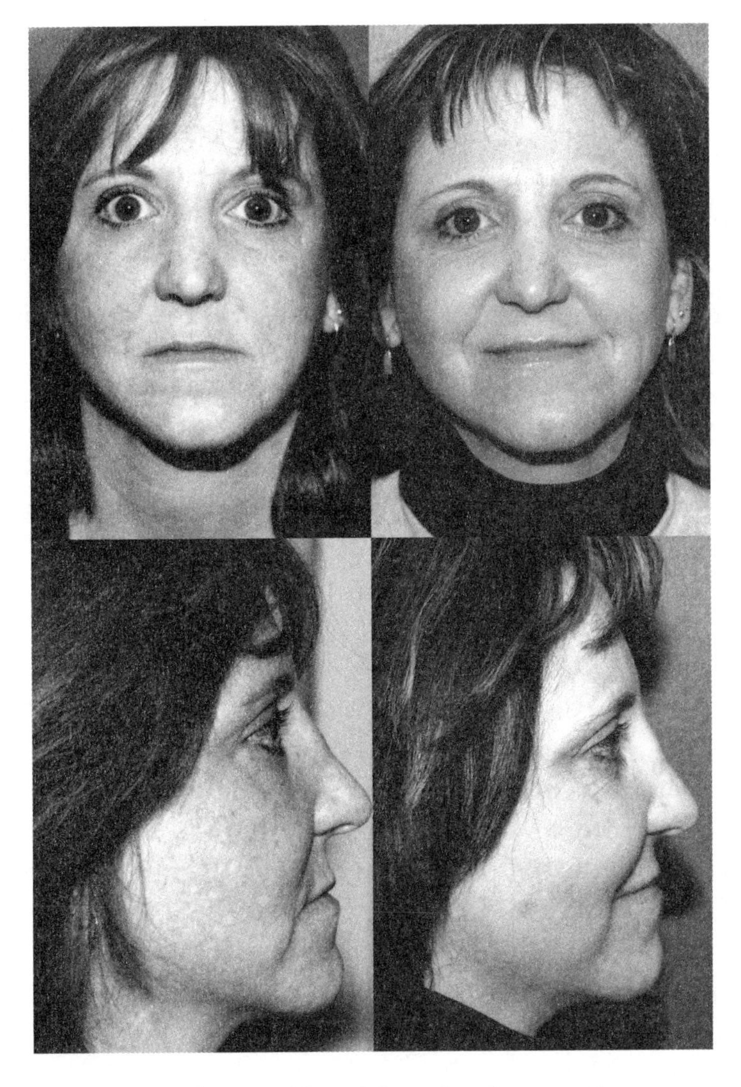

This woman underwent a dermal-fat graft to the upper lip, balancing the lower lip. A TCA peel also improved her skin tone and reduced the number of freckles. See color plates.

The dermal-fat graft is then shaped. For the lips, it is made into a narrow strip. The lips are numbed and tiny incisions are made in the corners on the inside of the lip. A space is made in the lip, under the red lip skin. With a long clamp, the dermal-fat graft is threaded through the lip and the incisions are closed.

This type of graft behaves just like any other skin graft. It gets its own blood supply within three days, and becomes part of the lip by three weeks. The percentage of graft that survives should be very high.

Not everyone wants this operation, however. It is most popular in people who already have a scar somewhere. Usually abdominal, back, or buttock scars work well. If your body is scarless, it is unlikely that you will choose to have a five-to-seven-centimeter scar created just for this purpose.

Fat Grafting Is a Much More Common Method of Lip Augmentation

Through tiny incisions inside the belly button or on the hip, the fat is numbed and liposuction is performed. Only a few teaspoons of fat are required. This fat is drained of blood and reinjected through tiny incisions in the lip. In actuality, the fat is not injected, but layered into the lip by lifting the lip skin with the injecting needle. Fat is extremely pressure sensitive and will not survive a true injection. To live, it must contact tissue but must not be placed under pressure. I use the instrument to create multiple tunnels not only through the tissue just under the lip skin, but also in the lip muscle.

After the procedure, the lips swell comically. Pleco fish come to mind. The profound swelling goes down rapidly, however, and within five days is almost gone. I graft extra fat, because much of it will not survive. Less survives in the lips than in the nasolabial folds. No one knows why, but I believe it is because the lips cannot remain at rest. Eating and talking probably disrupt some of the fat in the early days following the graft. My feeling is that about one-third of the fat survives long term. I tell my patients that in three to six months they will require a second augmentation. It seems that a higher percentage of fat survives during the second procedure. This makes sense biologically: for a time, the blood supply of the lips is increased after surgery, and grafts survive longer with a better blood supply.

If a Surgical Procedure Is Not Desired, Then "Lips from a Bottle" Can Be Created

I prefer Restylane for temporary lip augmentation. After the lips are numbed with lidocaine, Restylane is injected slowly and carefully. It will last between six and twelve months. One of the advantages of Restylane is that we can add another syringe the following week if even larger lips are desired.

Often a patient is not sure just how big her lips should be. In this case, I inject her lips with saline and have her look into the mirror. We record the amount of saline injected, and this is the amount of fat or Restylane that I use.

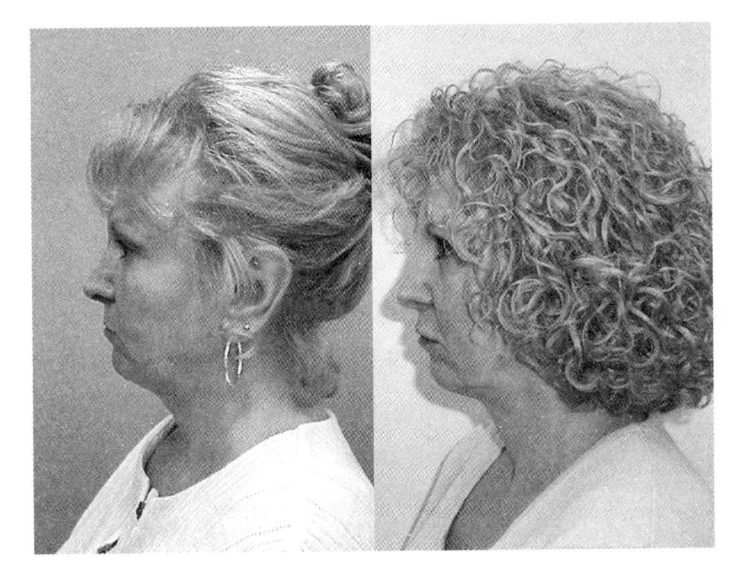

This woman had two fat-grafting procedures to augment her lips. Additionally, she underwent liposuction of the neck and jowls and a blepharoplasty. Preop on the left, postop on the right.

Other Procedures

Some surgeries that plump up the lips through incisions leave noticeable scars. For this reason, they are extremely unpopular. I have never met a woman willing to trade noticeable scars around the lips, requiring constant makeup, in return for larger lips.

Lip reduction is another rare procedure, particularly in this era of popularity of the large lip. Incisions are made inside the lips and a horizontal wedge is removed. Stitches are left in for about a week.

The Risks

Lip augmentations are safe. Infections and bleeding are rare. While portions of the lips can be destroyed when errant injections clog arteries, this outcome is rare. Loss of sensation occurs in lip augmentations, although it should be temporary, as these nerves grow back readily. The final cosmetic result is largely dependent on the artistic skill of the surgeon. As in most cosmetic surgical procedures, choose your surgeon wisely and ask to see photographs of prior patients. If too much Restylane is injected in one area, pressure can disperse it. With fat grafting, pressure will only work during the first three weeks; after that it must be surgically removed. Fat will go away unpredictably after injection of steroids. If asymmetry occurs with a dermal-fat graft, either some can be removed, or additional fat can be added with a fat graft.

Women should proceed slowly with a lip augmentation. It is easy to add more material later and make the lips larger. If the lips are made too large, Murphy's Law says that permanence may result.

The Cost

The average fee for a lip augmentation is $1,800. The statistic is a difficult one to compile, since some augmentations use your own fat (free) and others use expensive filling materials.

20

Ear Reshaping

Protruding ears are one of the most common abnormalities at birth. No statistics are available that fully define the occurrence of this deformity. It would be difficult to figure out, because there is no solid definition of what is normal and what is too much protrusion.

In 1881, the first operation to correct protruding ears was performed in New York City. In 2005, 27,000 otoplasties were performed in the United States, with 40 percent of the patients being men. Many men have this procedure because their hair doesn't hide their ears in the way women's hair does.

The framework of the ear is made of cartilage, a plastic-like material that can bend to the point of complete folding without breaking. During fetal development the cartilage of the ear folds into its final shape. In babies born with protruding ears, the portion of the ear known as the antihelix is not folded properly.

The ear cartilage of babies is very malleable. Just as we can mold the ears of a dog, it is possible to tape and shape human protruding ears. This molding must be performed within the first three months of life to be effective.[1]

The ears grow early in life, and 85 percent of their growth is completed by age 4. Children begin to be teased about their ears around this age, and they can undergo reparative surgery as early as 4 years. Some plastic surgeons and parents prefer to wait until a child is older before considering surgery. By about age 6, it may be possible to perform the surgery under local anesthesia with sedation in a properly motivated child. It may not be possible in other children until much later. The decision to have surgery takes into consideration the amount of teas-

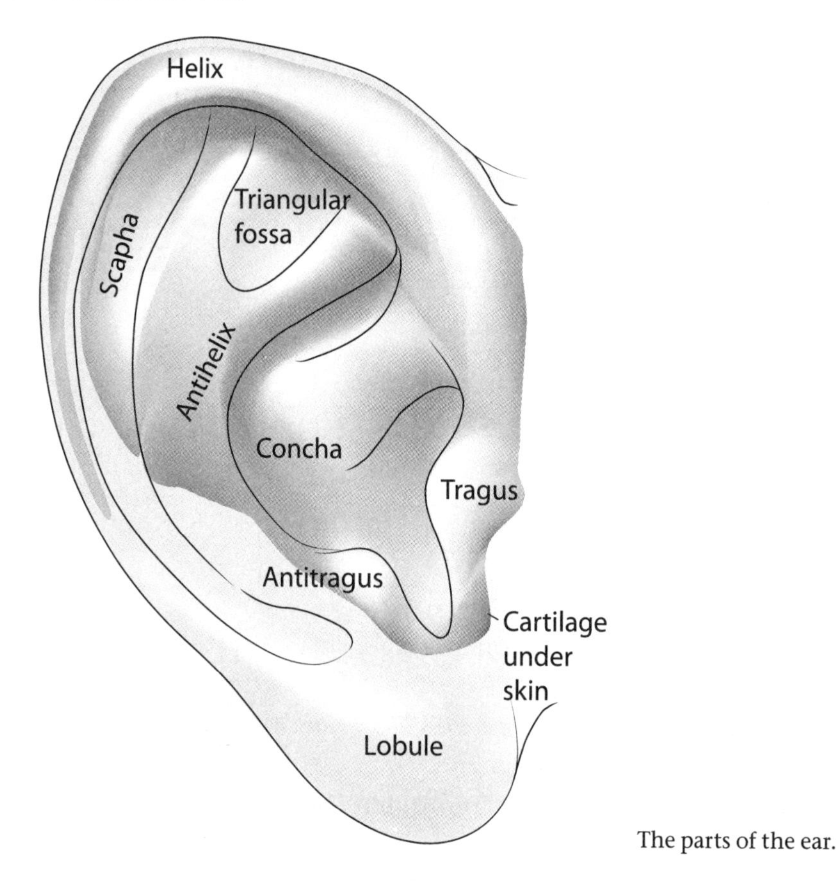

The parts of the ear.

ing and the emotional distress the child has to endure. You can buy time if the girl or boy can grow hair long enough to hide the ears.

Surgery on Protruding Ears Is Called an Otoplasty

This literally means "ear molding." An incision is made in the back of the ear. Through this incision, the antihelix is folded into the appropriate shape. Several permanent stitches hold the shape. The front of the cartilage is scratched to purposefully warp it into a more appropriate shape. Skin is usually not removed, although in some cases cartilage is removed from the concha, the main cartilage of the ear. Sometimes the earlobes are reshaped. As in all cosmetic surgery, individual problems are assessed and the surgeon generates a specific operative plan.

During an otoplasty I can shape the ear in a number of interesting ways, even creating "Spock ears." After reading this paragraph, someone will almost certainly call my office and request them. They are actually possible to create and will certainly create conversation around the office!

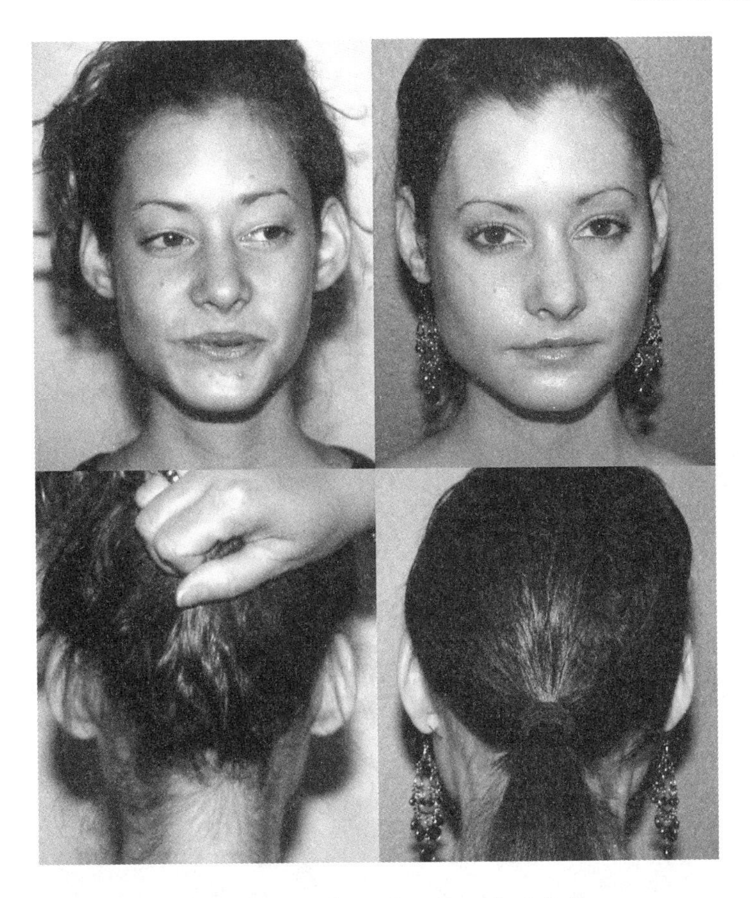

This woman underwent otoplasty. Preop on the left, five years postop on the right.

The Procedure

This surgery can be performed under local anesthesia, local with sedation, or general anesthesia. The hair is taped back and Betadine kills skin bacteria. The front and the back of the ear are injected with the anesthetic lidocaine and epinephrine, a chemical that stops bleeding. An incision is made in the back and possibly the front of the ear. Through these incisions the cartilage is shaved, bent, whittled, and removed, depending on the surgeon's preference and the particular problem. The skin is then closed, and a form-fitting dressing is placed. Each ear takes about an hour.

In an otoplasty, the surgeon often sets the ear back slightly more than the desired final result because a small amount of rebound will occur. The artistry of this operation includes not only creating a beautiful ear, but correctly predicting how much bounce-back will occur.

The look on this teen's face says everything. Preop on the left, postop on the right.

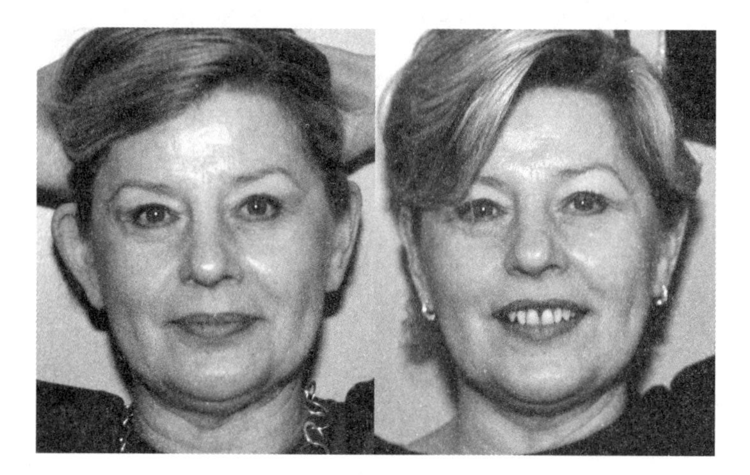

Sometimes only one ear protrudes. This woman underwent correction at middle age and said she only wished she had done it sooner. Preop on the left, postop on the right.

The Risks

Besides the risks that can occur with any surgery, if bleeding occurs with this procedure it can result in a "cauliflower ear." This deformity forms for fascinating biological reasons. The connective tissue next to cartilage can create new cartilage if it is lifted away from its neighboring cartilage. Whacking the ear of a boxer creates bleeding between the cartilage and the connective tissue. The body's response is to make more cartilage. And so it thickens.

Artistry is paramount in this operation. If the cuts in the cartilage are too deep, they may be visible through the skin. If not deep enough, the cartilage may not bend appropriately.

Infection is a dreaded complication of otoplasty and can result in a horrible but uncommon deformity. Sometimes the stitches become visible behind the ear and even poke through the skin. If this occurs, after about six months they can be safely removed.

The Cost

The average fee for an otoplasty is $3,000. Of course, this does not include the fees for the facility and the anesthesiologist.

21

Hair Restoration

Hair transplantation surgery has been around for decades. Interestingly, it is losing its popularity—virtually the only cosmetic surgical procedure to do so. The year 1997 saw 61,000 hair transplants performed, but in 2005 there were only 14,000. There was a 41 percent decrease in procedures between 2004 and 2005 alone. Typically poor results are the most likely reason. Now we have seen what twenty-year-old hair transplants look like, with obvious scarring when the transplanted hair follicles follow their genetic destiny and fall out.

Although the first hair transplants were performed in 1822, they did not become popular until the 1950s. It was the New York dermatologist Norman Orentreich, M.D., who showed that transplanted hair retains its genetic destiny when moved. That is, if it is taken from an area that will never become bald, the hair will survive no matter where it is placed.

At birth, there are many hair follicles all over the body that remain small until the male hormone testosterone causes them to grow. The same stimulus that causes hair to grow on the face also causes it to fall out from the scalp. Typical male-pattern baldness results.

Back in the 1960s, plugs of hair were taken from the back of the scalp and planted into the bald areas. The result looked somewhat like a cornfield. By the 1980s, procedures used smaller plugs, called minigrafts, containing two to five hairs. Results were better, but still not great. When single hairs were transplanted into the scalp, the results were more natural.[1]

Today most plastic surgeons use single hair follicles for the front hairline and fill in behind the hairline with minigrafts. Some use only single-hair grafts.

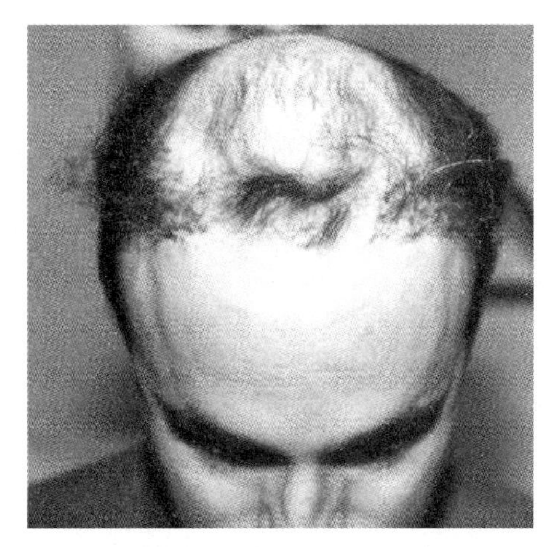

This man had undergone a poorly performed hair transplant by another doctor. He ultimately had all of the plugs removed, preferring scars to the corn-row appearance.

To decrease the bald area, scalp reductions (called scalp lifts) can be performed before grafting. Often several procedures are used to repeatedly remove bald skin, thereby making the bald area progressively smaller.

Older procedures that move large flaps of hair containing skin, or even microsurgical transplantation of hair, are becoming increasingly rare.

Many plastic surgeons do not perform hair transplants because the procedure is tedious and, frankly, boring. Nonsurgeons have filled the void. Chains of hair transplant clinics advertise heavily and act like factories, with "surgeons" who have often not completed any surgical residency. Technicians perform much of the operation. They cut up the hairs and even plant them into the scalp. In most states, performing a procedure in this manner is technically illegal, since by law the surgeon must be in the operating room supervising all aspects.

The Procedure

Local anesthesia is usually used, supplemented by oral narcotic painkillers and antianxiety medication. A strip of hair-bearing scalp is removed from the back of the head and sliced like a loaf of bread. The slices are divided into small pieces, which in turn are divided into groups of individual hairs or clusters. The donor area (where the hair was taken from) is closed, and tiny incisions are made in the scalp where the hairs are to be transplanted (the recipient area). Once the hairs are ready, they are chilled on ice until they are placed in the incisions. Behind the frontal hairline, clusters of more than one hair can be transplanted. Hairs cannot be placed closer together than one millimeter in a single session. The process is extremely tedious and takes several hours. Typically, five hundred to twenty-five hundred hairs are transplanted in a daylong session.

After the procedure, the hair follicles will not be sturdy for a few weeks. Most of the hairs will initially enter a resting phase for several months and fall out, but then will start growing again. Results cannot be evaluated before one year. At that time, a second procedure to further increase density can be considered.

The Risks

Infection and bleeding are risks with this procedure. The hair follicles must be handled gently or they will be destroyed. Typically, about 85 percent of the transplanted follicles will grow within a year. Scarring is a problem with hair transplant surgery, particularly when performed by amateur surgeons. Transplants should not be done in men who are expected to lose all of their hair. If the hair is destined to be lost eventually, transplantation will not alter its fate.

Poor cosmetic results are the most common risks of hair restoration. They are so frequent that many patients are reluctant to undergo the procedure. Problems can occur with scarring in both the donor and recipient areas. Cysts in the transplanted area occur in about 5 percent of people.

The hairline must be designed properly, not too low and with a natural design. The transplanted hair density may be inadequate. Sometimes the transplanted hairs fall out, following their genetic destiny. Newly growing hairs can be angled inappropriately, causing difficulty in combing the hair. The grafts can be depressed or can grow with a cobblestone appearance.

Hair Transplants in Women

Women usually do not lose the frontal hairline but, rather, thin throughout the front half of their hair. Transplantation in women involves filling in between existing hairs with either single-hair transplants or minigrafts.

Eyebrow Transplantation

This is an uncommon procedure. Women and men who lose portions of their eyebrows can have them reconstructed by moving single hairs from the back of the scalp or the nape of the neck to the eyebrows. The hairs must be angled precisely to match the existing hairs. Because they will grow faster than the slow-growing eyebrow hairs, they must be trimmed weekly.

Other Possibilities

Minoxidil (Rogaine)

This drug was originally used to control hypertension. A side effect was hair growth. Minoxidil can jump-start shrunken hair follicles and cause thicker hair to grow for longer periods. Now available without a prescription, minoxodil liquid is applied to the dry scalp twice a day. It takes about twenty minutes for the chemical to be absorbed. Most people do not have actual new-hair growth. Used in both men and women, it does cause a significant proportion of people to have their hair loss slowed or stopped.

Finasteride (Propecia)

This drug prevents conversion of the male hormone testosterone into dihydrotestosterone. By doing this, the testosterone cannot cause the hair follicles to shut down. The drug can, indeed, prevent ongoing loss of hair but cannot cause lost hairs to regrow. Propecia is taken daily; sexual side effects occur in over 3 percent of patients. These include less desire for sex, difficulty in achieving an erection, and decrease in the amount of semen. Propecia should be taken only by men.

Infrared Light

Luce LDS 100 is a new infrared light therapy that claims to cause hair growth. Patients are exposed to the light for a half hour, twice a week, for five weeks. Then the treatment is performed at progressively longer intervals. There have been no controlled studies of this system. The idea probably came from a research study in 2003 that used a similar light to effectively treat a hair-loss disease called alopecia areata.[2]

The Cost

The average fee for a hair transplant procedure is $5,000. Many procedures cost upward of this number, depending on the region of the country and the skills of the surgeon. Of course, the costs vary with the number of grafts implanted. Procedures as expensive as $10,000 are not unheard of. Scalp reductions would be expected to cost between $2,000 and $4,000 for each procedure.

22

Cosmetic Dentistry

It always amazes me that people can be worried about the smallest wrinkles on their faces but fail to address their yellow or crooked teeth. Patients who are interested in cosmetic surgery usually don't have chipped or missing teeth, however. According to my dad, Michael M. Perry, D.D.S., a practicing dentist for 60 years, the current crop of television shows has caused a definite increase in people having cosmetic dentistry. "In the last ten years, people became conscious of the color of their teeth. When the whitening materials came out, they started lightening the teeth. This caused an interest in porcelain crowns and veneers. Several television programs have shown before and after photos, including teeth with cosmetic surgery and physically fit bodies. A nice smile became important," my father told me.

Cosmetic dentistry started out as a specialty performed by oral surgeons and periodontists. Today, more and more well-trained general dentists perform cosmetic dentistry, which works hand in hand with cosmetic surgery.

What really is cosmetic dentistry? Much of dentistry *is* cosmetic dentistry. The accent now is not only on preservation, but on creating an attractive smile, too. Attention is focused on the bleaching and cosmetic restoration of teeth and on orthodontia, which allows crooked and malpositioned teeth to be moved into proper alignment.

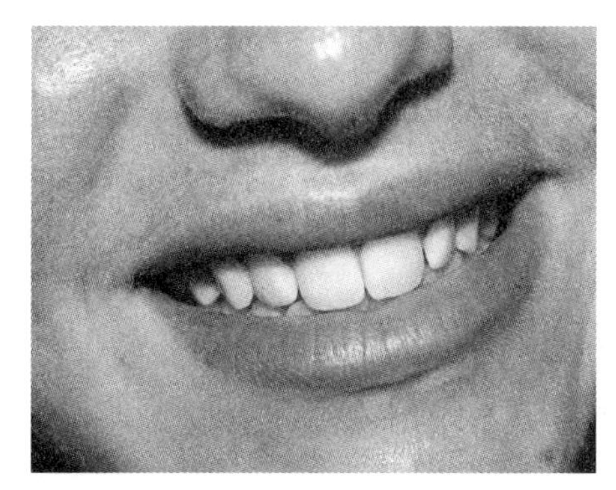

Beautiful teeth are critical to an attractive appearance.

Bleaching

Bleaching is the whitening of darkened teeth. A peroxide gel ranging from 7 to 30 percent concentration is put on a clean tooth and is left in place for a variable amount of time. Heat or special curing lights speed up the process.

Bleaching originally started as a home treatment. The teeth are first cleaned, then the peroxide gel is applied and held in place with a form-fitting tray for half an hour a day, for about two weeks. If the teeth are dirty, bleaching is less effective. Bleaching can irritate the gums and cause tooth sensitivity. It will not lighten fillings or crowns and can result in splotchy teeth. Tetracycline-stained teeth are difficult to correct with bleaching.

Dentists now perform bleaching in their offices, using a variety of different techniques and materials. The in-office procedures frequently employ a curing light, which supposedly increases their efficiency. Face and gums are protected from the chemicals and lights. The results are superior to results from home products.

Name-brand bleaching systems such as Brite Smile, Zoom, and Sapphire-Rembrandt light systems are available, but no one system appears to be markedly different than the others. Sensitivity is less of a problem with the newer techniques and improved chemicals.

Orthodontia

Orthodontia is the movement of malpositioned teeth into a more normal position. The correction has both functional benefits (better chewing of food and better mouth breathing) and cosmetic benefits. In addition, by improving the bite, the teeth don't bang into one another and last longer. Invisalign is a new technique that uses clear plastic mouthpieces. Patients wear the mouthpieces for six weeks and then have new trays made, sequentially moving their teeth. This is appropriate for simple orthodontia. Invisalign can be performed by a general dentist. Costs range from $2,500 to $5,000 depending on the number of mouthpieces required and the time involved.

Prosthodontics

This branch of dentistry deals with the replacement of teeth by removable and fixed dentures. The latest improvement is the use of a flexible material that replaces the ugly metal wires that hold partial dentures in place. The new material is tooth colored and snaps into place almost invisibly. The newer materials are lighter and thinner and stronger than their predecessors.

Veneers range from $500 to $2,000 per tooth, depending on the materials and the experience of the dentist. Porcelain crowns are a little more expensive, ranging from $750 to $2,500. There is much less tooth removal with a veneer: a crown must shave down the existing tooth structure to adhere properly. If the tooth structure is significantly decayed or twisted out of shape, a crown is required to produce an aesthetically pleasing result.

Implants are titanium metal anchors screwed into the bone. Crowns are later fastened onto these anchors. The surrounding teeth do not need to be ground down. If multiple implants are placed, a removable denture may not be necessary.

Changes in Operative Dentistry

Tooth-colored materials are starting to replace the more noticeable silver fillings. First used in the 1970s, the materials have been improved and now last much longer. "These types of materials will eventually replace silver fillings completely," says Dr. Perry. Gold is still considered the best material, "the gold standard," but is expensive, time consuming, and noticeable. It still is used in less visible teeth in the back of the mouth.

The new dental drills are much faster than the older drills, spinning at 350,000 rpm. They are air- and water-cooled, creating less heat and less pain. Air abrasion and laser abrasion techniques are used for drilling the teeth, theoretically with less pain. The use of lasers in dentistry is still in its infancy and is expected to increase with further research.

Bonding is used to attach filling material to the tooth without drilling large holes. The benefits are less tooth destruction and a stronger tooth after restoration.

Periodontia Deals with the Gums

When Shakespeare said a man was "long of tooth," he was referring to receding gums in the elderly. In actuality, the bone structure recedes and the gums shrink along with it. This process exposes the root and makes the tooth look longer. It can also result in sensitivity when the teeth are exposed to heat, cold, and sugar. Periodontia is the treatment of the supporting bone and gums. The periodontist may recommend baking soda and peroxide rinses to destroy the bacteria that cause receding gums. The gums can be moved to cover the bone, and gum grafting can be performed to lessen the appearance of periodontal disease.

After cosmetic surgery, your plastic surgeon may suggest a visit to a cosmetically oriented dentist who will complement your new appearance with an oral makeover. It doesn't make sense to have youthful eyes and a mouth full of bad teeth.

III

Lasers

23

Laser Hair Removal

Hair. Those who have it don't want it. Those who don't have it want it. Sometime in the 1990s hair became ugly. I'm not talking about hair on the top of your head. I'm talking about the rest of the hair on your body. Certainly, unwanted hair has plagued women for centuries. More recently, men too seem bothered by it.

Hair shows up in the usual places: the underarms and the groin. But it also appears in places we don't expect it: the nipples, the upper lip, even the chin in women; inside the ears and nose in men. Techniques of hair removal include plucking, waxing, bleaching, shaving, and electrolysis. In the 1990s, lasers that destroyed hair were invented. The early versions were not too powerful and resulted in only temporary hair removal. Newer, more powerful lasers chill the skin and kill the hair permanently. These lasers were breakthrough products. Some of the most satisfied patients I have are women who have had beards removed, or who are rid of chronic ingrown hairs in the bikini area. Over 1.5 million people underwent laser hair removal in 2005.

How Lasers Work

Lasers split light into one precise color. Think of light coming in your window, hitting your chandelier, and causing a rainbow of colors to be projected onto the wall. Now capture just one of those colors and intensify it millions of times. That is basically a laser. The laser light passes through substances that are not the correct color and is completely absorbed by one specific color. Scientists and doctors call this substance a chromophore—the target for a specific laser.

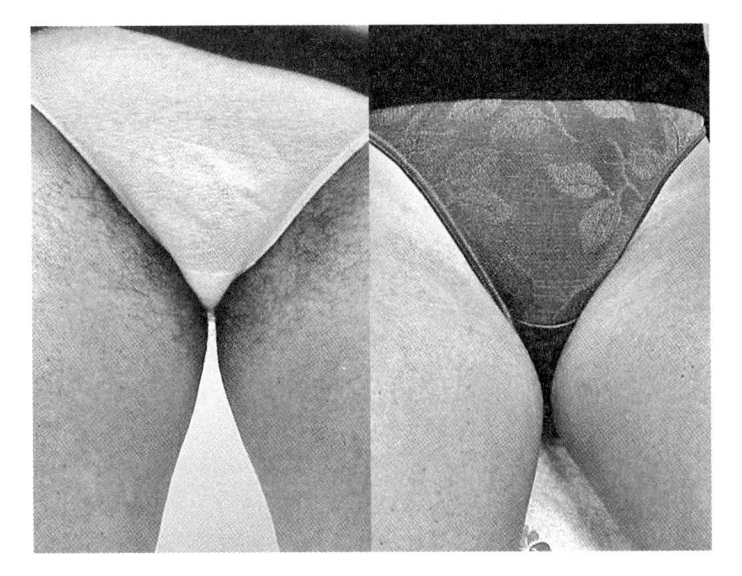

The bikini area is one of the most common areas for laser hair removal. On the left, prior to treatment; on the right, after three treatment sessions. See color plates.

When intensified light, specifically tuned for, say, dark brown color, hits the hair, it superheats and vaporizes it. The hair acts as a fuse, leading all the way to the hair follicle. When the heat absorbed by the hair hits the follicle, it fries and permanently destroys it. If you think about it, laser hair removal is strikingly like electrolysis. In that process, a metal electrode is placed down the hair shaft until the follicle is reached. Electric current is then passed along the wire, frying the follicle.

Laser hair removal is much more efficient than electrolysis. In a half second, a 9 × 9 millimeter square of skin is treated. All the growing hairs within this square are destroyed. The process is relatively independent of the skills of the practitioner; the laser finds the hairs.

At any given time, only a portion of the hair grows. Other hairs are resting. Laser hair removal kills only the actively growing hairs. And it kills only follicles that have actively growing hairs. That is why techniques that remove the hair, such as waxing or plucking, decrease the efficiency of laser hair removal. These techniques should be avoided for months before the laser procedure. You don't have to be hairy, though; you can shave the hair or use a depilatory cream such as Nair.

The laser destroys any brown it encounters. It doesn't know the difference between hair and tanned skin. The darker the skin, therefore, the more the blistering. When the skin is hit by the laser, it responds by making pigment. This can be a big problem. It turns out that people with olive complexions have the most severe problems with hyperpigmentation after laser hair removal. Southern Asians, Puerto Ricans, and southern Italians all have problems with hyperpigmentation. People with darker skin should use the lightening cream hydroquinone for at least two weeks prior to laser treatment to decrease the chance of this troublesome side effect.

Today's Lasers Really Do Work

Early hair removal lasers did not kill hair: they heated up the follicles and "stunned" the hair. It fell out but returned in a few weeks. I always wondered why someone would pay for temporary hair removal when they could wax at a fraction of the cost. When the stronger lasers were made, we saw the first permanent results.

The lasers available now can kill black and dark brown hair, but still have a hard time with light brown hair. None can kill blond or white hair. If white hair persists after a course of laser hair removal, electrolysis can be used.

In the average patient, each laser treatment kills between 20 and 40 percent of the hair. After three treatments, about 80 percent of the hair is destroyed. After four treatments, about 90 percent of the hair is gone. The darker your skin, the lower the power that must be used in order not to cause burns. But the lower the power, the less hair is killed. It can take a dozen or more treatments to decrease the hair on dark-skinned or tanned people.[1]

Hormonal Problems Can Cause Hair Growth

In women who have hormonal problems, many laser treatments will be necessary unless the problems are brought under control. Polycystic ovary disease (*Stein-Leventhal syndrome*) is one of the more common conditions that cause facial hair growth. Under the influence of hormones, each little blond hair of the face has the potential to become a large black hair. It's kind of like weeding a garden. Even though you pull all the weeds, the seeds are already in place for a new crop in a few weeks. The blond hairs are the seeds of larger hairs.

There Are Different Types of Laser Hair Removal Systems

Laser hair removal systems use various wavelengths of light to kill hair, and each system is a little different. Before beginning a laser hair removal program, be sure you know what kind of laser is being used. Find out the percentage of hair destroyed with each treatment. Make sure the hair removal is permanent. De-

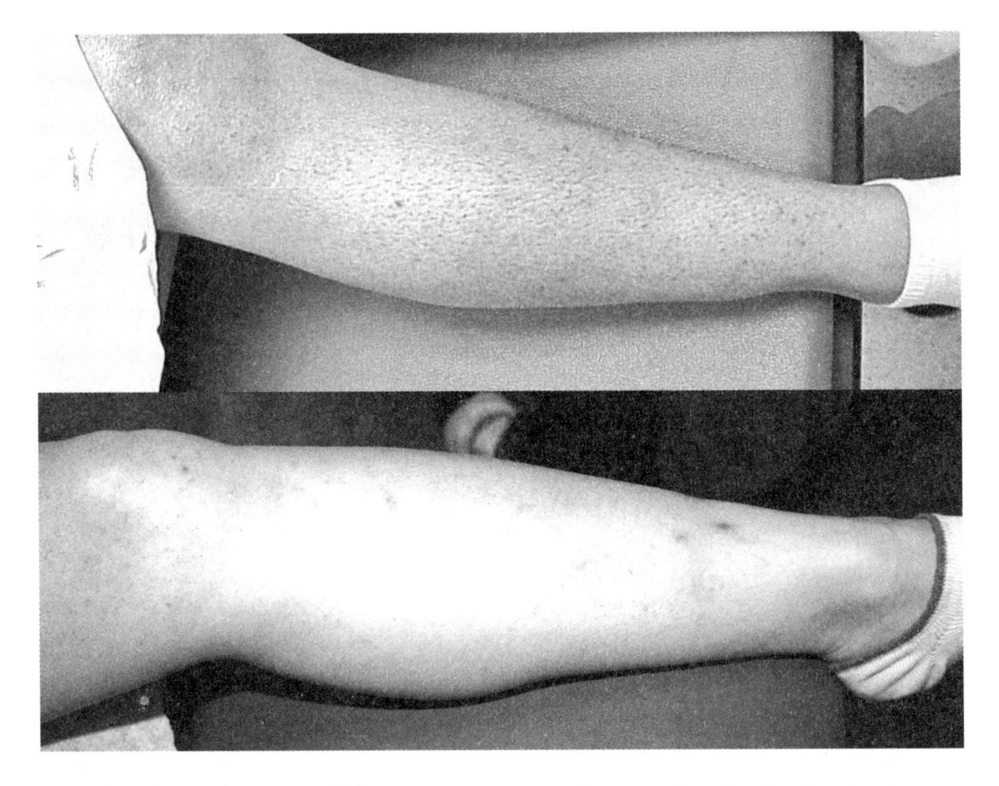

Top photo shows this woman's leg prior to treatment. Bottom photo is after four treatments. See color plates.

termine the chance of complications such as blistering, bruising, and increased or decreased skin pigmentation with your doctor's particular system.

Alternative Methods of Hair Removal

Techniques such as plucking or waxing the hair are temporary, but they do work. Depilatories such as Nair contain chemicals that dissolve hair. Shaving works, but creates stubble.

Electrolysis uses electric current to fry the hair follicles. It is painful and tedious, with weekly treatments that sometimes stretch on for years. The effectiveness is only as good as the electrologist. There are competent electrologists and there are those who go through the motions, plucking the hairs at the end of the treatment. Electrology cannot treat ingrown hairs, and it can result in dreadful scarring if not performed correctly. I believe that this technique will be relegated to the removal of blond and white hairs, since laser hair removal is so much more efficient on darker hairs.

The drug Vaniqua (eflornithine) slows the production of hair and causes it to form improperly. The result is an early breaking off of hair shafts. Again, with the advent of laser hair removal my opinion is that this drug is best used on blond or white hairs.

The Cost

The average cost for laser removal is $350 per treatment. Areas such as the upper lip might cost $200 per session, the entire back or thighs about $2,000 per session.

24

Tattoo Removal

One would think that tattoo removal would be a booming business. The tattoo craze of the 1990s left many baby boomers with embarrassing tattoos plastered on their bodies. But tattoo removal is still not that popular. Probably the overwhelming reason is the lack of an effective method of removal.

Laser light can vaporize the tattoo pigment. The light can be tuned to virtually any wavelength, corresponding to any color. But a specific laser can create only one wavelength. Tattoos are composed of many different colors of pigment, and because there is no standard for coloration of the pigment, laser tattoo removal is problematic.

The most common color in tattoos is a dark blue/black and the most widely used laser is tuned to destroy that color. When tattoo artists force pigment into the skin, it gets picked up by cells. The laser heats the pigment and explodes the cells, killing them and dispersing the pigment. A heat injury with blistering occurs. Nearby cells pick up the released pigment. A month after the first treatment, the tattoo merely looks a little blurred. Monthly treatments repeat the process. After many treatments, the tattoo is significantly decreased in intensity. I didn't say gone.

Too Many Colors, Not Enough Lasers

If there are many colors in the tattoo, multiple lasers are needed. Several companies manufacture boxes containing several different lasers. I have seen as many as four different colors in a single machine. If your tattoo has one of the chosen colors, you're in luck! If not, your tattoo will not be removed as efficiently. It will

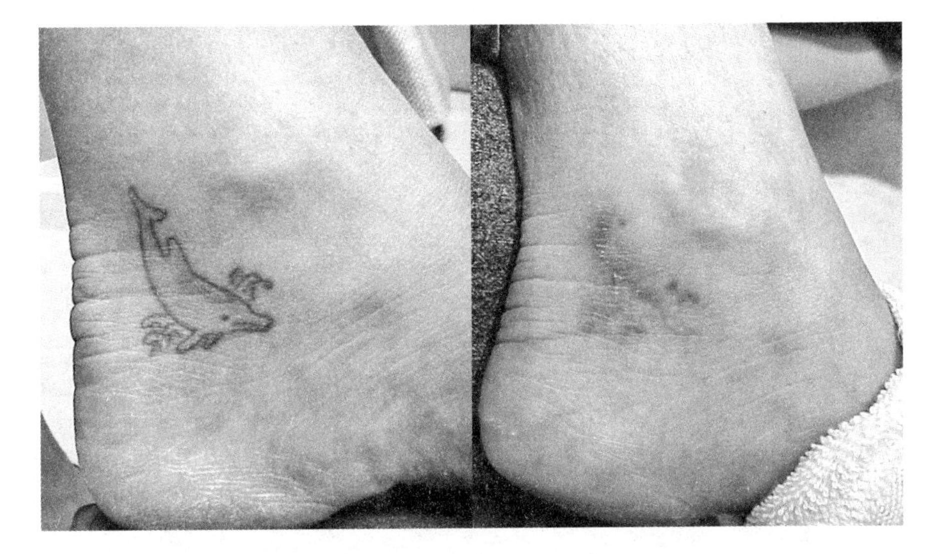

This woman underwent seven laser treatments to get rid of the pesky dolphin. Even so, some pigment remains. Left, before treatment; right, postop. See color plates.

take longer to remove and may not be completely erased. Until a laser is created that can measure the actual color of your tattoo and dial a particular wavelength (color), the tattoo removal process is doomed to be inefficient at best.

Not only are lasers not particularly effective in removing tattoos, they do so at a high price. Financially *and* medically. Because the laser creates heat when the tattoo is destroyed, it generates scar—usually white and often shiny. True, it is hard to know how much scar was created by the actual process of getting the tattoo, but I believe that most of the scar comes from the heat of the laser.

Lasers can unexpectedly change the color of tattoos. This huge problem causes the immediate and irreversible pigment darkening of white, pink, and flesh-colored inks. Ruby, alexandrite, neodymium:yttrium aluminum garnet, and pulsed dye lasers can cause this unpredictable change. If this happens, the only solution may be to cut out the tattoo. The laser can also release contained pigment, causing an allergic reaction.

When You Get a Tattoo, Open Up a Money Market Account to Save for Removal

A $75 tattoo can cost thousands of dollars to remove. There are usually fees for using the laser as well as surgeon's fees. Laser-use fees are in the range of $200–$400 per treatment and the surgeon's fee depends on the size and complexity of the tattoo. It could be as little as $200 or more than $1,000 per treatment.

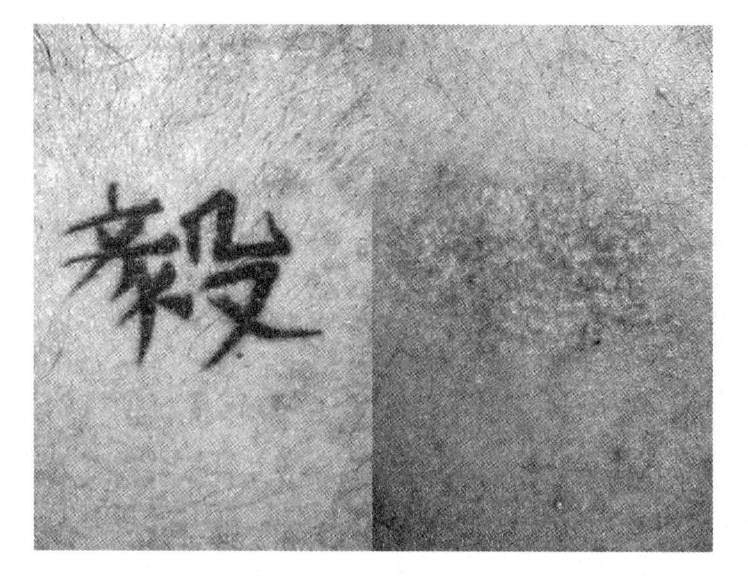

This tattoo had the most common color, blue-black. The photo on the right is after just three laser treatments. Notice the scarring left behind by the laser, or perhaps by the tattoo itself. See color plates.

The constellation of problems with laser tattoo removal makes it an unsatisfying procedure for most plastic surgeons. They usually are happy to pass the procedure to the neighborhood dermatologist. When a laser company comes out with a single laser that removes all tattoo colors without creating scar, the product will be a blockbuster.

Because of the problems associated with laser tattoo removal, many people with small tattoos simply opt to have them cut out.

The Health Risks of Tattoos

Aside from the permanence of tattoos, other problems can arise. Certainly infections are a possibility whenever the skin is penetrated. Although they can be serious, most are easily treatable with antibiotics. Still, viruses such as HIV and hepatitis B and C can be transmitted through tattoos. Even if the tattoo establishment looks clean and has a sterilizing autoclave, and even if disposable one-time use needles are used, you can still get sick. The pigments used by the average tattoo artist are not individually packaged, sterile pigments. If they *were* used, the price of the tattoo would triple or quintuple. In light of this downside, health departments should mandate sterile pigments as well as sterile needles. Until they do, once again, it is "buyer beware."

Permanent Cosmetics

Permanent eyeliner was introduced in the 1980s. It is a tattooing technique that permanently draws dark pigment on the upper and lower eyelids, simulating eyeliner. The eyelid must be anesthetized to apply this tattoo safely. Otherwise the patient may jump, causing eyeball puncture by the needle. To protect the eyeball, steel contact lenses must be used.

Anesthetics and steel contact lenses are the province of physicians. Occasional physicians perform the permanent makeup procedure, but most are done by nurses or specially trained individuals. Physicians should *supervise* the procedures and use sterile tattoo pigment and needles. Permanent cosmetics applied outside of physicians' offices are dangerous.

The tattooing pigment can cause all sorts of problems, including blistering, scarring, and chronic infections. Even severe disfigurement has been reported. At present, the FDA does not regulate permanent cosmetics. This stance is illogical, since these pigments function basically as microimplants and are capable of causing much damage. Their problems are escalating, and government regulation is imminent.

The Cost

Tattoo removal varies from $200 per treatment to over $1,000 per treatment. The cost depends on how large the tattoo is and how many colors it contains.

25

Laser Removal of Red and Brown Marks

Fortunately, the laser is better at removing red and brown marks than it is at tattoos. In particular, the laser readily destroys red marks on the face such as capillaries and hemangiomas. No anesthesia is required and, in fact, none can be used, for the anesthetics cause the red color to fade, decreasing the efficiency of the laser.

What Are Those Red Things on My Face?

As we age, facial capillaries become more visible. They predictably form around the nose and the cheeks. Several conditions are associated with these capillaries. The most common, acne rosacea, is associated with visible capillaries and acne on the cheeks and nose. Dermatologists usually treat this disease with drugs such as tetracycline pills or metronidazole (MetroGel) cream. When capillaries become visible, they can be destroyed with the Nd:YAG laser. The laser heats up the red blood cells and damages the capillary walls. The newest lasers do this without causing bruising. (Older versions caused two weeks of bruising.) Since the lasers destroy between half and two-thirds of the capillaries in one treatment, sometimes patients have several treatments.

Other red marks on the face are called senile angiomas. These benign collections of blood vessels look like red moles. A plastic surgeon or dermatologist should properly diagnose them before treating with the laser, since there will be no specimens to send to pathology.

Many different lasers can treat red marks. The first was the argon laser. I remember using this laser at Harvard's Beth Israel Hospital with laser pioneer the late Joel Noe, M.D., in 1981. The laser was so large that it took up a whole room and required a dedicated cooling system. Today's lasers are tabletop units. Different surgeons use different machines with mystifying names such as KTP (potassium titanyl phosphate), krypton, copper vapor, copper bromide, Nd:YAG, and flash lamp-pumped pulsed dye (FPDL) lasers. All of them treat red pigment slightly differently. Some cause bruising, some don't. All have a chance of scarring. All cause swelling that increases with the number of blood vessels that are treated.

The Risks

Any laser can cause blistering. Blistering can result in scarring—which is unusual. This treatment can be performed with minimal swelling if only a few vessels are treated. If a large area is treated, there can be a lot of swelling. Occasionally, huge swelling requires steroids to be controlled. As with any laser, the skin must be protected from sunlight for a few weeks while there is skin redness.

How About Those Brown Marks?

I prefer to remove brown marks on the face, arms, or hands with TCA peel. If that is unsuccessful, I use the laser. The laser heats up the brown pigment and fries the cells that make the pigment, leaving a blister. After healing, there is a light red spot that gradually fades. In the end, the brown mark is gone.

While the laser can remove brown marks, I believe it is overkill. I reserve its use for very difficult brown spots. The most popular lasers used to treat brown pigment have names like ruby, alexandrite, neodymium:yttrium aluminum garnet (Nd:YAG), and pigmented lesion dye laser (PLDL). Again, these different lasers use different types of light to destroy the brown pigment.

The lasers that treat brown spots are also some of the lasers that treat tattoos. Because they significantly heat the pigment in the skin, they usually cause blistering. Healing takes five to eight days and often leaves a reddened area that can result in a white area. Scarring and permanent pigmentation changes are possible.

The Cost

Sessions for capillary removal generally vary from $200 to $1,000, depending on the size of the treated area. The cost of brown-mark removal is similar.

26

New Lasers and Intense Pulsed Light

Since the mid-1990s, literally hundreds of new lasers and laser-like devices have been invented, each claiming superiority in performing one or another task. As we have seen, some lessen wrinkles, and some remove hair, tattoos, and red or brown marks. Each is a little different. By tweaking the physics of the high-energy light, slightly different effects are achieved. Many of these lasers are released onto the open market before valid studies are performed. Companies get FDA approval by claiming their machine can perform a limited function, such as removal of brown spots. Once the machine is approved, physicians can use the machine for off-label purposes such as wrinkle removal. Manufacturers place the lasers in the offices of prominent plastic surgeons or dermatologists, then arrange for television publicity. Patient interest develops and in turn generates demand by physicians—and therefore sales. In 2005 these new devices were used in 42,000 procedures.

Devices That Help Wrinkles and Signs of Aging All Injure the Skin

It is almost comical to see how manufacturers and researchers come up with new and unique skin torture machines. Some vaporize the surface and some heat the dermis with laser or electrical or radiofrequency energy. Some poke holes in the skin and some fry it with blowtorch-like devices. In the end, most simply injure the skin, starting a series of complex healing events that perhaps will ultimately result in smoother skin. Perhaps.

It is increasingly difficult for plastic surgeons to sort through the huge numbers of available machines. Most do not want to be "first on the block" to have a laser that turns out to be useless. On the other hand, no one wants to be left behind and not use new, proven technology. It's as difficult for the doctor to figure out as it is for the patient.

Intense Pulsed Light (IPL)

This relatively new method of facial rejuvenation was introduced around 1995. Treatments that have been called photofacials can decrease red and brown sun damage to the skin. Intense pulsed light (IPL) also lessens red and blue veins and capillaries, as well as rosacea in the face. The hands and the upper chest can also be treated. Results are better if the skin is not tanned.

Intense pulsed light can thicken collagen, creating a younger and fresher appearance. Many companies state that the technology can improve scars and wrinkles, but this claim has not yet been proven. IPL is supposed to improve the quality of the skin without the massive destruction caused by the CO_2 laser.[1]

Basically, this technology is a "poor man's laser." A laser is a very specific intensified color (wavelength) of light; IPL is light that is narrowed down to a specific range of colors by filters. The blend of different wavelengths causes a broad range of effects.

There are different types of intense pulsed light systems, all of which require a series of treatments. Most send high-intensity light through skin cooled to 10°C. Without this cooling, blistering is likely.

Anesthetic cream may be applied to the skin before the treatment, but most physicians do not do so. Some exfoliate the skin with microdermabrasion first. In order not to have any downtime (the lunchtime peel), low energy must be used. Therefore, three to six monthly sessions are necessary.

The real question is whether this new technology actually works. Numerous studies indeed show that IPL decreases wrinkles, skin roughness, mottled brown spots, and red blood vessels of the face. It takes about two weeks to see the effects of the treatment.[2]

Bruising and swelling can occur; side effects include whitening or darkening, and blistering and even scarring of the skin. These problems occur in 5 percent of patients.[3]

Costs for each treatment are fairly high when compared to peeling techniques, since the machine itself costs between $50,000 and $100,000.

Light *and* Electricity Systems

Syneron's Elõs electro-optical synergy technology combines pulsed light energy or a diode laser with radio-frequency electrical current. This complex system cools the surface of the skin while it heats the dermis, creating a "controlled injury." The hope is that this trauma will prompt the dermal cells to pump out collagen, and tighten the skin with less damage to the epidermis. The manufacturer claims significant improvement in facial wrinkles, as well as red blood vessels and brown marks. The technology also claims to remove hair of any color and treat any skin color to decrease acne. The system that combines the diode laser with radio frequency showed modest wrinkle improvement when evaluated six months after a series of treatments. Long-term results are not known.[4]

Thermage Radio-Frequency Tissue Tightening

Thermage uses radio-frequency electrical current to heat the undersurface of the skin, simultaneously cooling the anesthetized surface of the skin. Tighter skin with reduced wrinkles results, presumably from increased collagen production. One study showed that this technique raised eyebrows half a millimeter in 61 percent of people, *but* three of the eighty-six patients developed permanent scarring. Other studies showed better shrinkage of the skin. Results are visible in one week to six months. Often only one sixty-minute treatment is needed. The real question is whether the tightening of the skin will last more than a few months and what the long-term effects will be. This machine costs about $40,000 and is expensive to use. Treatment costs range from $800 to $3,500.[5]

Thermage has a troubled history, however. An Internet search reveals many articles about unsatisfied patients, some with scarring and some with loss of facial fat. In fact, the private section of the American Society of Plastic Surgeons website last year contained a lively discussion of Thermage disasters. A Boston law firm has filed a lawsuit against the company on behalf of two women who claimed to be damaged by the treatment.

The Dallas otolaryngologist Benjamin Bassichis, M.D., has reviewed fifty-three problems caused by Thermage, including scars, dimples, and depressions. Bassichis showed that Thermage could cause loss of the fat in the temple area. "One-third of patients had a good result, and one-third had no result," he told me. Because of this experience, he no longer uses the machine.[6]

Proponents of Thermage counter these claims by noting that their treatment protocol has changed. Newer techniques have better results and fewer complications, they say. Indianapolis plastic surgeon Charles Hughes III, M.D., and Dallas surgeon Jay Burns, M.D., both have had excellent results with the Thermage device. "There is definitely immediate and late tightening. The heat denatures collagen. It then remodels and tightens," Dr. Hughes said at a 2005 New York University conference. He believes Thermage can also tighten the underlying connective tissue. Some doctors use the machine to shrink arm, thigh, buttocks, and abdominal skin after liposuction; the FDA has approved these uses. Only time will tell whether this technology is *uniformly safe* or will result in long-term problems.

The GentleWaves Light Emitting Diode (LED) Photomodulation System

This is one of the new machines designed to decrease facial wrinkles. The company says that "specially coded arrays of light emitting diodes modulate the activity of living cells." That sounds terrific, but what does it really mean? Most doctors don't know—and few patients will. The real question is whether the system works. Patients expose their face, neck, and chest to two panels of more than two thousand yellow cascading LED lights for less than a minute. According to Dr. David Goldberg, director of dermatologic laser research at the Mount Sinai School of Medicine in New York, "the light can activate new collagen formation. There are at least two peer-reviewed published studies that show the machine is effective." Ninety percent of patients in one study had smoother skin with fewer wrinkles.[7]

The "Lumifacial" Device

This Silhouet-Tone Company machine combines blue, green, yellow, and red LED lights, along with what it calls micronized (electrical) and high-frequency polarized currents. The thirty-minute treatment allegedly improves acne, pigmentation, redness, and photoaging of the skin. This machine is marketed to aestheticians who have completed three hundred hours of skin care training, not just to physicians. Scary. I was unable to find any peer-reviewed controlled studies in the scientific literature showing that this machine actually did anything.

Reliant Fraxel Laser

The Fraxel laser is a new, exciting addition to the war against wrinkles. Approved by the FDA for treatment of wrinkles and age spots in 2004 and for skin resurfacing in 2005, it uses what is called fractional photothermolysis to decrease wrinkles. This erbium laser treats a tiny portion of the skin's surface at a time. Basically, the machine splits the laser beam into thousands of tiny beams, narrower than the diameter of a hair. The beams penetrate the skin, deep into the dermis. The dermis fries, causing shrinkage without removing the entire epidermis. The epidermis flakes off in a few days. The beam is so narrow that stem cells in the skin are not destroyed, a key factor in preventing long-term problems.

The company states that improvement is seen as early as a week following the treatment. One study showed about 2 percent shrinkage of the skin, which corresponds to 18 percent fewer wrinkles clinically. The response to the laser is new collagen formation. There is some pain with the thirty-minute procedure, so usually a cream anesthetic is applied. Swelling and redness resolve within a few days. Typically, three to five treatments are performed, spaced two weeks apart.[8]

This laser is billed as useful for all skin colors, including dark pigmented skin. However, it has not been possible to use any laser safely on dark skin because of the danger of pigment changes. Unless this laser is truly unique, some pigment problems in the mid-brown skin colors (such as Asian or Hispanic skin) are expected to surface as experience with this laser grows.

I am optimistic about Fraxel. If clinical experience in the next few years confirms the early studies, this laser will become widely used. Wrinkle removal with a low complication rate and use on all skin colors are big advantages. This laser shows promise in treating age-spots on the hands and arms. We need to learn just how deep a wrinkle it will treat, and how long the effect will be maintained.

The Fraxel machine costs about $100,000 and consumes $70 worth of disposable goods with each procedure. Expect a fee of $1,000–$1,500 per treatment. Patients will require a series of three to six treatments over a month to achieve wrinkle reduction. Other companies have picked up on the new technology; many competitors are already pitching their machines.

Rhytec's Portrait PSR^3 Technology

This brand-new technology is unique in skin care. The PSR delivers very short pulses of nitrogen plasma to the skin. A flowing lilac gas changes to a yellowish light as it flows from the handpiece. The plasma is an ionized gas that rapidly heats the skin. It is not a laser and supposedly does not damage collagen. My understanding of this machine is that it produces a controlled burn of the face, without creating char, since oxygen is swept away by the nitrogen gas.

The face is treated in a fifteen-minute session and requires an anesthetic cream. The skin swells and is red for about one week. Because the epidermis is not destroyed, healing is faster than with other lasers. The machine is supposedly effective in treating wrinkles, texture, and brown pigmentation by stimulating collagen and elastin formation. Early reports show a 50-percent improvement in wrinkles and a 10–15 percent skin tightening, which is as good as a CO_2 laser.

Patients will probably have a series of four treatments, spaced three to four weeks apart. I wonder how many patients will allow a week of down time monthly for four months. That requires great commitment and would require stupendous results. The machine was just cleared by the FDA for treatment of facial wrinkles.[9]

I look at this machine as simply another way to injure the skin, like peels and lasers and all technologies that reduce wrinkles. The body responds by making new collagen and shrinking the skin. It is too early to know whether this machine has long-term benefits and what the complications will be. Of course, if it doesn't prove to be useful in facial rejuvenation, it may be great clearing ice off sidewalks!

How to Choose a Rejuvenative Treatment

Few studies compare dermabrasion, chemical peels, laser procedures, and other rejuvenative procedures. The field is so confusing that even plastic surgeons and dermatologists are unsure which procedures to recommend to patients. It is a fact of life that when a surgeon purchases a $100,000 laser, he will use it. Given the choice of a less expensive peel or a more expensive laser procedure, the profit motive often will win out. This is just the way things work. Look at the marketing information that the laser companies disseminate: maximizing profit is often the goal. And once a surgeon has purchased the laser, it is unlikely to be put in a back room when a better one comes along. More than likely it will be used until a malfunction requires a repair that costs more than a new machine.

In general, the less invasive the surgical procedure, the more subtle the results. This is certainly true of laser and laser-like machines. Those that do not destroy the epidermis have less obvious effects and usually require multiple treatments.

Every patient has a different problem. Every surgeon offers a different set of procedures. Sometimes, several are suggested, such as microdermabrasion to clean the skin, followed by conditioning peels, and finally surgery to lift the sagging tissues. My best advice is to choose your doctor wisely, based on the criteria laid out in Chapter 2. Once you have chosen him, trust your surgeon to do the right thing. Read about the suggested course of treatment and ask to see examples. Ask to speak to other patients.

The Cost

The cost of these procedures varies enormously, depending on the type of machine the surgeon uses and what his fixed monthly costs are. Generally, laser procedures run between $350 and $2,500 per treatment.

IV
Body Contouring

27

Liposuction

No procedure exemplifies the Hollywood image of the plastic surgeon more than liposuction. The first such surgeries were performed in 1976 in Switzerland and Italy and in France in 1977. Since bursting onto the American scene in 1983, liposuction has perpetually been one of the most popular procedures. And its popularity continues to rise. In 1992, for instance, it was the third most common cosmetic surgical procedure, with over 47,000 performed in the United States. By 2005, liposuctions surpassed eyelid lifts in popularity, with 455,000 procedures. It had become the most common cosmetic surgical procedure in both men and women.

The Surgery

Liposuction is considered a violent procedure. A blunt-tipped, hollow tube called a cannula breaks up fat. At the same time, a fancy vacuum cleaner, aptly called a liposuction machine, suctions out the fat. Different-sized tubes are used, with a multitude of configurations of holes through which the fat flows. The tubes range from 1.5 to 6 millimeters in diameter.

The areas to be suctioned are first infiltrated with a "wetting solution" of sterile saltwater with or without the local anesthetic *lidocaine.* If the procedure is performed under local anesthesia, lidocaine is necessary; it isn't needed if general or epidural anesthesia is used. Most surgeons add a chemical called epinephrine (adrenaline) to the solution. This important additive enormously cuts down blood loss.

There are variations of this procedure, with different surgeons injecting varying amounts of fluid. Names such as the "dry," "wet," "superwet," or "tumescent" techniques flood the literature. Most plastic surgeons simply inject enough fluid to distend the tissues. The amount of fluid injected usually works out to be about the same as the amount of fat that is removed. The surgeon then suctions the fat, preferably using more than one incision. The incisions are hidden in creases of the body and are usually shorter than 4 millimeters. More are better, since the resulting scars are difficult to see, and extra incisions give a smoother result. Back in the dark ages of liposuction, the 1980s, fewer but longer incisions were made. We now know better.

> It is the surgeon's artistry that determines a good result. It is medical judgment that determines whether you will have a complication.

Removing Fat Is Easy; Sculpting Fat Is Not

The training of a plastic surgeon involves an understanding of the beauty of the human form. Artful liposuction is the sculpting, not the extraction, of fat. What matters is not how much fat is removed; it's what the remaining fat looks like that is important.

Different areas of the body react differently to liposuction. To look good, the skin must shrink after the procedure. Before pregnancy, in a person who has never been overweight, most areas of the body will shrink nicely following fat removal. On the other hand, if there are stretch marks or if the skin has poor tone, the surgeon must be careful to avoid creating contour irregularities. Rippling, dimpling, and hanging skin are the curses of liposuction.

The inner thighs have limited ability to retract following liposuction, whereas the hips and outer thighs have excellent ability to shrink. We call different areas more or less "forgiving" to liposuction.

The Procedure

On the morning of surgery, all modesty must be put aside. You'll stand up naked as the surgeon draws a topographic map on the area of your body to be suctioned. The marks are important because your fat will change shape when you lie down during surgery. Skilled surgeons are able to translate their marks into guidelines for the fat removal. Some surgeons have tried to operate on standing patients, but this method is dangerous, with a high chance of fainting reactions.

You will be brought into the operating room where the nurse will prep your

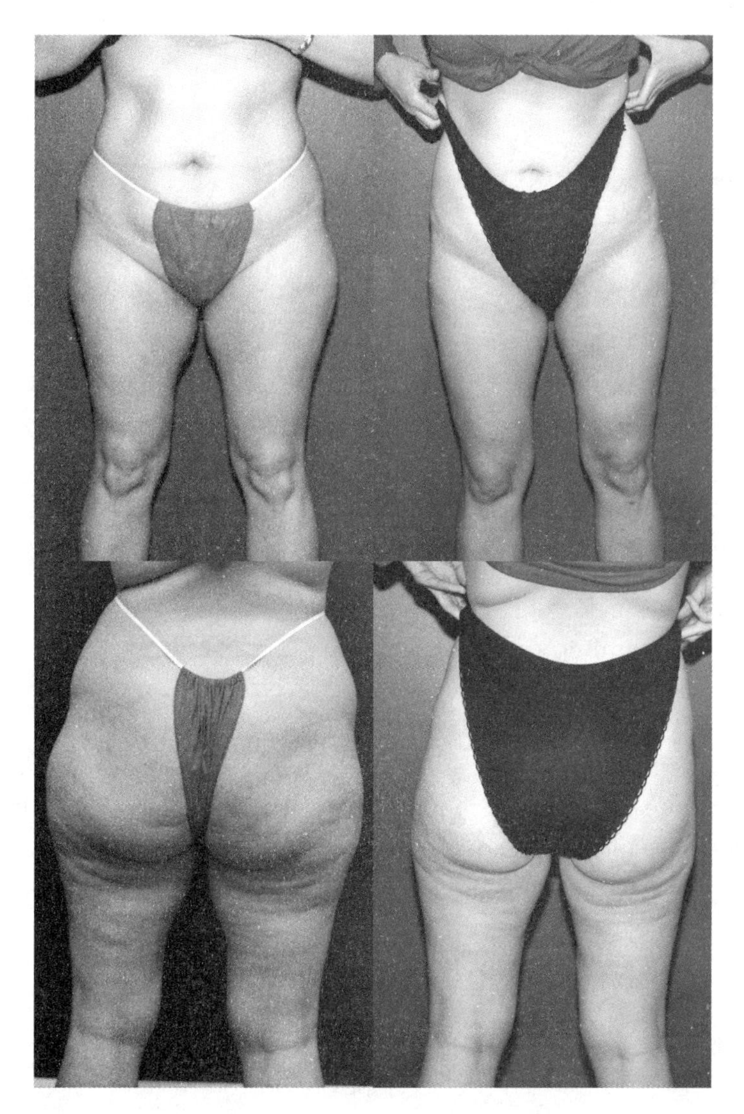

This is my typical liposuction patient. Some cellulite, some dimpling. Before suction of hips and thighs on the left; after, on the right.

body with Betadine, a germ killer. Again, you will stand, and Betadine will be painted from your collarbone to the tips of your toes. While the procedure is somewhat embarrassing, it is necessary. Liposuction is one of the few surgical procedures where both the front and back of the body are operated on. It is impossible to sterilize both sides of your body once you are under anesthesia. You'll then lie down on the sterile operating table, where the anesthesiologist will place an intravenous line in your arm and hook you up to EKG, blood pres-

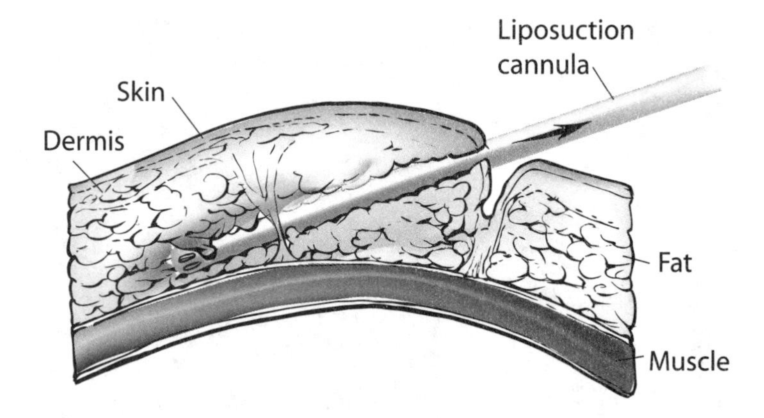

Liposuction is performed by sucking fat through a cannula, a hollow tube.

sure, and oxygen monitors. You will then be given general anesthesia and the operation will be performed. Some sort of blood clot prevention, such as compression boots or mini-heparin, should be used in every case.

Incisions are made, the wetting solution is injected, and the surgeon suctions the fat. You will be turned from side to side, if necessary. The surgeon keeps track of the amounts of fat removed from each location—the thighs, knees, and hips, for instance. If you are symmetrical, even amounts are taken from each side. If you have more fat on one side, then more will be taken from that side.

The surgeon judges the adequacy of fat removal by observing the amount of fat suctioned and by feeling the thickness of the remaining fat. Remember, the goal is not complete fat removal—it is a thinning and sculpting of the residual fat.

Your incisions will be closed with stitches and Band-aids and gauze will be placed. A specially designed girdle will be put on you while you are still asleep, to provide pressure where it is needed. It is washable and even has a hole in the crotch to avoid painful removal when you go to the bathroom. You will stay in the facility about two hours and then will be sent home.

It is not important to eat, but it *is* vital to drink after the surgery. A lot of swelling and some oozing will take place. Dehydration is possible and can lead to complications. If your legs or thighs were operated on, you will need to keep your legs elevated.

You'll be seen by your surgeon a day or two after surgery. You will be given a clean girdle and then you may shower. Your stitches will be removed in about a week.

Your bruising will be profound after surgery. *Think eggplant.* After two weeks, the bruising will be almost gone, and the swelling will have settled enough for your clothing to feel loose.

You may drive and return to work four or five days after surgery. Pain should be minimal. In a study I published in the journal *Plastic and Reconstructive Surgery,* the pain after liposuction was only 1, on a 0 to 10 scale. Interestingly, the pain was the same even if local anesthetic was injected into the suctioned area. For this reason, local anesthetic is not necessary and should not be used if the procedure is performed under general anesthesia.[1]

The final result shouldn't be assessed for at least three months following surgery. Subtle changes can occur for up to a year.

The Risks

Liposuction is an incredibly safe procedure—*when performed by people who know what they are doing.* The surgeon should be a true surgeon, trained broadly in surgical principles and specifically in liposuction. He should have been in a real residency program with graduated responsibility leading up to a chief residency. He should be certified by one of the ten *surgical* boards recognized by the American Board of Medical Specialties and should be operating within the anatomic boundaries of his specialty. *Remember, the T in ENT does not mean thigh; it means throat!* While it may seem obvious to the public, the American Society of Plastic Surgeons has taken the bold step of recommending that only surgeons perform liposuction.

Like any medical field, liposuction is dangerous when performed by someone who does not know what he is doing. It is a real surgical procedure, with real potential complications. Even in the best of hands, problems can crop up. You will stack the odds in your favor if you choose a board-certified plastic surgeon, with good credentials and with recommendations from other doctors or patients.

The risks include infection and bleeding. Infection is very rare with liposuction. But when it does occur, it can be aggressive and even life threatening. The average patient loses a pint or more of blood during and after the surgery, an amount that does not require a transfusion. If there is any question about the patient's ability to clot (such as if excessive bleeding has occurred after a tooth extraction or with menstrual periods) then more extensive clotting studies should be performed prior to the surgery.

Liposuction can cause very serious blood clots in the legs. If these travel to the lungs, they can be deadly. Symptoms of blood clots in the legs include swelling, pain, and fever.

The liposuction cannula is a weapon, and it has speared virtually every organ during liposuction. Death can occur if this complication is not immediately recognized. The cannula can also penetrate the skin, causing scarring. Suctioning too close to the skin causes blistering and even skin loss, resulting in severe scarring.

If local anesthesia is used, high doses can be toxic. While local anesthesia seems safer than general anesthesia, to an anesthesiologist the amount of lidocaine sometimes used is frightening. High doses can cause problems with the heart and the brain. Certainly, many of the deaths from liposuction in the last decade have been from overdoses of local anesthetic.

More common and less important complications include contour irregularities in the skin such as dimpling, waviness, rippling, and even hanging skin. Before surgery your surgeon assesses the chance that these problems will occur. If you have good-quality skin, the chance is low—about 5 percent. If you have poor-quality skin, perhaps already with some hanging, your chance could be as high as 100 percent.

As in all surgery, risks will be higher if you have diabetes, heart disease, impairment of the immune system, or other chronic disease. You and your surgeon need to fully assess whether surgery to improve your appearance is worth the risk of complications, even death.

Can Liposuction Make You Healthier?

The great liposuction debate burns in the minds of plastic surgeons at scrub sinks across the country. They argue about whether liposuction is a purely cosmetic procedure or whether it has medical benefits. Some plastic surgeons have stated that the way the body handles cholesterol and fat after liposuction is changed, improving diabetes and inflammation. Several studies support this theory, and others claim a reduction of blood pressure; still others show that even the removal of more than ten kilograms (twenty-two pounds) of fat does not alter the way the body handles insulin, and that there is no cardiovascular benefit.[2]

The fat removed during liposuction may be a source of stem cells, immature cells that can be coaxed to develop into different mature tissues. In fact, stem cells from fat have been found to create bone in experimental studies. It will be interesting to see whether other tissues can be created in this way.

The Techniques

Power-Assisted Liposuction

Surgeons seem to like to advertise "power-assisted liposuction." Promoting this technique is like a chef advertising the kind of pans he cooks on. Power-assisted liposuction does nothing for the patient. It simply makes the surgery easier mechanically.

With power-assisted liposuction, instead of the surgeon moving his arm back and forth to mechanically break up fat, the cannula itself reciprocates. It takes the physical effort out of liposuction. I occasionally use the technique, usually at the request of a patient. I suppose when I get older, I'll use it more and more and get my exercise in the gym instead.

Ultrasound-Assisted Liposuction

Ultrasonic liposuction began in Italy in the early 1990s. Late in the decade the American Society of Plastic Surgeons saw this technique as critically important and waged an effort to train members in it. Ultrasonic energy is the energy of sound waves, specifically, sound waves higher in frequency than humans can hear. A common use of ultrasound is in dental cleaning.

I spent thousands of dollars taking the course and operating on cadavers. Ultrasonic liposuction was billed as the next big thing in liposuction. It was supposed to be able to glide through not only traditional fat, but also fibrous fat such as in the breast and back—areas that were more difficult to suction.

Ultrasonic liposuction uses high-energy sound waves to blast apart and liquefy fat, leaving blood vessels, fibrous tissue, and muscle intact. Unfortunately, the technique has a higher complication rate than traditional liposuction in that over 8 percent of patients experience a complication. Burned skin, pigmented skin, collections of fluid (seromas), and longer scars can result. The seroma rate may be higher than 11 percent. Nerve injuries can lead to significant disability. Other problems include lumpiness; in the abdomen, it may ultimately require an abdominoplasty to correct.[3]

Some say ultrasonic energy generates free radicals and causes disruption of DNA, possibly even causing cancer. These free radicals can be reduced in number by adding the antioxidant vitamin C to the wetting solution. Because of these long-term safety questions, some advise not performing this procedure in either the male or female breast. Ironically, this is the area of the body in which ultrasonic liposuction works the best.[4]

Most plastic surgeons who were trained in ultrasonic liposuction no longer perform the procedure. But sometimes old technologies die hard, and in 2005 ultrasonic liposuction was again being touted as an effective technique for removing breast tissue and fat (*gynecomastia*) in men.

The Vaser

The manufacturers of this machine have marketed it as Lipo SelectionSM, whatever that means. This technique is supposed to be the next generation of ultrasonic liposuction, using pulsed (intermittent) ultrasound and a specially shaped cannula tip. The complication rate is said to be lower than for traditional ultrasonic liposuction. The real question is whether the Vaser is better than traditional liposuction. For a doctor to switch to a new technology, it must be better, cheaper, faster, or have a lower complication rate. I do not believe that this has been shown to be true of the Vaser. Yet when a plastic surgeon advertises that he has the new machine, some people are impressed.[5]

Expect the Vaser to increase in popularity. Sound Surgical Technologies Company is not selling its machines directly to plastic surgeons: it is placing them in offices free of charge but charging a $350 fee for each procedure. The surgeon must perform at least four procedures each month and sign an agreement for a certain number of years. It's a fine deal if the surgeon is busy—ten procedures a month yields $3,500 for the company. The problem is that if the surgeon decides he doesn't like the technique, he is stuck with the machine and its obligatory costs. Do you think it will just sit in a closet, costing $1,400 a month?

External Ultrasound

This technique applies ultrasonic energy to the skin prior to liposuction, or even instead of liposuction. There are no solid data to support its use. I tried external ultrasound on a few dozen cases and was happy to return the machine to the manufacturer. Not only did it not work, but it caused blistering of the skin.

High-Energy Focused Ultrasound

Two newer external ultrasound devices, Liposonix's Sonosculpt and the Ultrashape device, are being tested to see if they can noninvasively destroy fat. This focused ultrasound disrupts fat cells, sparing the skin. According to Dr. Ami Glicksman, plastic surgeon and president of Ultrashape, Ltd., "Ultrashape uses ultrasound that breaks up tissue but doesn't generate heat. So burns are not possible." The Ultrashape system uses a computer to guide the treatment. A typical

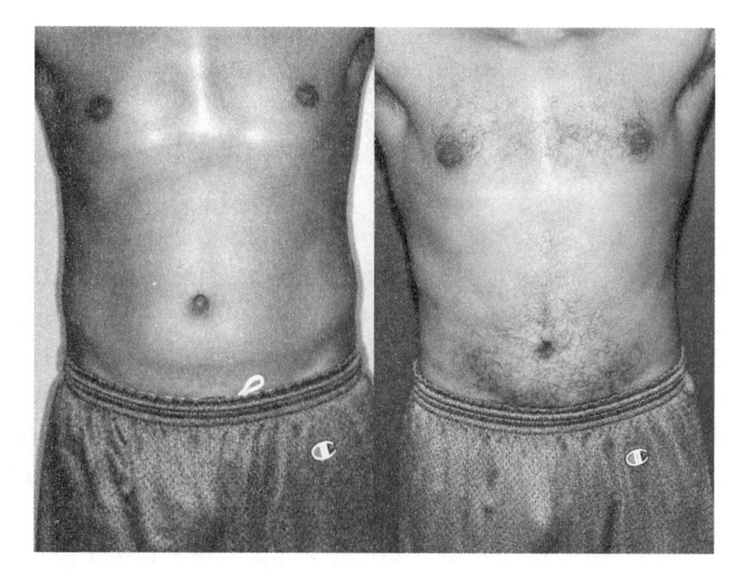

This patient underwent liposuction of the abdomen and flanks. On the left is his preoperative appearance; his postop result is on the right. These areas are the most common for liposuction in men.

session takes about an hour and a half and does not require anesthesia. Fat cells are damaged, causing them to spill out their fat. This fat is "recycled" by the body and lands somewhere else. Patients may have one or more treatments, depending on how much fat they want removed. The results from a single session are impressive but certainly are more subtle than with liposuction. But considering the risks and downtime of surgery, if the results hold up with time, this technique will revolutionize fat removal and possibly make liposuction obsolete. In July 2005, the machine received the CE Mark (the European Union stamp of approval) and is now available in Europe. FDA approval in the United States is pending. If this technique really works, it will be a blockbuster—the holy grail of reducing fat without surgery. Hold your horses, though—we don't know yet what are the risks or long-term consequences of ultrasonic energy. In 2004 the Liposonix Company raised over $27 million to market its product, so expect to see a big push for this technology. The first studies are showing its effectiveness.[6]

Low-Level Laser Liposuction

Erchonia Medical's new device has received FDA clearance. The laser is used prior to liposuction to liquefy the fat. The company claims that liposuction is then easier and quicker, and causes less pain, swelling, and bruising. However, a study

I performed in 1999 showed that pain should be minimal following liposuction, so I wonder how the new laser could improve on that. One other study failed to show that this laser actually disrupted fat.[7]

The Limits of Suction

As liposuction became safer and safer in the 1980s, there seemed to be a race among surgeons to suction the most fat. During this time, the death rate escalated. I am a strong believer in limiting the amount of fat suctioned in any one procedure. We must honor the philosophy that we perform surgical procedures that are uniformly safe—not what we can "get away with." The official recommendation by the ASPS is that no more than five kilograms (eleven pounds) can be safely suctioned from patients without admitting them to the hospital. If this number is exceeded, the procedure and the patient belong in a hospital overnight. It is safer to perform a series of small liposuctions, rather than one large procedure.

After the Surgery

Expect to see weight loss following liposuction. Not right away, because the swelling immediately replaces the removed fat. After about two weeks, your clothing will fit more loosely. Because of the inactivity following surgery, some

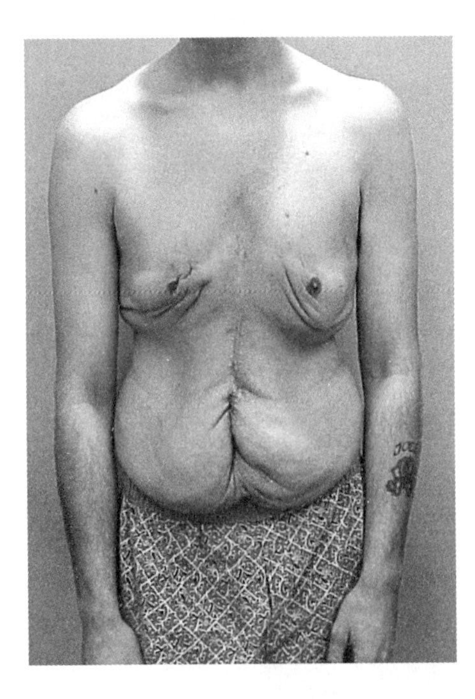

This 20-year-old man had a gastric bypass and lost over one hundred pounds. A dermatologist performed liposuction on his breasts and abdomen, with disastrous results. He was not an appropriate candidate for liposuction in the first place.

early weight gain is common. Once you start exercising at three weeks, that weight should come off. Your clothing size will decrease after surgery, although the sizing is more than a little arbitrary.

The Cost

Liposuction varies in cost but averages $2,700 per area, more if ultrasonic liposuction is performed. Generally, when more than one area is suctioned in a given operation, the cost of the additional areas is discounted. If specialized technology is used, expect to have further charges added to your bill.

28

Tummy Tucks and Body Contouring

Tummy Tucks

According to the American Society for Aesthetic Plastic Surgery, nearly 169,000 tummy tucks were performed in the United States in 2005. We're seeing more and more tummy tucks because baby boomers are nearly finished having children. Their bellies were overstretched as the developing fetuses pushed the rectus muscles (the "abs") to the sides of the body. Tummy tucks, formally called abdominoplasties, restore a youthful appearance to the midsection.

The belly muscles are similar to suspenders in a man who gains weight. At 150 pounds, the suspenders hang straight down the front of the body. By 250 pounds, the suspenders drift far off to the sides. As the rectus muscles drift from the midline to the sides of the body, the oblique muscles on the sides of the abdomen lose their tone. Their efficiency decreases. Since the abdominal muscles stabilize the spine, low back pain and poor posture result. The abdomen juts forward, causing the pelvis and the shoulders to shift backward.

With pregnancy and rapid weight gain, the abdominal skin permanently overstretches. The rapid expansion of the dermis causes stretch marks. After a baby is born, most women can't shrink back their skin. Hanging skin with excess belly and flank fat is the typical result.

Many women accumulate scars on their bellies. Modern Cesarean section (C-section) scars are horizontal in the lower abdomen. Because of the popularity of laparoscopic abdominal surgery, fewer women have long upper and mid-abdominal scars.

In the first tummy tuck in 1899, only hanging abdominal skin was removed.

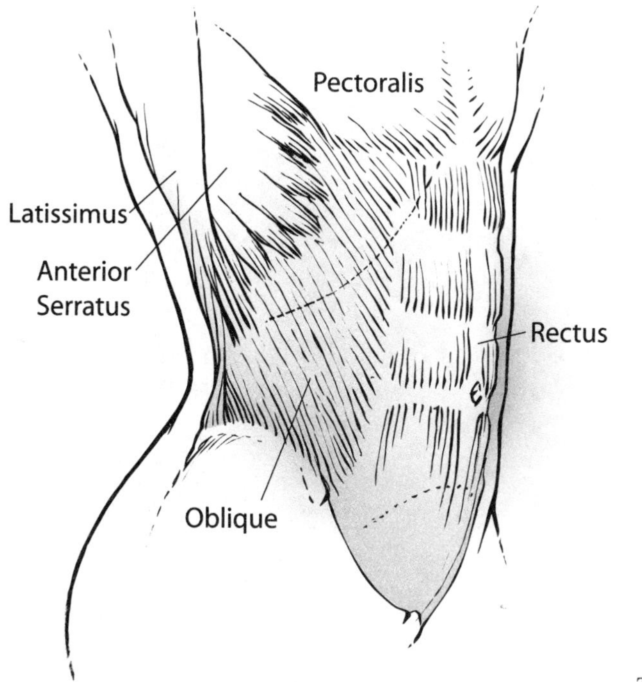

The muscles of the abdomen.

Today's tummy tucks are all based on this early operation. After the 1960s, upper-abdomen fat was also removed and the abdominal muscles were tightened.

The Techniques

The goal of a tummy tuck is a flat belly. Depending on the exact problem, a variety of different procedures are possible.

The Full Tummy Tuck

An incision is made in the lower abdomen. If there are no scars on the lower abdomen, a gentle W-shaped scar is made. This shape causes less distortion of the pubic area than a curved scar. Styles are changing, though, and I have recently had requests for purposeful pulling up of the genitals along with the tummy tuck. If there is a C-section scar, the new scar follows the gentle curve of the old one. In most cases, the old scar is removed.

After massive weight loss, there is usually a huge amount of extra skin. Sometimes vertical incisions are added to the belly in this situation. The skin of the belly is then lifted like the hood of a car, exposing the muscles. The belly button (*umbilicus*) is protected, and the skin is lifted all the way to the ribs and to the

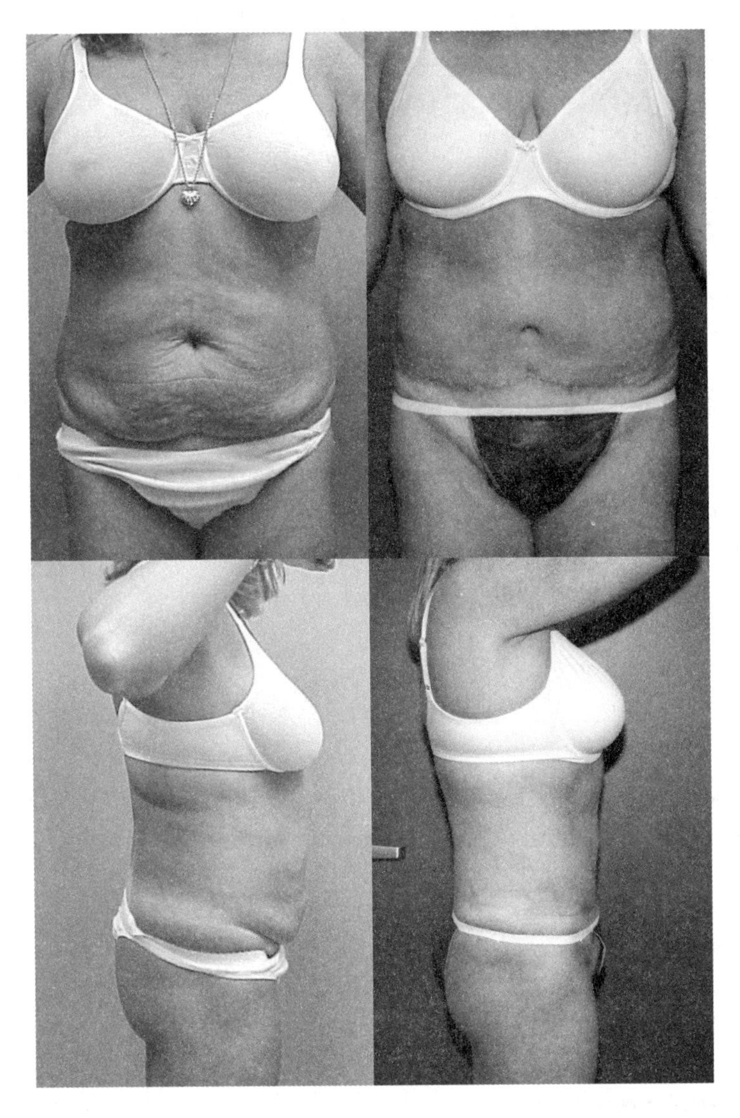

This 34-year-old woman underwent a tummy tuck and liposuction of the flanks.

bottom of the breastbone. In most cases, all of the skin between the belly button and the pubis is then removed. The rectus muscles are sewn together, using permanent stitches placed very close together. This "seam" absorbs tremendous stress, so the stitches must be strong.

Many people think the belly button is moved in a tummy tuck. Not true. The belly button stays put, but the skin slides over it. A cut is made in the skin and the belly button is popped through. Some surgeons remove it; except to save time in the operating room, I can't think of a reason why.

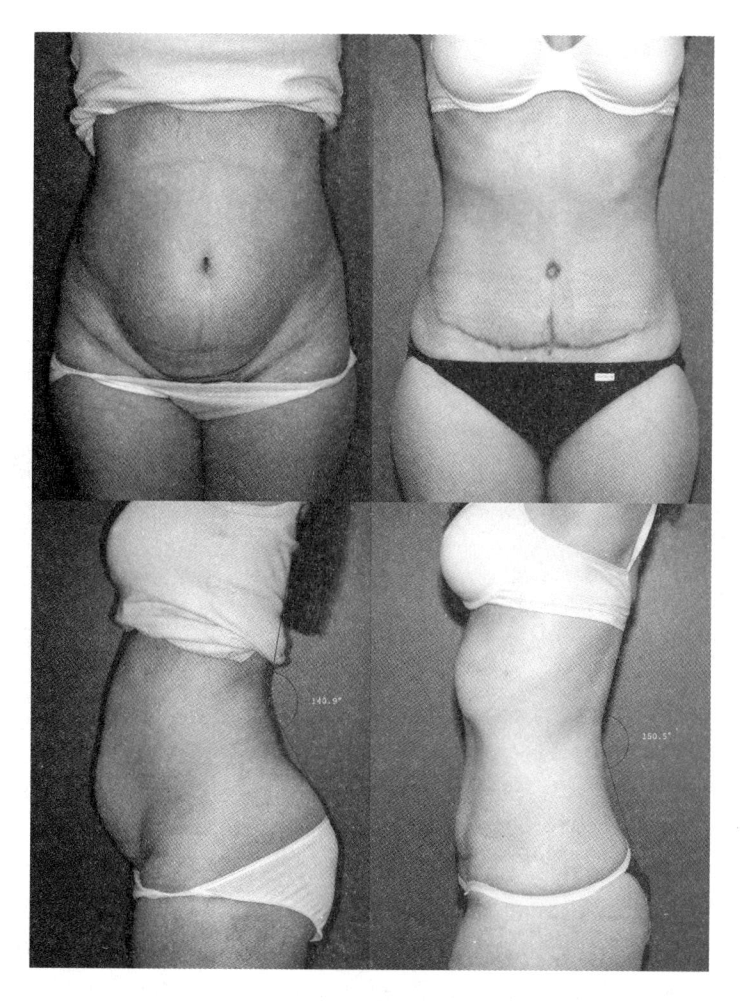

This woman had very weak abdominal muscles, without much extra skin. She underwent a tummy tuck with no liposuction and experienced an astounding ten-degree improvement in posture after surgery. Her pelvis and upper body, shifted forward prior to surgery, look as if they have been surgically repositioned. The penalty for this improvement, her scar, at six months is red and hard. It will settle with time.

Two drains are placed. This is the only operation in cosmetic surgery that drains are *required,* although they are optional in other operations. One of the biggest problems with a tummy tuck is a seroma, a collection of clear, blister-type fluid. Drains decrease the chance of a seroma.

The wound is closed, usually in two layers, with a deep layer of dissolving stitches and a superficial layer of either stitches or Dermabond.

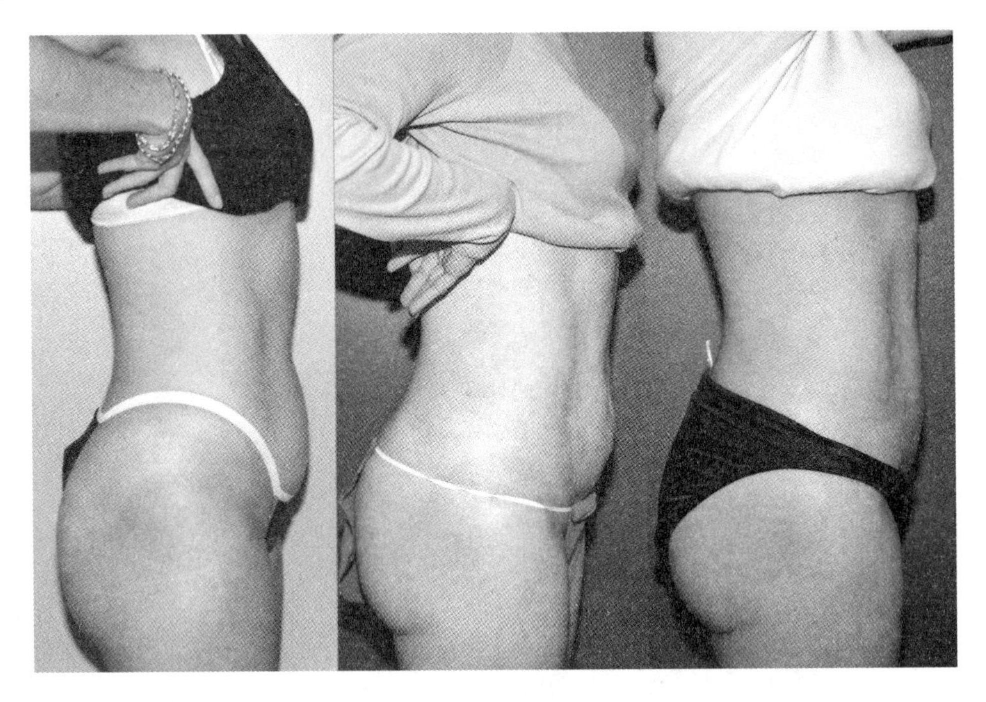

On the left, the patient preop. The middle photo shows her appearance after liposuction only. On the right, the same patient after mini abdominoplasty.

The Mini Abdominoplasty

If the muscle bulges only in the zone below the belly button, then a mini abdominoplasty might be an option. In this operation, a smaller incision is made and only the muscle below the belly button is tightened. Not much skin is removed and there is no incision around the belly button.

Other Types

Many other operations are possible. Some surgeons do the operation through an upper abdominal incision. Others use a midline incision. When there is not much extra skin, the belly button can be "floated" down about two centimeters, eliminating the surrounding scar. A panniculectomy removes only the extra hanging skin of the lower abdomen.

The Procedure

In the preop area before surgery, your surgeon will draw a line on the middle of your belly, from the bottom of the breastbone to the pubic bone. The ribs and

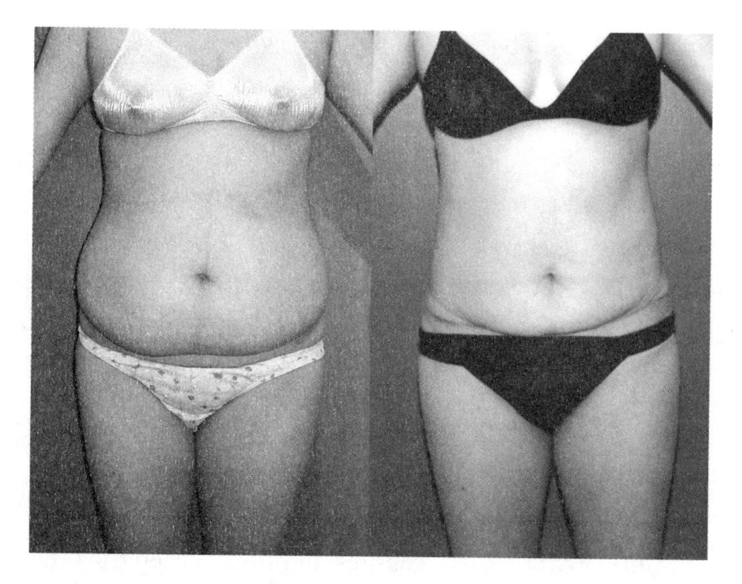

This woman underwent liposuction of the flanks, along with a panniculectomy (removal of the overhanging skin of the abdomen).

the skin to be removed will be marked. If liposuction of the flanks is to be done, a topographic map of the fat in this area will be drawn.

The operation will be performed under general anesthesia. Technically, it is possible to perform it under local anesthesia, but most surgeons prefer general anesthesia. A urine drain is often placed in the bladder. Compression calf boots will keep the blood flowing, lessening the chance of a blood clot in the leg. An injection of heparin, a blood thinner, further decreases the likelihood of clotting. The surgeon paints the skin with an antibacterial solution prior to the start of the operation.

Sometimes a hernia is found at the time of surgery. If so, either the plastic surgeon or a general surgeon will repair it at the time of the tummy tuck.

At the end of the procedure, the breathing tube is removed and the patient is moved onto a bed in a flexed position and rolled into the recovery room. A new technique is the continued infusion of anesthetic into the wound through a pump for a few days after surgery to help with the pain. After a few hours, the patient is brought to a hospital room, her home for one or two nights.

After leaving the hospital, you really need to take it easy for a few weeks. The abdominal muscles will spasm when they flex, causing pain. Most patients need narcotic pain medicine for about a week after this surgery. The drains will be removed by the surgeon two to five days following surgery, depending on the amount of drainage.

You cannot lie on your belly or exercise in any way for a month following this surgery. No sex for about two weeks. No football for six weeks.

The Risks

Tummy tucks are the riskiest of all cosmetic surgical procedures. Besides the typical risk of infection, the risk of bleeding following this procedure is high. There is even a small risk of requiring a transfusion. Because the belly is tightened, pressure is placed on the big veins of the lower body, causing a significant chance of a blood clot in the legs. This deep venous thrombosis, or DVT, can break loose and travel to the lungs, causing a potentially deadly pulmonary embolism. Deaths have been reported with tummy tucks. It is probably wise not to combine this procedure with other complex procedures, in order to keep the total operating time down. Plastic surgeons have different opinions about this, but my philosophy is to do what is uniformly safe.

Infection is a risk with any procedure. If it occurs with a tummy tuck, it can spread over an alarmingly large area, especially if liposuction is performed. Intravenous antibiotics in the hospital are required if infection occurs.

Seromas are common with tummy tucks. Drains and tight binders decrease their chance of forming. "Fibrin glue" may decrease seromas, but it carries the same risks as a blood transfusion.

All patients temporarily lose sensation in the lower abdominal skin. Usually it returns in several months, but the loss may be permanent. Until sensation returns, patients must be careful not to burn the skin with a heating pad and are prone to develop sunburn or frostbite. Ten percent of people develop loss of sensation in the thigh after a tummy tuck, a number that increases if the incision is made low on the belly. This procedure is particularly dangerous for smokers, *most* of whom develop wound problems. Diabetics also have a high rate of wound problems after tummy tucks.

Tummy tucks are usually performed in the hospital with an overnight stay. There is a trend toward sending patients home the same day. But most patients feel as if they were hit by a truck after this procedure. Considering the significant amount of nursing care and the real risks of the surgery, this procedure belongs in the hospital.

Lower Body Lifting and the Belt Lipectomy

Massive weight loss is becoming more common now that gastric bypass and other types of bariatric surgery are so popular. Plastic surgeons are seeing more patients with huge amounts of drooping skin. In 1997, only 2,000 body lifts were performed; by 2005, there were 12,000. I would have expected the number to be higher, but in fact a 21-percent decline in lower body lifts occurred between 2004 and 2005, possibly owing to the risks of the surgery and the extensive scarring.

Skin hangs off the neck, arms, chest, abdomen, buttocks, and thighs after massive weight loss. A procedure called a belt lipectomy removes excess skin of the abdomen and back by making scars completely around the waist. The buttocks and abdomen are lifted in a single operation. This operation may be combined with an inner thigh lift, in what I call a womp. Another procedure is called a body lift. It is more appropriate for people who have loose skin of the buttock and thigh areas. Plastic surgeons disagree on what to call these two procedures; some consider them to be synonymous.

Albert Cram, M.D., F.A.C.S., is professor emeritus of surgery and former chief of the plastic surgery section at the University of Iowa. With his partner, Al Aly, M.D., F.A.C.S., he has performed over a hundred belt lipectomies, probably more than anyone else in the country.[1] "In a belt lipectomy, we make an incision completely around the body," Dr. Cram says. To do this, he and his partner take nearly four and a half hours to remove the excess skin and fat from the abdomen, back, and sides. "We move the patient three times in this procedure, operating first on the abdomen. Then we place the patient up on each side to remove skin from the sides and back before returning the patient to the supine (face up) position."

In a belt lipectomy, typically seven kilograms (fifteen to sixteen pounds) of skin and fat are removed. "People are always disappointed with this number. They have a lot of empty skin that doesn't weigh too much," says Dr. Cram. Body lifts may remove only two kilograms (four to five pounds). In a belt lipectomy, as much as forty centimeters of skin are removed in front and twenty-five centimeters in the back. A tummy tuck is a part of every belt lipectomy, and many patients have hernias that require repair at the same time. The incisions are high on the waist with a belt lipectomy. Drains are left in for more than a week, to decrease fluid collection under the redraped skin. These procedures are obviously performed under general anesthesia and require a hospital stay of several days or longer.

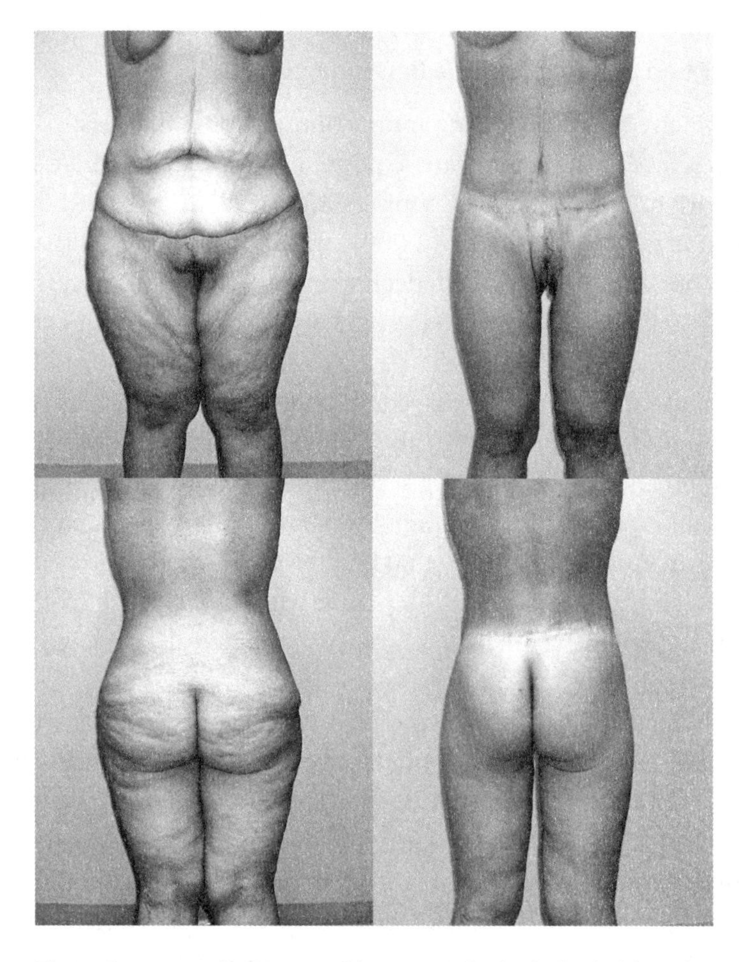

These photos are of a 33-year-old woman who had a body lift and a thigh lift after gastric bypass surgery. The pictures on the left are prior to surgery and those on the right are two years after the body lift and eighteen months after the thigh lift. Photos courtesy of Dr. Joseph F. Capella.

There Are Many Complications with Belt Lipectomies

Belt lipectomy patients have suffered from all the ravages of morbid obesity, usually with diabetes, arthritis, and even heart disease. "The complication rate is dependent not only on how much the patient weighs on the day of the operation, but also what she weighed prior to weight loss," explains Dr. Cram. Patients who were in the 450-pound range have so much extra skin that 40 percent of them develop collections of fluid under the skin. The tension on the incision lines is so strong that wounds can open when the patient bends over as late as one month after surgery. Diabetes usually goes away once the patient has lost the weight. Some physicians routinely have diabetic patients take exercise stress tests prior to surgery. Other complications include infection, bleeding, and blood clots in the legs. The belt lipectomy is a serious procedure, with the rare possibility of death.

Patients who became morbidly obese often have depression and other psychiatric issues. The depression does not necessarily go away with the weight loss. A psychiatric evaluation is therefore important prior to surgery. Psychologically, patients must be able to accept possible complications and the huge scars that are created.

Dr. Cram does not lift the inner thighs at the same time as he does the belt lipectomy. He feels that the lipectomy is a major operation and the inner thigh lift would add even more operating time. Some surgeons, however, will do the thigh lift and the belt lipectomy at the same time. Such procedures are usually performed at academic medical centers, where large teams of plastic surgeons and residents can participate.

Joseph Capella, M.D., a New Jersey plastic surgeon who specializes in body contouring after bariatric surgery, also does not perform an inner thigh lift at the same time as a body lift. His four-hour-long body-lift procedure improves the abdomen, outer thighs, and buttocks. He waits at least three months before tackling the inner thighs. Besides a lower complication rate, the aesthetic result is better with separate procedures

Other plastic surgeons break up the body-lifting procedures, performing a more traditional tummy tuck first and then a type of "back-lift" six months later. This more conservative approach sounds safer than the big womps.

Patients who have been morbidly obese and have had a gastric bypass or a lap-band procedure often lose weight rapidly and are motivated to contour the body once the weight loss is complete. It is the ideal time for body contouring. Surgery is safer at this point and the results will be better. Dennis Hurwitz, M.D., a Pittsburgh plastic surgeon, waits until weight loss following gastric bypass has stabilized for four months before removing excess skin. In his burgeoning body-lifting practice, the wait is about sixteen months after the bariatric surgery.

Thigh Lifts

The first thigh lift was performed in 1956. Twelve thousand were performed in the United States in 2005, only nine hundred of them in men. Thigh lifts can

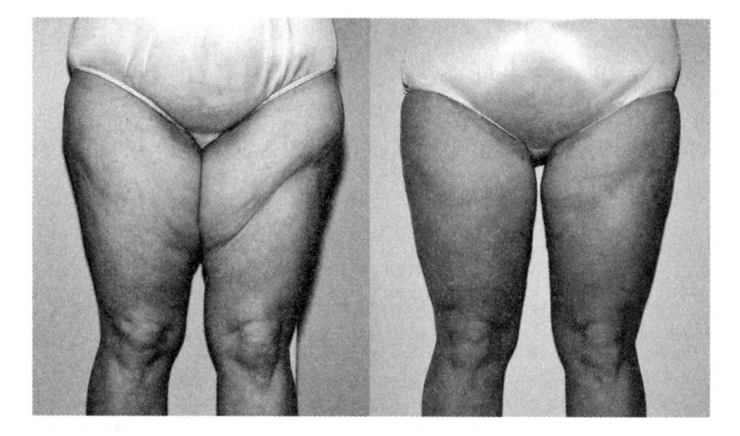

This woman had a deformity of her thigh from a childhood injury. Preop on the left. On the right, after undergoing liposuction of the thighs followed by a medial thigh lift during a separate procedure.

be isolated to the inner thigh or, less commonly, to the outer thigh. Complete thigh lifts are becoming more popular, with circumferential lifting.

Candidates for thigh lifts may have lost a significant amount of weight or may have achieved a poor result from overzealous liposuction. A body lift combines an inner thigh lift with a lower-body lift and abdominoplasty.[2]

The inner thigh lift is a tailoring procedure that removes excess skin and fat through incisions in the groin crease and the thigh. Trimming of the skin leaves significant scars that usually can be hidden underneath shorts or a creative bathing suit. The weight of the thigh skin can pull down and distressingly open the labia. To prevent this, the skin is firmly attached to stiff, deep structures.

General anesthesia is used, and patients generally stay in the hospital at least one night. The procedure begins with the patient lying on her back. Liposuction debulks fat, and an incision begins in the groin crease and extends to the buttocks crease. The patient is then turned on her side. Extra skin is removed and the wounds are closed. A final turn allows the opposite side to be operated on. A liposuction-type girdle compresses the thighs after surgery.

While this operation can have spectacular results, it does leave huge noticeable scars. The surgery is very involved and is fairly disabling for weeks. Exercise must be avoided for three weeks. The risks include wound healing problems, infection, bleeding, and blood clots in the legs.

While some surgeons promote operating on the belly, back, and thighs all in the same sitting, the procedure is enormous. A womp would be expected to

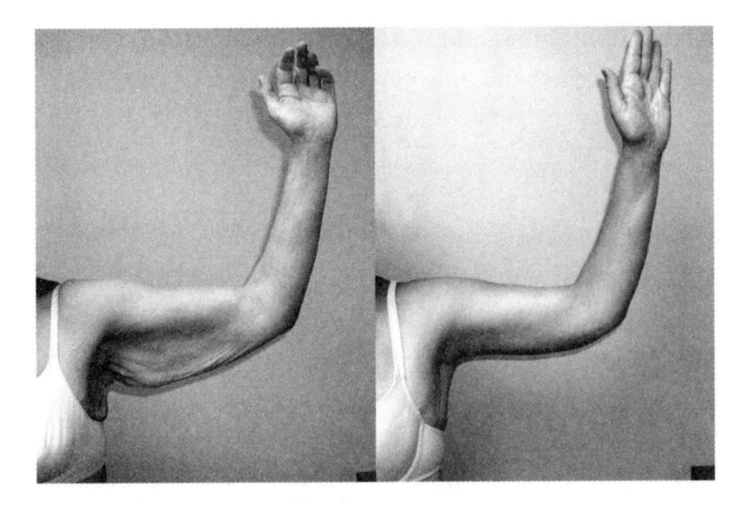

This 38-year-old woman lost 140 pounds following gastric bypass surgery. She is shown before (left) and seven months after (right) an upper arm lift. Although the contour improvement is excellent, the scarring can sometimes be thick and raised. Photos courtesy of Dr. Joseph F. Capella.

have a higher complication rate than several smaller procedures performed at separate times. Longer anesthesia time, more blood loss, and a larger body area operated on are simply more dangerous.

Upper Arm Lifts (Brachioplasty)

Originally described in the 1920s, sixteen thousand upper arm lifts were performed in the United States in 2005. The giant scars that are created are the main reason this procedure is not overly popular.

In an upper arm lift, incisions are made from just above the elbow and continue to either the armpit or even onto the chest wall. Some surgeons perform a zigzag at the end of the incision, to decrease webbing in the underarm area. Others do not feel this is necessary, but extend the scar into an L-shape on the chest. Sometimes liposuction is performed at the beginning of the procedure and sometimes it is a separate procedure, a few months prior to the arm lift. The upper arm lift takes an experienced surgeon just under three and a half hours to perform.

While a brachioplasty effectively removes hanging skin from the upper arm, it can leave conspicuous scars. However, in patients who have lost a huge amount of weight and have a lot of hanging skin, the arm lift may be an appropriate procedure. Recent operations have tried to move this scar from the arm to the armpit area. The risks include nerve damage as well as wound healing and contour problems. Scars can become red and raised for a long time.

Buttock Lifting

One of the new fads in plastic surgery is buttock enlargement, reshaping, or lifting surgery. Techniques include placement of silicone implants, injection of the patient's own fat (fat grafting), and lifting. Most plastic surgeons agree that implants do not work well in the buttocks; these are not popular procedures. Fat grafting is probably reasonable, but there are no good studies verifying how long the fat lasts. Buttock lifting is being popularized by Brazilian plastic surgeons, who state that "there is no technique for all seasons." The procedure makes the most sense when performed as part of body lifting after massive weight loss. But huge incisions are made at the top of the buttocks. This defeats the aesthetics of the operation when the patient is naked. I suppose that if the shape of the buttocks is important in clothing, the benefits might outweigh the risks and scars. When the dust settles, however, I don't see this procedure as being particularly popular.

The Cost

Abdominoplasties average $5,200 in cost. However, mini abdominoplasties would be expected to be in the $4,000 range, full abdominoplasties in the $7,000 range. The inclusion of flank liposuction adds about $1,000–$2,000 to the bill. Thigh lifts average $4,700 and upper arm lifts $3,600. Buttock lifts cost $4,900, while lower body lifts cost $7,800. Belt lipectomies are complex procedures, usually with teams of two surgeons. Typical fees will vary from $10,000 to $16,000, depending on which variation of the procedure is performed. Hospital and anesthesiologist fees add another $13,000.

Health insurance will sometimes pay for the procedures, although at a level too low for many plastic surgeons to accept. While the figures quoted above are from surveys, the fees involved tend to be significantly higher in major metropolitan areas. The entire field of body contouring after massive weight loss is so new that the procedures are still not well defined. A five-year "shake-out" period is needed for the plastic surgeons to figure out which procedures work best.

A warning for consumers: Body contouring procedures after massive weight loss are long and difficult for the plastic surgeon. A team of at least two surgeons is required. The hospital must be equipped to handle this type of surgery. While reputable plastic surgeons adapt to new situations and learn new surgical procedures, make sure your surgeon has previously performed these procedures or has trained with experienced surgeons. The learning curve is short, but you don't want to be part of it.

29

Breast Enlargement

Breast enlargement, also called breast augmentation, is a procedure held in awe by men and women alike. In 1997, some 101,000 women in the United States underwent breast enlargement surgery. By 2005, that number had risen to a staggering 365,000, a 260-percent increase in spite of imperfect breast implants and more than a decade of bad publicity. Plastic surgeons are always amazed that simple lumps of fat and milk glands can create such a sensation. During surgery, after the breast is disassembled into tissue components, the reassembled pieces come together in a sensuous way that stimulates the primitive recesses of the brain.

In my two decades in plastic surgery, I have never met a woman who was happy with both the size and the shape of her breasts. True, I see a selected population of women. But breasts either are too small, too large, too saggy, or poorly shaped, according to my patients. The pigmented area around the nipple (the *areola*) is often too large. The size of the breasts is usually asymmetrical. Most women have one breast that sags more than the other and one nipple that is lower than the other.

Many women (and most men) have no knowledge of this asymmetry. I suppose it is different in the south of France, where breasts are on display constantly. But in more conservative countries such as the United States, women see other naked women briefly in the gym, and men see only a few women naked beyond those in the pages of *Playboy*. (And *Playboy* doesn't show average women. The perfectly symmetrical, usually augmented breasts there are not typical of the general population.)

The History of Breast Augmentation Reads Like a Novel

In 1895, the first augmentation moved fat from the buttocks into the breasts. The first implants in 1930 were made of glass balls or elephant ivory. Imagine the surprise when someone unknowingly touched a glass-ball–filled breast for the first time! In the 1940s, dermal-fat grafts were used. By the end of World War II, plastics such as the sponge-like Ivalon, Polistan, Etheron, Teflon (good for pans, bad for breasts), and hydron were used to make breasts larger. Early results were often satisfactory, except for scarring that caused the breasts to become rock hard. Paraffin injections and preserved human skin also were tried. Later, silicones and silicone plastics were in vogue. In 1949, a Japanese company manufactured a silicone "artificial fat" that was injected into breasts with disastrous results. Medical-grade injectable silicone, discovered around 1960, was widely used to lubricate suture material and needles to decrease drag and pain. Surgeons used high pressure to inject this liquid into breasts. Irritants such as olive or peanut oil, even snake venom, were added to the silicone to keep the liquid in place. These irritants incited a scar capsule to be formed that was supposed to contain the silicone within the breasts.

Silicone injections into the breasts were certainly one of the low points in the history of plastic surgery. They were injected blindly and sometimes killed people, after ending up in blood vessels. And the lumpy, cystic breasts were impossible to examine for cancer. Silicone breast injections were banned in Japan in the 1950s and deemed criminal in the United States by the 1970s.

The modern era of breast augmentation began in the early 1960s. The Dow-Corning Corporation invented silicone rubber (Silastic) in 1945 for use in airplane motors. Silastic combined silicone with carbon, hydrogen, and oxygen. In 1953, the company produced medical-grade Silastic tubing for gallbladder surgery. Silicone could be made as a liquid, gel, or solid by stringing its molecules together in chains of different lengths. The longer the chain, the more solid the material.

In 1962, Houston plastic surgeons Drs. Thomas Cronin and Frank Gerow performed the first silicone gel breast augmentation, paving the way for millions of women. Early breast implants had Dacron patches that allowed attachment to the surrounding tissue. To keep implants soft, their walls were made very thin. This structure ultimately led to the downfall of the implants, because the thin walls allowed leakage. Different types of implants were tried, some with two layers of silicone, some with a layer of silicone surrounded by a layer of saltwater, some inflated with saline, and some with plastic on the surface.

Silicone gel implants—textured on the left, smooth on the right.

Soon after implants were introduced, it became apparent that they were not the panacea everyone had hoped for. In as many as half of women, the implants hardened. Or the implants ruptured, spilling a gooey material into the chest. The human body walled off the implants with dense scar and, in some people, actual bony material. Truly, some women had breasts as hard as rocks and as strong as bone!

The operation soared in popularity, however, aided by the likes of Hugh Hefner and *Playboy.* Plastic surgeons pored over the photos in *Playboy,* some even taking their subscriptions as tax deductions. (They didn't really, but not a bad idea.)

Current silicone gel implants have the gel inside a silicone rubber outer shell. The gel and the shell are made from variants of the same chemical.

Silicone is a safe material that is virtually everywhere in our modern society. It is used to coat needles to decrease the pain of injections. It is a part of hundreds of medical devices, from pacemakers to artificial joints, to the Norplant device and, of course, breast implants. The material in breast implants is also used to make silicone nipples for bottle-feeding. Silicone is part of Simethicone, used to treat upset stomachs, and is a part of many other medications. It is allowed as a food additive, to prevent excessive foaming of liquids like soft drinks. Silicones are part of lipstick, hair spray, and cosmetics. If these compounds caused disease, hospitals would be filled with silicone-diseased patients; few living humans do not have some silicone in their bodies.

Breast augmentations are conceptually very simple operations. Bags of silicone gel are placed above or below the *pectoralis major* muscle. A pocket is made and filled with the implant. But the body sees silicone as foreign, like a splinter, and walls it off with scar. The scar can be so intense that the soft gel implants can become rock hard. In fact, that happened to about half of the early silicone gel implants. With implants placed in the 1980s, about 10 percent of people grew so much scar that the breasts became deformed. A woman lying on the beach might find that one breast looked south while the other looked north. In a small percentage of people, the scarring was so intense that the implants caused pain. And so the implant companies developed a thinner outer layer. Softer until scar formed, but definitely more fragile.

Please Don't Squeeze the Implants

Implant rupture was another common problem. Even today, no one knows exactly what percentage of gel implants will rupture, but my guess is that *they all will* if left in long enough. A European study showed that more than 15 percent rupture by ten years after surgery. When the implant ruptures, the gooey gel spreads out of the implant into the surrounding tissue. It can spread into the armpit and down the arm. It can spread to the abdomen and around to the back. When pressure is placed on the chest, the silicone travels farther. In a mammogram, the silicone is surrounded by scar tissue. A breast exam can be difficult, with each and every bead of spreading silicone feeling like a nodule.[1]

I used to tell women how strong breast implants were, and I repeatedly squashed them to impress patients. Eventually one broke on my desk, oozing a material strikingly similar to the silicone caulk I've used to seal my windows.

In the 1970s and 1980s, the height of the gel implant era, plastic surgeons told women to massage their breasts to keep the implants soft. If breasts hardened, we often performed "closed capsulotomies," pushing hard on the breasts until a pop was felt, causing immediate softening of the breast. Doing this ruptured the scar around the implant, also called a capsule. But this was probably the event that also ruptured the implant. Proving this point are the seventeen documented cases of gel implant ruptures during mammograms! Today most plastic surgeons don't advise massage after augmentation.[2]

Oy Même

In the 1980s, thousands of women received the Même, a polyurethane-coated implant. Although the Même created far fewer hard breasts, the polyurethane degraded rapidly into a mushy material. I remember reoperating on a woman a few months after her Même augmentation, only to find a gooey mess. I always

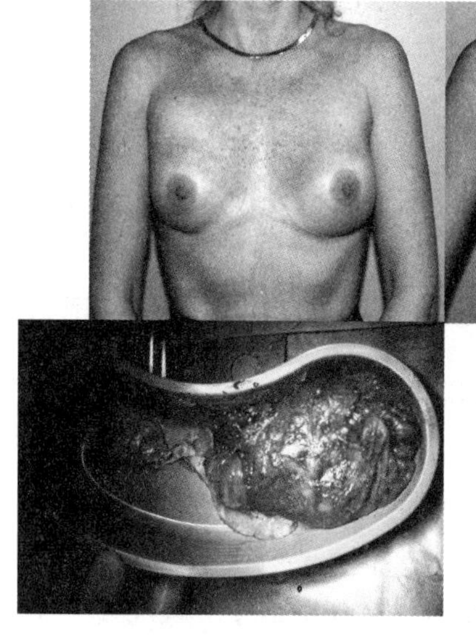

This woman had silicone gel implants for ten years. The implants ruptured, and she could push the silicone into her armpit (above). At left, the ruptured implant is largely contained within the thick scar capsule.

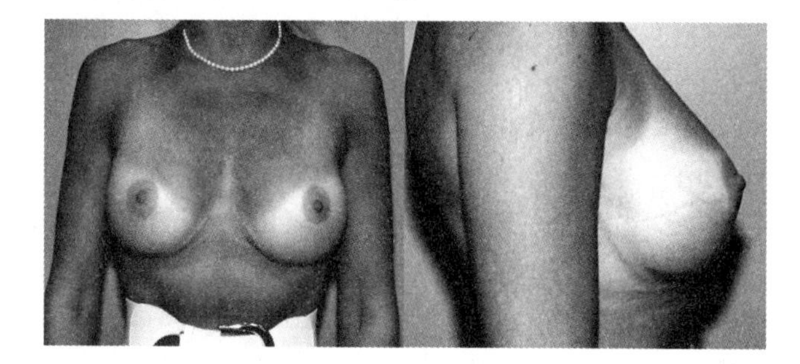

The same patient four months after removal of the ruptured gel implant and replacement with saline implant.

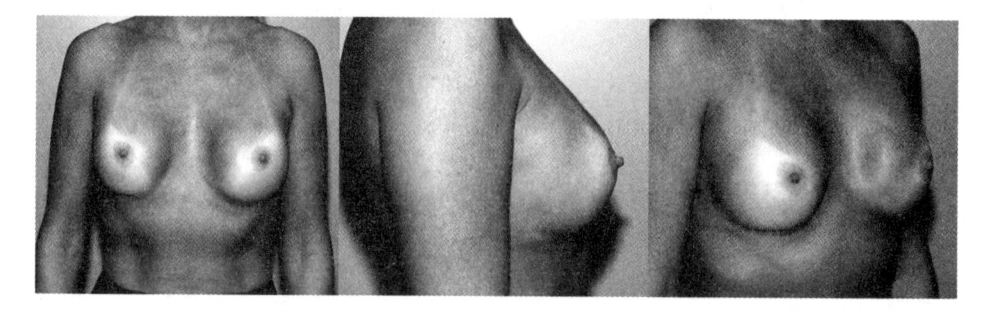

Eight years postop, sagging and noticeable ripples occurred.

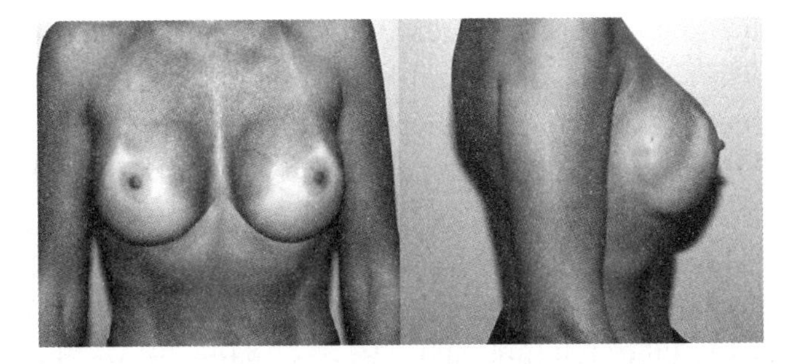

The implants were replaced with larger implants, at the urging of the patient and to the dismay of the surgeon.

wondered what had happened to the polyurethane, considering it was no longer around the implant. It turns out that 80 percent of women with these implants have toluene, a breakdown product of polyurethane, in their urine. It is possible that these chemicals may cause problems such as cancer later in life. While there is no documented increase of cancer in these women, and the FDA is not yet recommending removal of polyurethane implants, I personally do recommend removal.

The polyurethane implant was taken off the worldwide market in 1991, but was later reintroduced. Because polyurethane implants have the CE mark (the European Union stamp of approval), they can be marketed in the twenty-five European Union and three European Free Trade Association member states, despite the objection of the Medicines and Healthcare Products Regulatory Agency, or MHRA in the United Kingdom (the FDA equivalent). This agency in 2005 warned doctors that there was solid evidence that the implants, made by Polytech Silimed Europe, were carcinogenic.

The polyurethane implant did result in softer breasts than smooth gel implants. So the late 1980s saw textured silicone gel implants that combined the desirable properties of polyurethane with the presumed safety of silicone gel. Around this time, the implant companies used a thicker-walled implant with a firmer gel. It was called a low-bleed gel, since less of it leaked through the implant wall. Most plastic surgeons saw fewer hard breasts with the new implants.

The Breast Implant Controversy

By the end of the 1980s, countless implants placed in the 1960s and 1970s had ruptured, many because of vigorous massage or even forceful breaking of the hardened breast implant capsules. Silicone gel spilled out of the thin membrane of the implant and progressively spread, farther and farther. The year 1977 saw the first successful lawsuit against ruptured breast implants.

Plastic surgeons were told that silicone gel breast implants have between 1.4 percent and 6.4 percent chance of rupture during the first three years, depending on the brand. Data supplied to the Food and Drug Administration show that 9 percent of gel implants rupture in each of the first four years. The FDA estimates that 74 percent of implants will rupture during the first ten years. Since most ruptures are silent, it is difficult to determine the precise rate.

Because breast implants existed prior to the Medical Device Act of 1976, they were neither approved nor disapproved by the FDA. They were grandfathered in. In 1988 the FDA classified them as Class 3 medical devices (see box).

Connie Chung's Crusade

By the early 1990s countless women who had problems with breast implants began to band together. The infamous "Face to Face with Connie Chung" tele-

The FDA Medical Device Classification System

- Class 1 devices are the simplest and have the least chance of harm to patients. They must be appropriately manufactured and appropriately labeled. The facilities must be registered with the FDA. Examples of Class 1 devices are wound dressings and scissors used in surgery, as well as microdermabrasion and electrolysis machines.
- Class 2 devices are potentially more hazardous to patients. They require guidelines, performance standards, *postmarket* surveillance, clinical data, labeling, and tracking. Examples are sutures, silicone chin implants, and lasers and intense pulsed light devices.
- Class 3 devices are life sustaining or life supporting, prevent impairment of human health, or present unreasonable risk of illness or injury. These devices usually require controlled studies to establish safety and effectiveness and *premarket* approval. Examples are special wound dressings and silicone breast implants.
- Exempt devices do not need FDA clearance, but the manufacturer and the product must be listed with the FDA. Stethoscopes and mercury thermometers are examples.
- Some devices are not classified. Silicone sheeting and gel for scar management are examples. Companies may obtain Section 510(k) clearance, or Premarket Notification, for "copycat devices" (products that are basically the same as other already approved devices). This approach is a shortcut to FDA approval.

vision show in December 1990 discussed the medical hazards of breast implants and dramatically affected the procedure. The year before the show, 120,000 breast augmentations took place in the United States. By 1992, half that number were performed.

Some of the claims were real: the implants became hard in many women and many, if not most, would rupture. Women with breast implants who had symptoms such as arthritis, headaches, or hair loss, found like-symptomed women. Lawsuits mounted against the breast implant companies and against plastic surgeons. Awards against manufacturers included $1.7 million in 1984, $5.4 million and $7.3 million in 1991, $25 million in 1992, and $27.9 million in 1994. A class action suit filed against the implant companies in February 1992 enrolled 410,000 women. Many patients who told me they were thrilled with their surgery and with me still wanted to cash in on the lawsuit. Fortunately, I was never named in these actions. The nameless, faceless big corporations became victims. It's part of the reason for today's skyrocketing medical and drug prices.

The Effect of the Ban

In 1990, congressional hearings about silicone breast implant safety were held. In 1991, manufacturers were told to provide data supporting their use. In January 1992, the FDA issued a voluntary moratorium on breast implants. The next month, an FDA panel determined that breast implants did not cause autoimmune disease. Two months later, the moratorium on implants was lifted, but gel implants were only allowed for breast reconstruction and were made part of carefully controlled studies. In 1992, the gel implants were banned except in experimental trials. Lawyers and ill-informed women's groups applauded this action. By that time, between one and two million women already had received silicone gel implants.

As this story played out, the ban was based on emotions, greedy attorneys, dishonest co-conspiring doctors, and shaky science. There is no question that the implants should have been studied and improved. But the ban scared women and injured the medical device industry and physicians.

After the lawsuits, studies looked at tens of thousands of women with breast implants. No problems were unearthed. Non–plastic surgeons, with no financial interest, performed much of the research. By 1995, more than twenty studies showed that implants did not cause *any* disease. In fact, *no* study ever showed that women with breast implants have a higher chance of any disease than women without implants. In 1995, the American College of Rheumatology said that implants do not cause disease. In 1997, the American Academy of Neurology said the implants do not cause neurologic disease. Later that year, the *Journal of the National Cancer Institute* stated that implants do not cause breast cancer. Finally, in 1999, the Institute of Medicine, part of the National Academy of Sciences, released a four-hundred-page report that concluded that implants do not cause disease. But because they harden and rupture, the class action lawsuit stuck.[3]

This mounting evidence did not stop runaway juries. When they were faced with unfortunate women dealing with bona fide medical problems, albeit not caused by implants, juries found it easy to give away the money of some faceless corporation or insurance company. Despite overwhelming scientific evidence to the contrary, in 1998 a Nevada jury awarded $41 million against Dow Chemical to a woman who said she developed multiple sclerosis from breast implants. In 1999 a jury gave $10 million to a woman who claimed her implants caused scleroderma.

As a result, the Dow Corning Corporation went bankrupt and most other manufacturers stopped making implants. Connie Chung and the commissioner of the FDA lost credibility, as did citizen advocate Ralph Nader and Dr. Sidney Wolfe, whose Public Citizen Health Research Group stirred up the public and upset many women needlessly. The whole experience taught physicians and medical device companies a lesson: Be more up front with your products. If anything, play it safe and *overstate* the risks. The experience has probably hurt society as well, because many investors are reluctant to risk their money in the medical devices industry after this fiasco.

The Current Situation

The only implant available in the United States from 1992 to 2006 was the silicone saline implants. Saline implants have advantages over silicone gel implants. Ruptures are easier to deal with. Incisions are smaller. Unfortunately, saline implants still block mammograms.

In November of 2006, the FDA approved silicone gel implants, allowing their return after a fourteen-year absence. The newer gel implants have a thicker "cohesive gel." If it leaks, it is not expected to travel as far. However, there still is a question of whether the new gel is really a semisolid at body temperature, or whether it turns into a runny goo. I believe that the gel will spread out, similar to the older material, perhaps at a slower rate. Breast implant expert Dr. V. Leroy Young of St. Louis says this will not happen. In these new implants, the silicone gel is dense enough to rebound to its original shape after it is touched. This allows different shaped implants to be created, opening up new possibili-

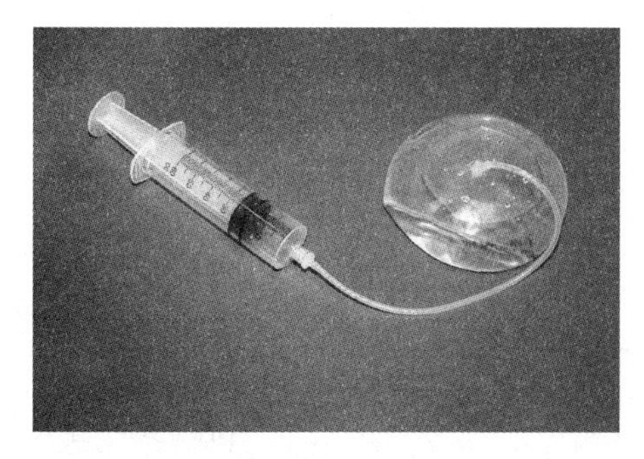

A textured saline implant, with the fill tube inserted.

ties for women. For instance, if a woman has very little tissue in the bottom of her breast, then a shaped implant can add more volume to this area.

These implants are allowed in women ages 22 and older, presumably because their breasts are fully grown. Gel implants require special consent forms, training courses for board-certified plastic surgeons, and long-term patient tracking. Most important, patients will have to have an MRI (magnetic resonance imaging) three years after the surgery and every two years thereafter. Dan Schultz, M.D., F.A.C.S., director of the FDA's Center for Devices and Radiologic Health, told me that "the FDA now has enough information about the behavior of breast implants to be able to write a label, giving women and their physicians enough information to make an informed decision." If the implants rupture, the FDA advises removal with possible replacement.

Unfortunately, there is no perfect breast implant. Saline implants are really just water balloons. And they feel like water balloons. I tell patients that the implant procedure is for looks, not feel. All women feel the implant—in particular, its folded edges. As breasts sag, many women can see these ridges. Troublesome while naked, breasts supported by a bra or a bathing suit rarely show these wrinkles. On the positive side, they are positioned uninflated, through much smaller incisions, and inflated once inside the pocket. When they rupture, there is no leakage of gel. Gel implants have the advantage of a more natural feel with less wrinkling.

Both saline and gel implants are available with either smooth or textured surfaces. They come shaped or round and have different degrees of projection. Only time will tell if this new generation of implant is better than the four previous generations.

The entire breast implant saga is wonderfully recounted in the book *Science on Trial,* written by Marcia Angell, M.D., former executive editor of the *New England Journal of Medicine.*[4]

Comparison of Saline and Gel Implants

Saline		Gel	
Advantages	**Disadvantages**	**Advantages**	**Disadvantages**
Smaller incision	Less natural feel	More natural feel	Longer incision
Rupture less troublesome	Wrinkles more visible	Fewer wrinkles	Rupture causes spread of gel

Other Implants

Only silicone saline and silicone gel implants are currently marketed, although other types have been proposed. Hydrogel implants are made of a complex mix of water, sugar, and salt encased in a silicone rubber shell. The NovaGold polyvinylpyrrolidone (PVP) is a hydrogel implant with a textured silicone shell, filled with PVP hydrogel and guar gum gel. We don't yet know if these implants are safe in the long run. They are not available in the United States and were taken off the market in the United Kingdom in 2000 because of safety concerns.

For a while, it looked as if soy oil implants, known as Trilucent implants, would be a viable option for breast augmentations. A few dozen patients in the United States received these implants in an experimental study. Thousands in Europe received them. Unfortunately, the soy oil can leak out and cause inflammation and swelling. The implants were taken off the market in the United Kingdom in 1999, and women were advised to have them removed.

A hydrogel implant containing a chemical called carboxy-methyl-cellulose (CMC) shows promise as a new implant. The implants feel natural and are purportedly not toxic if they leak. The implant is available in Europe but not in the United States.[5]

Implants of Native Tissue

Many women wonder why they can't just move fat from their buttocks or belly into their breasts. In fact, this has been tried. In the 1980s, fat retrieved with liposuction was reinjected into breasts. Some of the fat survived, as with any fat grafting procedure. Unfortunately, in the breasts, scar and even calcium deposits formed. When a mammogram was performed, this scar could be mistaken for cancer and require a biopsy. Accordingly, the American Society of Plastic Surgeons issued a ban on this type of procedure.

Technically, tissue from the abdomen or back ("flaps") can be moved into the breasts to increase their size. Microsurgery can also move large pieces of tissue from the abdomen or buttocks into the chest. These techniques are usually reserved for reconstruction after cancer, for they are long, difficult, and expensive procedures. They also leave huge scars.

Some surgeons augment the breast tissue with upper abdominal fat and dermis. This procedure is performed as part of a special tummy tuck in which an incision is made under the breasts and the upper abdominal skin is lifted to smooth out wrinkles. Instead of throwing out the excess fat, the surgeon moves it upward into the breasts.

The Consultation

In order to detail all the risks, benefits, and issues surrounding this operation, the consultation should take a full hour. Many decisions must be made prior to the surgery.

Smooth or Textured Implants

Implant surfaces are smooth or textured. Textured implants stay where they are put—they do not move around. That is an advantage, since a smooth implant scooting around within its pocket can be extremely disconcerting, not to say unnatural. Some believe that textured implants have a lower chance of hardening. The rate of significant scarring is decreased from 15 percent to 9 percent. But this advantage comes at a price; it seems that the rupture rate is higher with textured implants. Tissue that grows into the textured surface may cause tethering of the implant, weakening it over the years. Smooth implants, on the other hand, offer less chance of visible wrinkles.

Round or Shaped Implants

Breast implants come round or shaped. Round implants are preferred by most plastic surgeons because there is less chance of error. Because they are shaped like a circle, the surgeon doesn't have to worry about the orientation of the implant. With oval-shaped implants, if a few degrees of rotation occur, the implant may look perfect on the operating table, but after swelling has settled for a few weeks after surgery, the breasts may have a peculiar shape.

Proponents of the shaped implants say that they are more natural, that there is more volume in the bottom of the breast. In fact, the round implants *become* shaped when in the body. Better yet, when the patient lies down, the implant flattens in a natural way. The shaped implant retains more volume in the lower portion of the breast, which is *not* particularly natural.

Finally, the shaped implants cost more. I use them only if the patient wants very large breasts. In this situation, the round implants may be too large for the chest, touching each other in the middle of the chest and overlapping the outer chest wall. The shaped implant shifts the volume from the sides to the front of the implant.

The Spectrum Implant

This is an implant for the unsure. Its size can be adjusted for six months following surgery. The adjustment is made through a small valve placed under the skin to the side of the implant. The valve is a double-edged sword—it can potentially leak or be felt. The implant is more expensive than standard implants. Besides, size issues really should be hammered out *before* the procedure, not after.

Possible Locations for the Incision

The implants can be placed through incisions made under the breast in the crease, around the areola, in the armpit, or through the belly button.

Each incision has advantages and disadvantages. I prefer the incision in the breast crease. It is impossible to see unless the woman is lying down nude. If done properly, even when she is naked the incision should be hard to see. The operation is simplest in this location. Simple is good for the doctor, and good for the patient. Simple translates to a lower chance of complications. Specifically, this location is associated with the lowest chance of a misplaced implant.

An incision made around the areola will be visible when the woman stands up naked. This location is no longer popular. While the armpit incision waxes and wanes in popularity, it has many associated problems. It eliminates scars from the breasts—but it moves the scars to a more public location, under the arm. The scar is totally visible when the woman wears a bathing suit or a sleeveless shirt or dress. In 5 percent of people, the scars are red and raised. Even if the scars are perfect, little hair will grow along them, creating a noticeable rift in the stubble. A few women develop a numb area on the inside of their arm, and a small number will sustain a more serious nerve injury. Surgery performed through these incisions is harder to make symmetric; one breast could be higher than the other or off to the side. Finally, to assure a safe surgery, the scar must be longer than if made under the breasts.

The incision through the belly button is ill founded. Besides adding unnecessary difficulty to the operation, this Trans-Umbilical Breast Augmentation (TUBA) requires the implant to be pushed through a tunnel into the breast pocket. The necessary steel instruments could injure the implants, creating

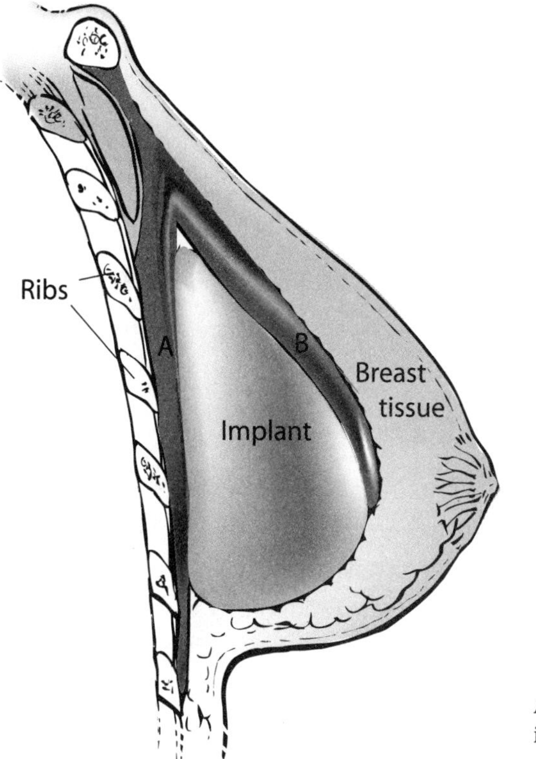

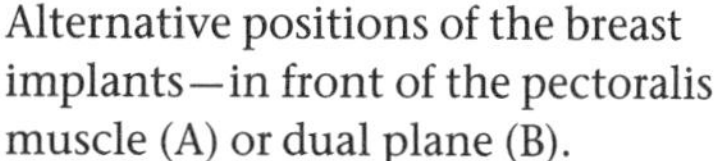
Alternative positions of the breast implants—in front of the pectoralis muscle (A) or dual plane (B).

weaknesses that could lead to early rupture. Implant companies do not encourage this technique. Yet there always seem to be plastic surgeons who swear by it. They claim that complications are rare. What if furious bleeding starts in a blood vessel over a third of a meter away from the incision? It is only a matter of time until this type of disaster claims a life.

A new technique involves placement of the implants through a tummy tuck incision, at the time of a simultaneous abdominoplasty. It is technically a difficult procedure, with longer operating time and more potential for complications.[6]

Under or Over the Muscle

The implant can be placed under or over the pectoralis muscle. The space made for the implant is called the "pocket." Since breast tissue normally is over the muscle, this location is more natural. However, with saline implants, particularly *textured* saline implants, wrinkles in the implant are more likely if it is placed in front of the muscle. Since the nipple receives its blood supply through blood vessels that penetrate the muscle, if the implant is placed in front of the muscle, any future breast-lifting procedure becomes more difficult, if not impossible. When the implant is placed under the muscle, the nipple blood supply is preserved, allowing a lift later in life. If the implant is placed completely under the muscle, however, bulging high up on the chest can be a problem. The implant can be squeezed when the muscle contracts, causing an embarrassing flattening of the breasts.

Playboy *Influences Plastic Surgery*

I always find it interesting when women *request* a breast augmentation with high bulging of the implant. Since the only nude people that many individuals see are in pornography magazines, and since most of the models have had breast augmentations, this look has been interpreted by our society as "normal." A breast augmentation that bulges up high is a complication of surgery, something undesirable. Remember: in cosmetic surgery we want to make the normal better. We don't want to create a look that never existed in nature. In breast augmentation, as in rhinoplasty and eyebrow lifting, plastic surgeons have succeeded in creating an appearance that society then mimics.

Dual-Plane Breast Augmentation

In 2001 the Dallas plastic surgeon John Tebbets, M.D., described his dual-plane procedure. Conceptually, it is a complex procedure. Basically, the implant is placed above the muscle in the bottom third of the chest. The muscle is then cut widely and the implant is placed behind the muscle in the upper portion of the pocket.[7] The muscle is not repaired, allowing the implant to fall through it, held by breast tissue and skin. The implant position is more natural, with less upper bulging. It also does not disrupt the blood supply to the nipple, allowing the patient to undergo a lift in the future. Finally, wrinkling is lessened, since the upper portion of the implant is behind the muscle.

Since the dual-plane procedure was introduced, I have switched from 80 percent of my patients having the implant placed in front of the muscle to 80 percent receiving the dual-plane procedure.

The AMBRA Procedure—Breast Augmentation Without Implants

This acronym stands for "augmentation mammaplasty by reverse abdominoplasty." Not many plastic surgeons perform this procedure, but it will find a niche. Tummy tucks usually are performed through lower-belly incisions. In this variant, the incision is made across the chest, in the creases below the breasts. The abdominal skin is lifted and the epidermis is removed. The fat and undersur-

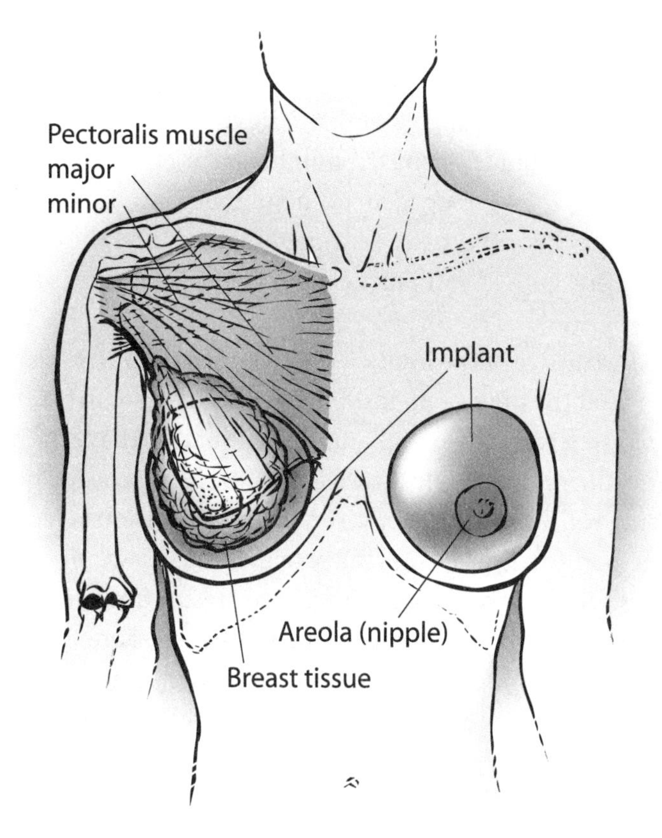

The dual-plane breast augmentation. The implant is partially behind and partially in front of the pectoralis muscle.

face of the skin are placed inside the breasts before closing the skin. Two operations for the price of one. Pretty cool.

Funny-Shaped Breasts Make Augmentations Challenging

There are a variety of deformities of the breast. One of the most common is the so-called tuberous breast. Plastic surgeons also call this the Snoopy deformity, because the breast looks like Snoopy's nose. The breast tissue herniates through a weakened areola. It is fixed via an incision around the pigmented areola. The skin is tightened as the areola is made smaller. The bottom part of the breast tissue is cut like a loaf of bread to spread it out, and then an implant is placed.

Now Comes the Hard Part—The Size

Once the type and position of the implants, and the location of the incisions, have been chosen, the remaining decision is *size.* While initially it is the least important decision, once the decision to have surgery is made, it becomes extremely important.

An average woman has a large B-cup or small C-cup breast. Bra sizing is not government regulated, but basically the number is the chest wall circumference just above the breast, rounded up to an even number. Cup size is measured over the apex of the breast. For every five centimeters more than the circumference, size increases by one cup. For all practical purposes, a cup size is about 150 cc, or 5–6 ounces.

To determine the size of the implants, I have patients purchase the bra they want to fit into after the operation. It should be a rigid full-cup bra. In the exam room, I place progressively larger implants into the bra and have the patient look at herself in the mirror. Eventually, she will like the size. I coach and advise her if she is choosing too large a size, but ultimately the decision is hers.

Despite the plastic surgeons on reality television shows, *bigger is not better.* There is no question that the long-term aesthetic result from an augmentation is worse with larger implants. Problems such as droop, abnormal shape, and the visibility of wrinkles increase with larger implants. I strongly believe that women should have average-sized breasts—not so large that they are stared at or that they begin to have the physical problems associated with excessively large breasts.

The Procedure

As with any surgery, aspirin should be stopped a week prior to the procedure. Birth control pills should be stopped one month prior to surgery, to decrease the chance of a blood clot in the legs. A mammogram should be performed. A second one will be taken a few months following the surgery. This second mammogram will establish a new baseline, because the implants will block mammograms and may cause calcifications that could be mistaken for cancer in later years.

The morning of surgery, the surgeon marks the position of the implant pockets. Leg compression boots are placed and the surgery is performed. I prefer general anesthesia, because of the rare chance of a puncture of the lung that has been associated with breast augmentation. This complication is more likely under local anesthesia, because the needle used to numb the chest wall comes perilously close to the lung.

Following surgery, sterile gauze is held in place by a restrictive-type bra. Most plastic surgeons do not use drains for this procedure, so showering may begin the next day. Stitches are removed in one week.

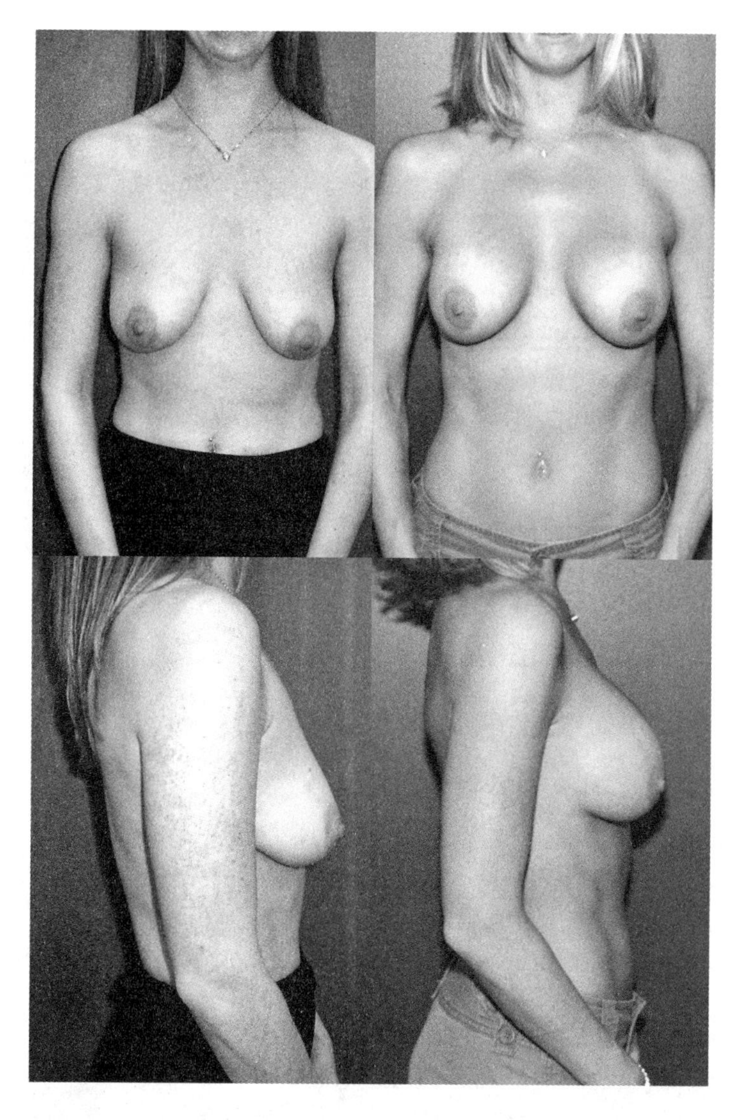

If the breasts droop and start high on the chest wall, the crease can be dropped to give the appearance of less drooping.

If the implants are placed over the muscle, there won't be a lot of pain after surgery. If the muscle has been penetrated, pain is severe for about three or four days.

Depending on the patient's occupation, she may return to work after two to seven days. She cannot exercise for three weeks after surgery, because bleeding can occur if the blood pressure or pulse rate increases. To let the tissues heal, refrain from upper body exercises for six weeks following surgery.

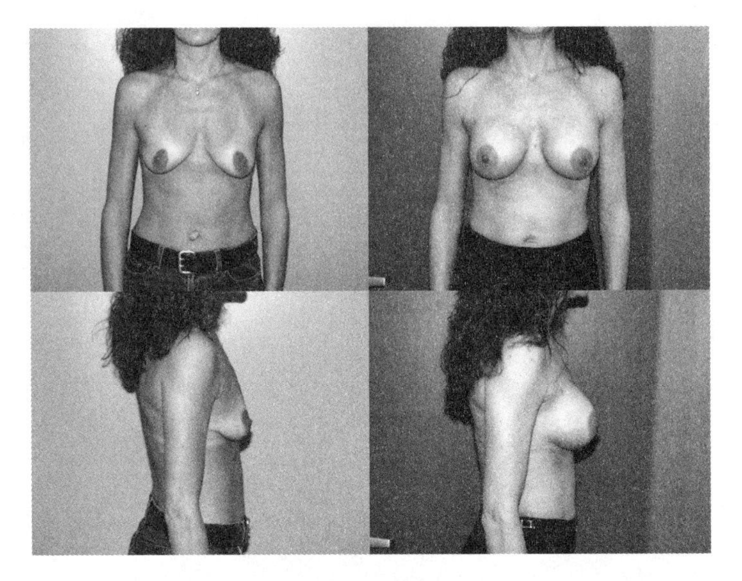

Typical appearance after dual-plane breast augmentation. This woman's breasts drooped following pregnancy because of a mismatch of breast volume and skin. When reinflated with an implant, the breasts drooped less.

The Risks

Breast augmentation has the standard risks of infection and bleeding. Infection is rare but when it does occur, it may require removal of the implants. The body simply cannot fight infection efficiently if a foreign object is present. Many plastic surgeons are reluctant to replace implants in a patient who has developed an infection. Bacteria lie dormant in scar tissue for a long time, and a second infection may be provoked by another operation. Most plastic surgeons will not replace the implants for at least six months.

Bleeding is not common after this surgery, but it can be dramatic if it does occur. Blood clots in the legs, clots spreading to the lungs, punctured lungs, and death have been reported with this procedure.

Implant-related problems include scarring around the implants, called capsular contracture. Scarring is the body's attempt to wall off the implant, to isolate it from the rest of the body. We simply don't know what causes the hardness or how to stop it.

In 15 percent of women, the breasts harden because of the scarring. About 5 percent develop deformed breasts, and 2 percent develop painful breasts. I will reoperate once, and once only, if one of these conditions arises. If the condition recurs, the patient should either live with the problem or have the implants re-

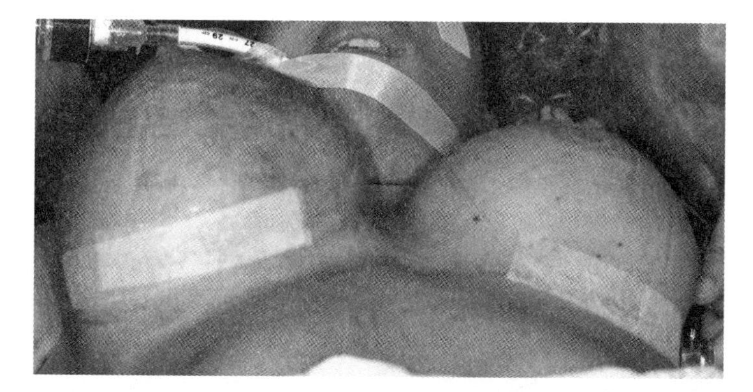

This patient underwent breast augmentation by another surgeon two and a half weeks prior to this photo. She developed acute swelling and pain of the right breast (at left in this photo). Nearly a quart of blood was drained and a large blood vessel was tied off. She recovered and has a satisfactory appearance following the procedure.

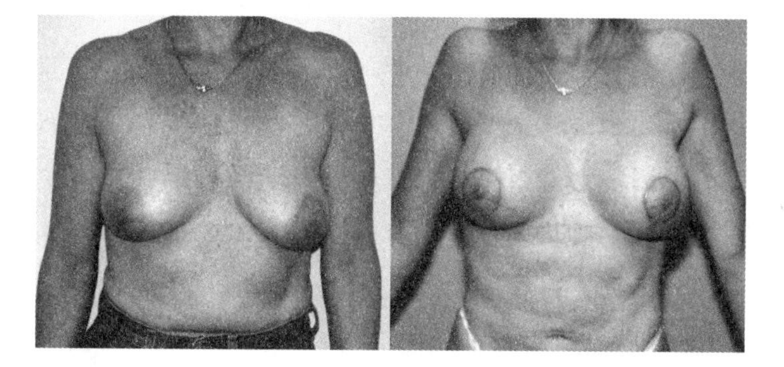

Bad scars occur in 3–5 percent of patients. Assuming your plastic surgeon has closed the wound properly, this hypertrophic scarring is genetically predetermined and largely unavoidable. This woman had small, droopy breasts. On the right, her appearance after a combined lift/augmentation procedure. She developed red, raised scars, although she had soft-feeling breasts. See color plates.

moved. Some women fall into the trap of repeatedly having their implants and scar tissue removed, and new implants placed. Some seem to make a career of it. Some plastic surgeons use the asthma and allergy drug Singulair for two months after breast augmentations, to try to decrease capsular contractures. There is no evidence that this drug works.

Scarring around implants has nothing to do with skin scarring. In fact, many women with hard breast implants have excellent scars and women with soft implants sometimes have red, raised scars.

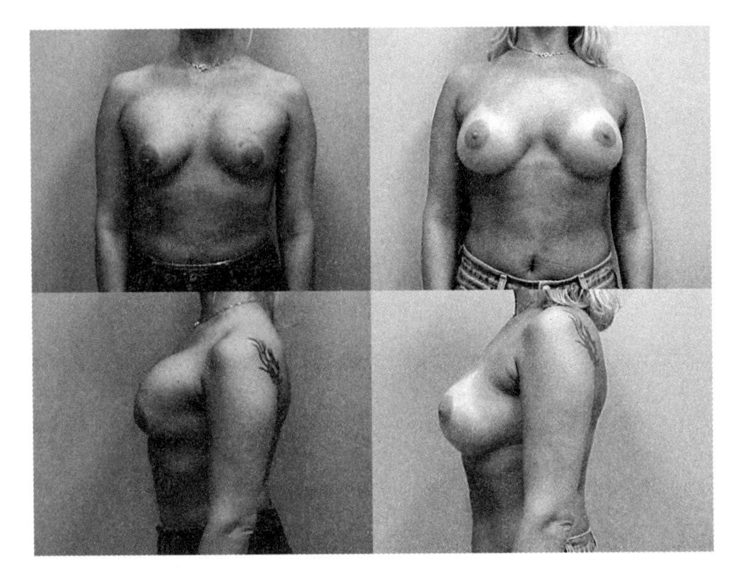

This woman underwent a prior breast augmentation by another doctor (on the left). Her implants were placed too high and were too large. On the right, her appearance after repositioning of smaller implants to a more appropriate position.

Implant wrinkles are common after augmentations, more so in droopy breasts, very small breasts, and when very large implants are used. The problem is most common if textured saline implants are used.

Saline implants can deflate after surgery. About 1 percent per year will do so. As a cumulative risk, this means that after ten years there is a 10-percent chance of deflation. The Mentor Corporation, one of the two implant companies, quotes a 3.7-percent deflation rate for its smooth, round saline implants after seven years. Inamed Aesthetics, now part of Allergan, counters with a 2-percent deflation rate for its similar product. This contrasts to Mentor's 12.4 percent rate of deflation for its textured variety. And larger implants may deflate sooner than smaller implants. If deflation occurs, wait until morning and then schedule an appointment with your plastic surgeon right away. Sometimes replacements can be performed under local anesthesia. The implant companies often will pick up part of the bill. Gel implants can leak. If this happens, the leak may go unnoticed for months or years. Sometimes there is an obvious change—a softening or hardening of the breast. But sometimes there is no change, and the leak is found on mammogram or MRI. When gel implants leak, they should be removed or replaced.

Problems with the position of the implants may be related to a poor choice of implant or to the surgeon's skill. The implants may simply drift, usually downward but sometimes toward the sides or the center of the chest. Reoperative surgery is challenging and associated with a high rate of recurrence.

Breast-feeding should still be possible following augmentation, perhaps with more problems if the incision is made around the areola. Certainly if poor technique is used or if infection occurs, breast-feeding may be impossible. If you are the leader of your La Leche League chapter, you might not want to have a breast augmentation until after your last child is born.

Nipple sensation should be normal, but may be decreased or absent. I have had many women whose nipples were overly sensitive after surgery, probably because the nipples were forced forward and rubbed on their clothing. These women wore Band-Aids over their nipples for months; the problem eventually disappeared.

Risk of disease does not increase after silicone breast implants. Tens of thousands of women have been examined in many studies. No increase has been found in rheumatoid arthritis–type connective tissue disease, nor in any other disease.[8]

The single most important issue is that all breast implants block mammograms. This blockage could be as low as 3 percent or higher than 60 percent. Doctors disagree. Radiologists try to take special views of the breast tissue, but there is no question that some of the tissue will be blocked. X-rays cannot penetrate silicone (either the gel or saline) implants.[9] The argument is academic as to just how much tissue is blocked by the implants. Even if 1 percent of tissue is blocked, if your cancer is in that 1 percent and the mammogram doesn't pick it up, *you lose.* Breast cancer is visible on a mammogram two years before it can be felt on examination.

MRIs are the solution to this problem. An MRI can see through breast implants. It is much better at imaging breasts and does not use harmful radiation. Unfortunately, an MRI costs at least ten times what a mammogram costs. Most insurance companies will not pay. I have had patients with obvious ruptures who were first required by health insurers to have useless breast ultrasounds to try and save money. If a woman has a breast implant placed inside her chest, blocking future mammograms, she must undergo MRI studies to remain safe. If you can't afford the expense, you shouldn't have the surgery.[10]

How can health insurance companies get away with not paying for MRIs of the breasts when they are mandated to pay for mammograms? Women's groups should realize that if they demand coverage of breast MRIs, insurance companies ultimately will have to provide them. As the demand increases, more MRI facilities will become available, decreasing costs. In twenty years, radiation-emitting mammograms will seem archaic; all women will have MRI mammograms.

Breast examination is usually *easier* with implants. No increase in breast cancer has been noted in women with implants. Despite the fact that mammograms are less useful, there is no difference in the size of breast cancers discovered in women with implants and those without them. Cancers are easy to miss. Implants provide a platform to push against and actually make breast exams easier (unless, of course, scarring causes rock-hard breasts).

Breast Implant Removal

One of the most difficult decisions a woman may face is whether to remove her implants. If problems such as severe hardening, distortion of the breast, or pain occur, or if repeated biopsies become necessary because of mammogram or MRI changes, sometimes the best decision is removal of the implant. For financial and emotional reasons this may be difficult to do. If implants must be removed, the breasts do not simply revert to the shape they had prior to the augmentation—they look worse.

The breasts stretch rapidly with implant surgery. It is like a nine-month pregnancy occurring in one afternoon. Stretch marks are common. The implants also put pressure on the fat of the breast, causing it to melt away. Drooping is more common in larger breasts, whether the size is a result of genes, pregnancy, or surgery.

For all these reasons, when the implants are removed, the breasts usually are droopier and smaller than before the initial surgery. Of course, surgical scars will also be present. When implants are removed, most surgeons remove as much of the scar tissue as possible, to minimize changes to future mammograms and to remove microscopic amounts of silicone that have leached into the tissues.

Nonsurgical Breast Augmentation

Get ready for this one. The Brava device is a $2,000 Rube Goldberg–looking machine that applies suction to the breasts ten hours a day for ten weeks. The breasts actually do increase in size by about 60–120 grams. In one study, however, a quarter of patients could not tolerate the device.[11]

It appears that breasts grow and stay larger for more than a year after the Brava. The real question is whether anyone will tolerate a device such as this for a small gain in breast size.

Dietary Supplements

There's not much to say about dietary supplements with names like Quick Bust, GroBust, Contour, and Natural Curves. They simply do not work.[12]

The Cost

The average fee for a breast augmentation is $3,600, with additional expense of about $1,000 for the implants. Gel implants cost several hundred dollars more.

30

Breast Reductions and Lifts

One of the most satisfying cosmetic surgical procedures is not cosmetic at all. Huge breasts cause back and neck pain, bra straps that dig into the shoulders, and chafing with recurrent yeast rashes. A breast reduction addresses all of these issues.

A full century since the first breast reduction, most plastic surgeons use a variant of Dr. Robert Goldwyn's 1977 technique. By 2005, the number of women in the United States who had their breasts reduced was 161,000. They have the procedure for medical reasons, and to be able to participate fully in sports. Interestingly, by removing the mass of tissue on the chest wall, breathing is improved.[1]

In this surgery, breast tissue and fat are removed. The skin is then tailored around the new breast volume. Most people want average-sized breasts—a large B or a small C cup. When breast size is altered, the bra number doesn't change: it depends on the circumference of the chest just above the breasts. The cup size, of course, will change.

The challenge of breast reduction is to maintain nipple sensation and blood supply. Almost fifty years ago, a plastic surgeon named Dr. Robert Wise cut up bras, removed cloth, and sewed them back together to figure out this operation. We still use Wise's pattern. Most surgeons leave breast tissue attached to the underlying muscle, and remove tissue above and to the sides of the nipple. The nipple and the skin are then repositioned with stitches.[2]

The major drawback of breast reductions and breast lifts is the scarring. All procedures require scars around the areola, extending to the breast crease. Tradi-

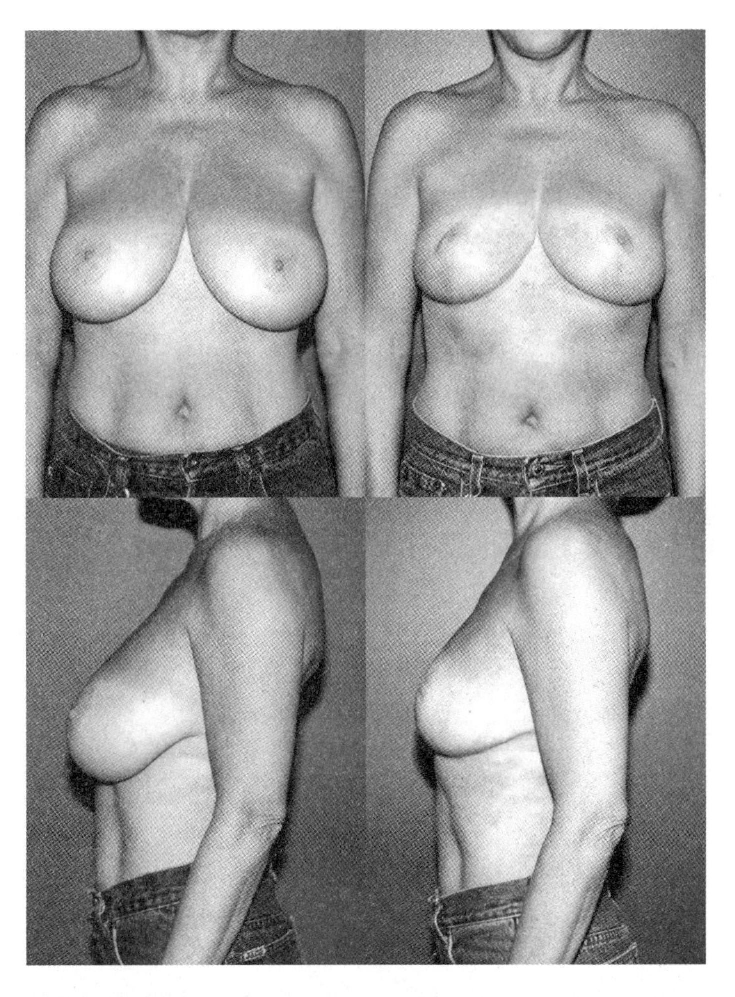

This woman underwent a breast reduction procedure. On the left, before; on the right, after the procedure.

tional breast reductions also make incisions along the length of the breast crease. A newer, controversial procedure, called the vertical mammaplasty, eliminates the horizontal portion of the scar. Many plastic surgeons do not consider it reliable. Some have seen severe breast distortions, and some will use the procedure only for small reductions. Interestingly, plastic surgeons can't agree on which scar they want to eliminate. An even newer procedure eliminates the vertical scar, leaving the horizontal portion.[3]

The Procedure

In your consultation, the plastic surgeon will examine your breasts both for aesthetics and for lumps. If your breasts have a pleasing shape, chances are that your final result will be good. However, many women have odd-shaped, boxy, unattractive breasts. After surgery they will look better, but they still will not have a perfect appearance. A mammogram should be performed before surgery. Since the operation can cause scarring and calcifications, unless a new baseline mammogram is obtained shortly after surgery, the changes may later prompt a doctor to recommend a biopsy. A postoperative mammogram that clearly attributes the changes to surgery allows these calcifications to be safely watched over time.

Before your surgery, your plastic surgeon will draw the correct nipple position and mark the skin and breast tissue to be removed. Skilled plastic surgeons agree that these planning marks are the most important part of the surgery. Just cut on the dotted line, so to speak.

You'll be brought into the operating room and placed under general anesthesia. The surgery takes between three and six hours. You will wake up wearing a surgical bra. It opens in the front and provides support without the underwires that can cut into new incisions.

Blood transfusions are extremely uncommon with this surgery, and most plastic surgeons no longer have patients donate their own blood ahead of time.

The Technique

Most plastic surgeons use a technique that keeps the nipple attached to the underlying breast. About 15 percent use shorter incision techniques, although most believe this to be an inferior procedure, best used in small reductions. Some surgeons remove the nipple, shape the breast tissue, and sew the nipple back on as a skin graft. If the surgeon understands the basic science behind the procedure, however, nipple removal should almost *never* be necessary. He removes breast tissue from either side, and above and sometimes below the nipple, using artistic judgment and experience. The weight of the removed tissue is also a factor. Often, more tissue will purposely be removed from one side to even out asymmetrical breasts, a determination made prior to surgery. Some physicians sit patients up during the operation, even while under anesthesia, to determine symmetry of removal. Because such movement is dangerous, it is avoided by many surgeons.[4] Sometimes liposuction is used to contour the fat under the armpit.

All tissue removed from the breasts is examined under the microscope to be sure that no cancer is present. Once an appropriate amount of breast tissue and skin is removed, large (key) sutures temporarily orient the new breast. The surgeon closes the skin in two layers, using dissolving stitches in the dermis, and stitches, Dermabond, or a combination of both to close the surface layer. Drains should not be needed. Dressings are placed under a restrictive surgical bra while the patient is still under anesthesia.

The patient stays in the recovery room for about two hours. About half of my patients go home the same day; the other half stay the night in the hospital.

After the Surgery

After surgery, there should be surprisingly little pain. The muscle has not been cut, and breast tissue has few nerves. Narcotics such as Percocet or Tylenol #3 are used for a few days. A visit to the doctor within a day or two of surgery to check the wounds is standard.

You can then begin showering. Your incisions are waterproof after twenty-four hours. The old myth that stitches should be kept dry arose in the days when silk sutures were used. When surgeons switched from braided silk to single-strand nylon material in the 1970s, they should have realized that wounds could now get wet after surgery. But it wasn't until the late Joel Noe, M.D., published a simple but landmark study in 1988 that the instructions changed. Noe showed that there was no difference in the infection rate or quality of the scar if the wounds were allowed to get wet. Showering allows more normal functioning in the postoperative period. It is preferable to bathing, since bacteria can contaminate the water and get into the wounds.[5]

Use a Bra That Would Make Your Grandmother Proud

The special surgical bra should be used continuously, except when showering, for six weeks. Other bras can be substituted, but must open in the front so that your arms do not have to contort, and it must not have underwires. The wires redistribute the weight of the breast and create greater pressure on the skin. Since the wires sit over the incisions, they can cause the wounds to open.

Minimal physical activity is allowed after this procedure, and no sexual activity for about two weeks. Basically, you should not do anything that will cause your blood pressure or heart rate to rise. You should not sleep on your breasts for about a month after surgery. During the first three weeks, doing so could cause your stitches to break and create major healing problems. After three weeks you can exercise, but no contact sports for six weeks. At that point, the scars have half their normal strength.

The Risks

In addition to the usual risks of infection and bleeding, breast reduction surgery is notorious for resulting in poor scars. There is a tremendous amount of tension on the incisions: the more the tension, the worse the scars. Most plastic surgeons place stitches in the deep layer of the skin; they stay in for months before dissolving. On the surface, fine sutures or glue is used. Sutures are removed within a week or two of surgery. In areas where the skin flaps are joined together (the T), the stitches are left in longer.

Small openings occur in almost 20 percent of patients. These heal within a few weeks, but lying on the breasts during the first six weeks makes them worse. Remember that the stitches are not even as strong as fishing line. I had one 240-pound patient who slept on her breasts two weeks after surgery and opened all of her incisions.

Red, raised scars are common after breast reductions. The wounds are subject to a great deal of tension, with breasts hanging in several directions during sleep. Use of a supportive bra for months may help the scarring. Or the scars may benefit from a thin piece of silicone sheeting after surgery, although there is minimal evidence to support its use. Patients receive a measure of psychological relief when *something* is done. Covering the scars with paper tape may improve them, as do steroid injections and laser treatments.

Numb nipples have been reported in a third of women, although my patients rarely have this problem. The disastrous complication of nipple loss is uncommon, except in people with poor circulation. It is the reason we prefer not to perform this operation on diabetics, on women with blood vessel diseases such as rheumatoid arthritis, or on smokers. If the nipple is lost, it turns brown and shrivels up. It won't be a fun year while waiting to have a nipple reconstruction.

Breast-feeding may not be possible after reduction surgery. It depends on how much breast tissue was removed and whether scarring interrupts the ducts. Of course, if the nipple was removed and replaced during surgery, breast-feeding will be impossible.

In most women, one breast is larger than the other. During a reduction the surgeon makes a judgment and removes more tissue from the larger breast. Until that three-dimensional imaging is widely available, there is no scientific way to judge how much extra tissue to take, although we try and weigh the removed tissue. Dense breast tissue weighs more than fat, so the volume of the breast asymmetry is difficult to predict. The surgeon's experience aids in this task. Even so, residual size differences are frequent.

Imperfect shape, too, is common after breast reduction. Nipples can be positioned too high and look as though they are "star-gazing,"

Finally, remember that this is *real* surgery that can result in blood clots in the legs, blood clots that travel to the lungs, and even death.

A Few Words about "Scarless Breast Reduction"

Those words are really very simple: *there is no such thing.* It is impossible to remove breast tissue without leaving a scar. The so-called scarless breast reduction uses liposuction to remove breast fat through small incisions. The scars may heal well and be hard to see or, in 5 percent of women, may become red and raised. Hardly scarless! In addition, the fat removal makes the breasts sag more. Ads for scarless breast reduction belong on the last page of *Consumer Reports,* along with the other fraudulent ones.

Liposuction of the breast is not advisable. It can cause "fat necrosis," which results in calcifications that mimic cancer. A biopsy is then needed.

Breast Lifting

Breast lifting is similar to breast reduction, except that no breast tissue is removed. The incisions are similar to those in breast reductions. If there is just a little drooping, a "donut mastopexy" can be performed. Skin around the areola is removed, hiking up the breast. Because two circles of tissue of unequal circumference need to be sewn together, any seamstress knows that the result will be bunching of the tissue. Fortunately, this bunching usually settles in a few months.

When more extensive drooping occurs, the next level of lifting requires a vertical incision between the areola and the crease below the breast. This is a hot topic: many plastic surgeons believe that most patients require a full lift. An incision is made along the length of the crease, together with vertical incisions and incisions around the areola. As in breast reductions, the search for different, better scars seems never-ending. Horizontal, vertical, around the areola—all incisions have been proposed. My suggestion is to choose the surgeon, not the technique.

Patients with small, drooping breasts can benefit from a combination of a lift and augmentation. Technically, this is a difficult procedure. It is hard to judge how much skin to remove prior to surgery if the breasts are also to be enlarged. For this reason, some surgeons prefer to stage the procedures: the augmentation is performed first and the lift is done several months later. Alternatively, if the patient wants implants placed in front of the muscle, the lift should be performed before the augmentation.

At the 2000 meeting of the ASAPS, a panel concluded that it is safer to perform these procedures separately. When done at the same time, the complication rate is 9 percent and many patients later desire additional lifting.[6]

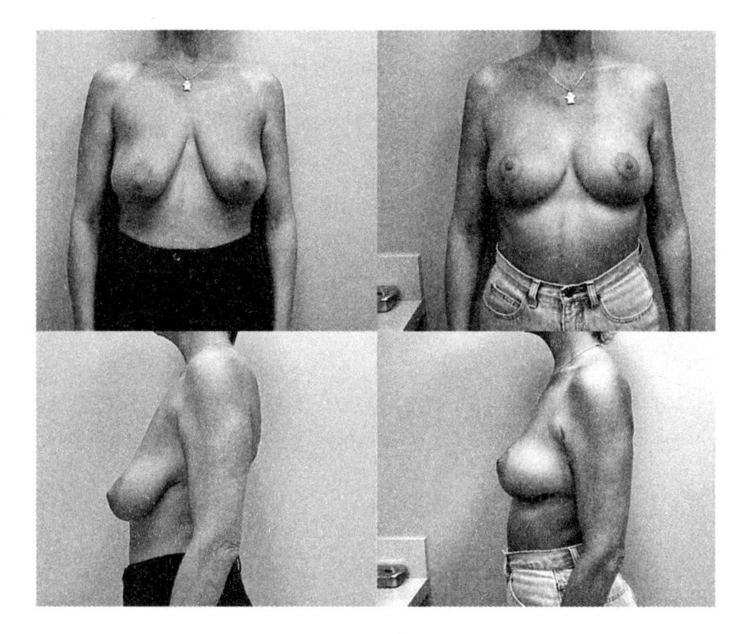

This 55-year-old woman underwent a breast lifting procedure. Preop on the left, postop on the right.

When a lift is performed along with an augmentation, it is much safer to place the implant beneath the muscle. This location preserves the nipple blood supply and decreases the chance of dreaded nipple loss.

Inverted Nipples

Many different methods have been tried to correct this problem. Devices that provide suction on the nipple can be tried before surgery, although I am not sure the correction holds up over time. Surgical techniques cut tethering breast ducts and support the nipple with sutures. Breast-feeding is impossible afterward, so the operation shouldn't be done if you may want to breast-feed. Sensation also may be ruined. A new technique doesn't cut the ducts but requires a skin graft on the areola. With this operation, you're trading one deformity for another.[7]

The Cost

Breast lifts average $4,300, while breast reductions average $5,600. The range for both procedures will go as high as $10,000 in some parts of the country.

31

Male Breast Reduction

Breast growth in men is more common than one might expect. It is a hidden, embarrassing deformity. Men with large breasts wear shirts all the time, even on the beach. About 18,000 men underwent surgery to reduce the size of their breasts in 2005.

Male breast growth, technically called gynecomastia, occurs in half of men at the time of puberty. It can be humiliating and prompt teasing. At a time when teenagers are beginning to grow beards, the last thing a boy wants to grow is a breast. The condition is due to an imbalance between estrogen and testosterone, the female and male hormones.[1]

Most Men's Breasts Spontaneously Disappear in Time for College

But they can start up again. Breast growth can occur with declining testosterone levels in aging men. And many drugs cause breast growth: hormones for prostate cancer, prescription drugs such as spironolactone and Propecia (finasteride), the antifungal ketoconazole, digoxin, lovastatin, verapamil, cimetidine, respirodone, methyldopa, melatonin, and the AIDS drug efavirenz. Marijuana and anabolic steroids are notorious for causing breast growth; half of steroid users develop gynecomastia. Thyroid disease, liver disease, alcoholism, AIDS, and kidney failure are associated with male breast growth. Overall, a third of adult men have some degree of breast growth.

If the breasts remain large for two years, the condition is likely to be permanent and may require surgery. After surgery, the men's social lives improve, and for the first time they will take off their shirts at the beach.

A medical evaluation will rule out diseases that cause breast growth. An internist, pediatrician, or endocrinologist should evaluate any male patient with recent breast growth. If a cause is found, appropriate medical therapy can be used. If no cause is found, then surgery may be the answer.

The Technique

Liposuction revolutionized the treatment of gynecomastia. Prior to liposuction, the only way to remove male breast growth was to perform a mastectomy-type operation. The huge scars traded one problem for another. In the 1980s, liposuction was first used to remove much of the fat of the breast through small incisions around the bottom of the areola. The breast tissue was removed through this small incision on an outpatient basis. The patient would need about a week off from work.

Ultrasonic liposuction in the 1990s was supposed to treat gynecomastia without excision of breast tissue. Unfortunately, it cannot remove all of the breast tissue by itself, and so an incision and tissue removal are still required.

Incisions outside the areola are needed only in giant male breasts. It is probably better to remove as much breast tissue as possible through small incisions and possibly perform a second procedure later to remove any skin that doesn't shrink. In most cases, the extra skin disappears after the breast tissue has been removed, without the need for larger scars.

The Procedure

While I prefer to do this procedure under general anesthesia, it is technically possible to use local anesthesia with sedation. The all-important topographic map of the breast and fat tissue of the chest is drawn while the patient is sitting. An incision is made running from the three o'clock to nine o'clock position on the areola, and liposuction is performed in the periphery of the breast. After maximal tissue is removed with suction, the actual breast tissue is removed after separation from the skin above and the pectoralis muscle below. The bleeding is controlled; drains suck down the skin and decrease the chance of fluid collection. A dressing is placed, and a tight-fitting vest without cups is used (we never say "bra" to men). After a two-hour recovery, the patient goes home.

The stitches are removed a week after surgery. Bruising and swelling take two weeks to settle down. The drains are removed three to seven days after surgery, then daily showering may be started. The vest is worn for at least a month.

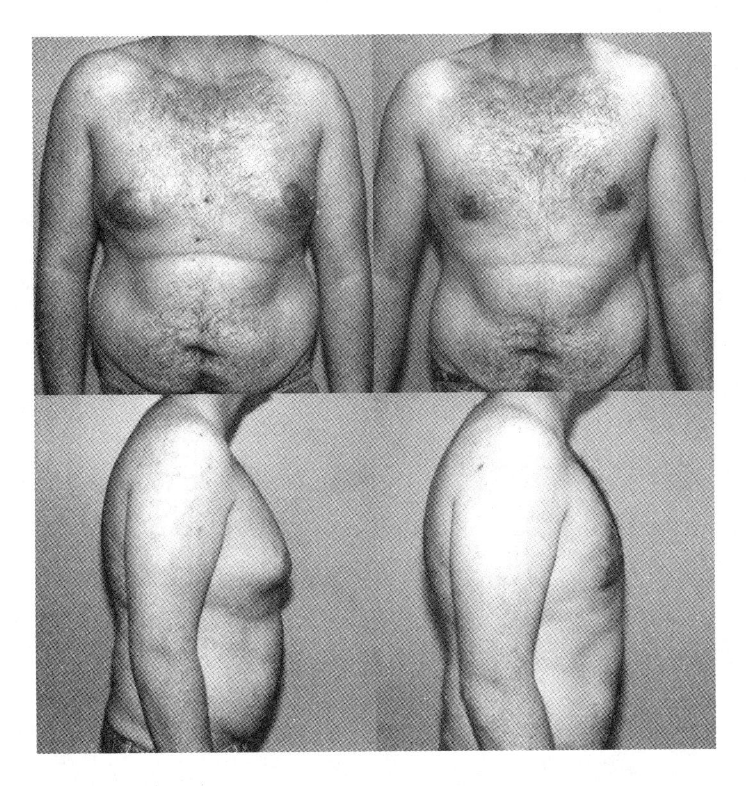

This typical 20-year-old underwent gynecomastia excision. Preop on the left, postop on the right.

The Risks

The risk of bleeding is higher than average with gynecomastia excision, because a large area is being operated on directly over an active muscle. Also, men are typically much less likely to follow instructions after surgery. There can be no sports or sex for at least two weeks, for we don't want the blood pressure or pulse rate elevated during this period. Bleeding can occur for as long as three weeks following surgery; it is more likely if the man lies on his chest.

Other risks include blistering of the nipple, even loss of the nipple. If not enough tissue is left behind the nipple, the nipple could invert. The overall complication rate is 20 percent, outstandingly high. As many as a third of patients need touch-up procedures to remove a little more tissue or smooth out the result.[2]

The Cost

Typical fees for this procedure range from $3,000 to $7,000, with $3,300 the average.

32

Calf, Buttock, and Pectoral Implants

Despite the hype about calf, buttock, and male pectoral implants, most plastic surgeons do not perform these procedures. In 2005, only 263 people in the United States underwent calf augmentation, 2,361 had buttock augmentation, and 172 underwent pectoral augmentation. Pectoral implants declined 76 percent in one year and appear to be headed for extinction. These soft solid-silicone implants are conceptually different from the more popular breast and chin implants. The latter increase the size of areas of the body that cannot be made larger by any natural means. The buttocks, pectoral muscles, and calves can be enlarged by three weeks of focused exercise.

A silicone gel implant is placed underneath the muscle through an incision in the back of the calf. Besides the visible scar, potential complications include infection, bleeding, and nerve and muscle injury. It may be possible to see and even feel the implant.

Buttock implants have similar problems. Patients commonly complain that the implant hardens and is painful while sitting or lying down. To avoid this, some surgeons have changed to fat grafting to increase the size of the buttocks. A typical patient receives 0.5 to one kilogram (one to two pounds) of fat, transplanted from the thighs or abdomen into each buttock cheek. There is less pain, a faster recovery, and a lower complication rate than with implants. No one knows how much fat will survive, but probably a minimum. The implant is basically a few pounds of dead fat that the body struggles to clean out for about a year. How can it be good for you?[1]

In light of the problems with these augmentations, my advice is to augment the buttocks, chest, or calves naturally—with exercise.

The Cost

Calf augmentations average $4,000, pectoral implants $3,600, buttock augmentations $4,300. Fat grafting to the buttocks is in the same range.

33

Weight Loss and Cosmetic Surgery

In this era of an obesity epidemic, interest in dieting is higher than ever before. Obesity is the second most common cause of death from behavioral risks; 400,000 people died of obesity-related illness in 2000. Only smoking killed more.

Robert Atkins has become a household name. Gyms and weight loss centers are as popular as the dry cleaner's. How does the consumer know what really works?

In a 2005 paper in the *Journal of the American Medical Association,* four common fad diets were compared: Atkins (low carbohydrate, high fat), Ornish (low fat), Weight Watchers (calorie restriction), and Zone (low glycemic foods). The more extreme the diet (Atkins and Ornish), the lower the success rate. People lost about four to five kilograms (ten pounds) with each diet. They all worked, but the longer the person was on the diet, the *more* calories were eaten. This "diet fatigue" results from deprivation. Cardiac risk was lower with all diets, corresponding directly to weight loss.[1]

In fact, in a typical person, diet, weight loss, exercise, and medication result in a reduction of no more than 10 percent of body weight. Unfortunately, virtually all diets fail in the long run, with a gradual gain of whatever weight was lost.[2]

A person who *tries* to diet loses weight. Diets that work identify the number of calories needed per day. Counting calories and weighing food are useful until a person has an accurate idea of what size the portions should be.

Why Diets Fail

Deprivation is the reason most diets fail. The longer you deprive yourself of a favorite food, the greater your desire to violate the diet. My concept of a workable diet is one that allows a weekly "holiday"—one dinner a week where anything is allowed. You will have something to look forward to all week, and the sense of deprivation and anger will diminish.

Dr. Michael Roizen has analyzed various diets and lifestyles. Comparing the Ornish, Zone, Weil, Pritikin, Atkins, and a variety of other diets, he found that no "bestseller diet" resulted in successful long-term weight loss. His own RealAge diet emphasizes nutritious, low-calorie foods, but with a little healthy fat at the *beginning* of the meal. The foods that he recommends slow your aging process and can make you healthy.[3]

Help from Drugs

Meridia (sibutramine) and phentermine are appetite suppressants. Xenical (orlistat) blocks fat absorption in the intestine. It is sold without a prescription in Australia and may soon be "over the counter" in the United States. Wellbutrin (bupropion) is an antidepressant that has the side effect of suppressing the appetite. Topamax (topiramate) and Zonegran (zonisamide) are antiseizure drugs that also have weight loss as a side effect. Finally, the diabetes drug Glucophage (Metformin) can cause weight loss. Unfortunately, all of these drugs are limited by the patient's motivation. They boost weight loss when a patient exercises and diets.

Surgery for Weight Loss

Bariatric surgery is better than medications for the severely obese. Bariatric surgical procedures limit the amount of food that can be eaten or absorbed. Like the jaw wiring of the 1970s, a sort of forced weight loss results. Surgical procedures designed in the 1980s limited food absorption in the stomach and intestines. Gastric bypass and gastric banding ("lap-banding") are the two most common bariatric surgical procedures performed today. In 1993, there were 20,000 bariatric procedures in the United States. Ten years later, 120,000 were performed and in 2005, there were 170,000.

Gastric Bypass Surgery

This surgery has been around for over two decades. Only in the past few years has it been popular. It reroutes the plumbing of the stomach and intestines, creating a tiny stomach and causing intentionally poor absorption of nutrients in the intestine. The procedure can be performed through a large incision or through small incisions, with the aid of an instrument called a laparoscope—basically a camera on the end of a tube. Weight loss is typically 100 pounds in a year. Side effects and problems (vitamin deficiency, diarrhea) are possible after the surgery. There is a 5 percent significant complication rate, with a death rate of 1 in 200.

Lap-Bands

Laparoscopic adjustable gastric banding (lap-banding) is a less invasive and less dangerous technique that restricts the size of the stomach but does not limit absorption of nutrients. It was introduced in the early 1990s. A silicone band is placed around the upper stomach and can be tightened after the surgery, further limiting how much food can pass through. The death rate with this surgery is ten times lower than with gastric bypass, but the complication rate is about the same. As the procedure becomes more common, complication rates are expected to decline. About half of the excess weight is lost during the first two years and then the weight loss continues at a slower rate; it is possible to lose as much weight as with gastric bypass.

New York University general surgeons Christine Ren, M.D., and her partner, George Fielding, M.D., have performed more lap-bands than anyone else in the United States, maybe in the world. According to Dr. Ren, "The lap band procedure will be as popular as gastric bypass surgery in five years and will be the predominant procedure in ten years." She argues that the procedure is simpler than gastric bypass and has fewer complications. "And it works," says Dr. Fielding, who himself has had the procedure.

The effectiveness of bariatric surgery cannot be argued. Each procedure successfully reduces between 60 and 70 percent of the excess weight. Medical problems such as diabetes, high blood fat (hyperlipidemia), high blood pressure, and sleep apnea go away in most patients who have the surgery. The benefits, including weight loss and improved lifestyle, are still present ten years after surgery. It is no wonder that this is the most rapidly growing field in general surgery. Worldwide, 120,000 of these procedures have been performed. Fielding has done 4,500 and can perform a lap-band procedure in about half an hour![4]

The Gastric Pacemaker

Performed experimentally in the United States but available in Europe, a compelling new procedure uses Transneuronix's gastric pacemaker to electrically stimulate the stomach. The electric current acts as a type of acupuncture, telling the brain that the stomach is full. At this point in time, the device is being used by only a few research surgeons. One of them is Tufts–New England Medical Center surgeon Scott Shikora, M.D. It holds great promise.

Plastic Surgery Is the Dessert after Weight Loss

The most rapidly growing field in plastic surgery is called postbariatric surgery body contouring. We have seen a meteoric rise in procedures such as gastric bypass surgery and lap-band surgery. "There have been over 100,000 bariatric procedures in the United States over the last three years. Of these, 35,000 were lap-bands," comments Dr. Fielding. The proliferation of bariatric surgery has resulted in tens of thousands of people who have lost over a hundred pounds.

While the health benefits of weight loss are life saving, unfortunately the cosmetic damage done to the body during the gaining of weight is usually irreversible. During weight gain, the epidermis expands rapidly; the dermis, however, cannot grow as rapidly. It actually splits, creating a stretch mark. A new stretch mark looks reddish or blue, because the translucent epidermis allows the underlying blood vessels to be seen. The new stretch mark is inflamed and red for the first year.

Stretch marks are permanent. No treatment other than surgical removal is known. Overstretched skin has limited ability to retract. Rapid weight loss affords the skin even less opportunity to shrink. With the growing obesity epidemic and the burgeoning surgical procedures, hundreds of thousands of people will be carrying around a lot of hanging skin. However, people who have lost a massive amount of weight care less about scars and more about looking good in their clothing.

Organized plastic surgery and medical product companies are gearing up for this influx of patients. Tummy tucks, body lifts, thigh lifts, breast-lifts, arm lifts, and face-lifts are procedures that will be needed by patients who have lost massive amounts of weight.

34

Spider Veins

The bane of the existence of many women, spider veins cause them to wear dark stockings or even pants. They keep women off the beaches. Even the name conjures up something evil. Spider veins, technically known as telangiectasia, are capillaries and small veins that become visible through the skin of the legs and thighs.

Spider veins are most likely to appear during pregnancy and are present in most women by age 30. They are caused by the weight of blood that dilates the capillary walls. As the skin thins with advancing age, the veins become more visible.

The Procedure

Sclerotherapy

Sclerotherapy is the most common method of treating spider veins. Over half a million people underwent the procedure in 2005. It is an office procedure, performed without any anesthesia. A chemical is injected into the capillaries, injuring the wall. The cells inside the capillary swell, blocking the blood flow. The body is quick to fix itself and will try to repair the blood vessels. Compression stockings flatten the inside of the capillaries (squashing them like a garden hose), allowing them to heal together, creating a permanent seal. The blood can no longer flow through the capillaries.

Sclerotherapy does not destroy capillaries. The remaining collapsed capillaries are unable to hold blood, however. It is the red (or blue) blood inside them

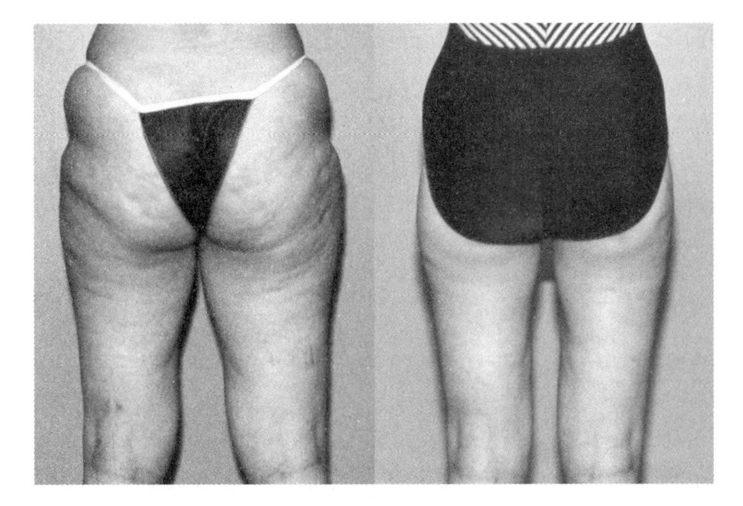

This woman underwent two rounds of sclerotherapy after having liposuction of the hips, thighs, and knees. Postop on the right. See color plates.

that is visible through the clear walls. By blocking the blood flow, the capillaries become invisible.

A variety of chemicals can destroy capillaries. The most common is a strong saltwater solution. This hypertonic saline has no chance of causing an allergic reaction, a major advantage. It does burn when injected, however. Other chemicals cause less pain, but all have risks of allergic reactions. Since the solutions are injected straight into the blood vessels, a deadly type of allergy called an anaphylactic reaction can result. With this in mind, the decision to use the hypertonic saline solution is an easy one.

Lasers

Lasers have also been used to destroy spider veins. Honest surgeons will tell you that the lasers are a less efficient technique than sclerotherapy. Only small numbers of capillaries can be treated in each session. And the technique hurts. Lasers are more useful for the capillary "blush" that sometimes remains after the larger capillaries are destroyed with sclerotherapy.

One of the new devices touted for leg spider veins is the Syneron Corporation's Polaris LV. It uses laser energy to fry the blood in the capillaries and radiofrequency energy to heat up the outside of the veins. No solid proof is available to support this laser, although Neil Sadick, M.D., dermatologist at Cornell–New York Hospital, has an unpublished study that shows 75 percent of capillaries cleared with three treatments. But even Sadick says, "No laser or IPL (intense pulsed light) is a substitute for sclerotherapy."

In the face, lasers *are* the treatment of choice for visible capillaries. The difference is the pressure of the blood. It is much lower in capillaries above the level of the heart. Also, these capillaries are closer to the skin and have a smaller diameter.

The Sclerotherapy Technique

Before the treatment, patients hydrate by drinking half a liter of pure water. If they are anxious, they are given oral Valium and codeine prior to the procedure. No anesthesia is used, since a large area is treated in each session. The skin is cleaned with alcohol, which also dilates the blood vessels, making them more visible.

A tiny needle is placed inside an individual capillary. Usually, magnification helps visualize the blood vessel. The technique requires great skill and an extremely steady hand. A small squirt of saline is injected inside the capillary, causing a dramatic clearing of the spider vein. I know I've done well when my nurse says "wow." If the right portion of the capillary is entered, several square centimeters of connecting capillaries can be treated with a single injection.

I work my way around the front of both thighs and legs and then turn the patient on her belly. Next, I treat the backs of the thighs and legs. The most noticeable spider veins are treated in the first session, then progressively smaller vessels in subsequent sessions.

After a capillary is injected, my nurse applies a small piece of foam tape. I inject a maximum of five milliliters of hypertonic saline, giving the patient as much salt as she would get in three and a half portions of large French fries at McDonald's. After the procedure, the patient drinks another half a liter of pure water to dilute the salt.

After the Surgery

Compression stockings, purchased ahead of time, are immediately applied to the legs and thighs. The stockings place pressure on the capillaries and increase the effectiveness of the sclerotherapy.[1]

The stockings are worn for forty-eight hours following the procedure. The longer the stockings are worn, the better the result. They are removed for showering. At the first shower, the foam strips are removed from the legs.

The Risks

Sclerotherapy with saline can cause pain, particularly if the solution is injected outside of the capillaries. The solution can cause blistering and even scarring. Bruising results if blood leaks out from the capillary. This leakage can cause brown skin pigmentation that lasts for weeks or months, or indefinitely. There is a tiny chance of the capillaries clotting and the clots extending into the larger veins to cause a painful superficial phlebitis. If so, the required treatment is warm soaks, elevation, and anti-inflammatory drugs. In a worst-case scenario, the clot can spread to the deep veins, causing a deep venous thrombosis (DVT). This serious but rare medical problem requires aggressive treatment. Symptoms are pain, swelling, and a hot leg. A noninvasive sound test of the leg and thigh are required to rule out DVT if these symptoms occur.

The results of the treatment are not known for about a month after the procedure. Some capillaries are successfully treated but contain blood that takes a month to be absorbed by the body. Others are successfully treated, but will be repaired by the body. In this case, "repair" is counterproductive: The body doesn't know the goal is to destroy the capillary. Its job is to fix it.

At the end of the month, we assess the results. A typical treatment destroys between half and two-thirds of the capillaries. Depending on the patient's wishes, sometimes a second or even a third treatment is performed.

Remember, spider veins are an ongoing process. The trouble started when women got up on their hind legs and stood. Higher pressure was placed on the leg veins and was increased by pregnancy and obesity. The treatment of spider veins is like weeding a garden. Every few years, we need to go back and clean up the new capillaries.

Spider veins are not varicose veins. While the two may occur together, varicose veins are due to diseased valves in the large veins of the legs; they are removed by vascular surgeons. These doctors evaluate the legs and either inject, strip, or destroy the veins. Spider veins are a cosmetic problem only, having no medical effects. Both develop in people who have occupations that require long periods of standing.

Treatment of Varicose Veins

Varicose veins are not just cosmetic problems. In fact, they are medical problems. True varicose veins can cause pain, swelling, ulcerations—and painful blood clots. Most plastic surgeons do not deal with varicose veins. Yet some dermatologists, who are not trained in vascular surgery, *will* treat them. This is foolhardy, since the potential complications from the surgery are dangerous. A full examination of the leg and thigh veins is necessary before beginning any treatment. Often there is an underlying problem with the vein that connects the deep and superficial veins. It must be dealt with in order to prevent recurrence.

Traditional approaches to varicose veins include vein stripping and injection. Stripping is a formal surgical procedure that is associated with significant disability and scarring and a two-week recovery. When veins are injected, a chemical called sodium tetradecyl sulfate damages the interior of the blood vessels. The technique is effective but can result in phlebitis, deep and dangerous clots, and allergic reactions.

Nonsurgical Treatment of Varicose Veins

The Endovenous Radiofrequency Obliteration Closure is a new technique that is rapidly growing in popularity. Its most common variant is the VNUS system. Ultrasound first precisely maps the legs and thigh veins. The culprits in varicose veins, broken valves that allow backflow of blood, are identified and treated with heat, basically frying the veins along their length. This technique is as effective as vein stripping in the long term, according to one study. Los Angeles surgeon Dr. Christine Petti is one of the few plastic surgeons in the country performing this procedure. "The nice thing about the VNUS is that there's no incision," Dr. Petti says. "My patients go back to work the next day." Other vascular surgeons do not believe that this technique will stand the test of time. If the connector veins are not dealt with, recurrence will be high. In addition, nerve injuries have been reported with the obliteration technique. Dr. Petti, who is a clinical assistant professor of plastic surgery at the University of Southern California and is trained in general surgery as well, says that the technique rarely causes problems because the ultrasound lets the surgeon see and avoid the neighboring nerves.[2]

The Cost

Spider vein injection averages $326 nationally. Typical fees in the Northeast are $900 for a forty-five-minute session.

V

Procedures That Don’t Work

35

Fraud in Cosmetic Surgery

When my patients spend their hard earned dollars on cosmetic procedures, they expect results. And *I* expect results. Subtle changes that require a statistical analysis to determine whether they work might be okay in a scientific laboratory, but they do not cut it in real life. This and the following chapters discuss procedures that leave patients shaking their heads after spending thousands of dollars and not seeing dramatic changes.

Let the Buyer Beware

Why do surgeons perform procedures that do not work? The reason is pressure. It's why otherwise normal, caring, plastic surgeons cross the line and offer questionable procedures—procedures with no scientific basis, procedures that simply do not work.

Plastic surgery has always been a competitive field. Only the best and brightest are accepted into plastic surgery residencies. That competition selects for the best doctors, and continues as the doctors enter clinical practice. Not only is there pressure to earn a sizable income, to be able to afford an elegant house and a luxurious car, but also to be regarded as the premier plastic surgeon. And that means offering the newest and the best procedures.

Plastic surgery is a unique field in that it is largely consumer driven. When a surgeon devises a new procedure, often his next stop is not the lab but the television station. An appearance on local or national television can catapult the surgeon's career, transforming a little-known curmudgeon into a celebrity surgeon. A flurry of new patients may follow and, often, financial reward. After a television story about some new procedure, plastic surgeons will receive many inquiring calls by the next morning. This prompts the surgeon to learn the new procedure or purchase the new machine—no surgeon wants to be left behind. Surgeons are haunted by stories of urologists who didn't learn the noninvasive kidney-stone crushing technique of lithotripsy in the 1980s, or general surgeons who did not learn laparoscopy in the 1990s, resulting in the demise of their practices. And so the new procedures are learned, the new machines are purchased, and the new procedures are performed, whether they work or not.

Many techniques in plastic surgery are introduced without controlled studies. The scrutiny of the news reporter seems to be all that is needed. The problem occurs when a plastic surgeon purchases a $100,000 laser, uses it for a few months, and realizes that it doesn't perform as advertised. Does the machine get returned to the company? Not much chance of that. Does it serve as a hood ornament on the doctor's Porsche? Unlikely. In fact, the machine will be used over and over again until bad publicity renders it useless, or until the doctor sells it to another unsuspecting and naive colleague. Now it is the consumer who has the problem.

Insurance Fraud

With a wink, the surgeon tells the patient that her surgery won't cost anything; it will be covered by insurance. Despite the fact that the surgery is cosmetic, the surgeon tells the insurance company that the surgery was reconstructive, and it is covered.

A crime with no victim? Hardly. The costs of insurance fraud are borne by the entire population. Your insurance premium pays for someone else's cosmetic rhinoplasty—but not yours, if you and your surgeon are honest. Society decides what is considered cosmetic and what is considered reconstructive. When devious doctors and patients defy the rules and defraud insurance companies, every person who buys health insurance pays for their cosmetic surgery. When your insurance company fails to pay for legitimate medical treatments because it is limited financially, remember the cosmetic surgery that you paid for.

Rhinoplasty is the most common procedure that is fraudulently billed to health insurance. Surgeons typically charge for repair of a "deviated septum" and then bill for rhinoplasty as well. Most people have deviated septums, but that doesn't mean that they need to be operated on. Only when the septum blocks airflow and interferes with breathing does it need surgery. If the septum is twisted, it may be operated on for cosmetic reasons.

Tummy tucks are also fraudulently billed to insurance. Insurance companies may approve the procedure for problems created by pregnancy, but most will no longer do so. In an attempt to gain insurance coverage, separation of the abdominal muscles is sometimes fraudulently called a hernia. As any general surgeon will attest, this is not a true hernia and should not be covered by insurance.

Upper eyelid surgery is another area for game playing. Surgeons fraudulently claim that the eyelid skin blocks vision. There have even been conspiracies between ophthalmologists who have documented impaired visual fields and plastic surgeons who have performed the insurance-covered eyelid lifts.

Even more heinous are the surgeons who perform liposuction and bill insurance companies for excision of "multiple lipomas." Some even try to bill insurance for breast augmentations, claiming the surgery was a breast reconstruction.

In my decade of experience as a member of the New Jersey State Board of Medical Examiners, I have observed that doctors who commit insurance fraud often have other quality-of-care issues. *Honesty is a character trait.* If a surgeon lies to an insurance company, would he not lie to a patient? When a surgeon tells you that your cosmetic surgery can be covered by insurance, run hard and fast!

36

Endermologie and the Quest to Cure Cellulite

Cellulite is a natural result of aging. It can hardly be called a disease, since over 90 percent of women have it. Cellulite of the thighs and buttocks is similar to the wrinkles of the face. Human skin is tethered to the underlying muscle by collagen connective-tissue strands. These connections prevent excessive motion of the skin. Animals such as dogs do not have these connections, and their skin can be pulled many centimeters. But because of our unique anatomy, when human skin is stretched, as with weight gain or in pregnancy, slight sagging will occur. The collagen connections hold up the skin, creating the typical appearance of cellulite.

As if this situation were not bad enough, dimples commonly accentuate cellulite in the buttocks and thighs. These depressions are acquired after a significant bruise occurs, actually fracturing the fat. Some of the fat loses its blood supply and is digested by the body. The bruise gradually subsides and the area of fat destruction settles in about six weeks, creating a depression. Often the original injury is subtle and the subsequent depression is not noticed until months or years later.

I always thought cellulite was not a real substance—that it was just the way the skin looks after being overstretched. In fact, cellulite *is* an anatomic entity. If you slice up the skin and look at it under the microscope, you will actually see cellulite. Scientists at the prestigious Rockefeller University in New York did so, and they found that cellulite was simply fat that extruded into the skin's dermal layer. They did not find any difference in structure or metabolism between cellulite and fat. Cellulite is not a pathologic condition, so no medicine can get rid of it.[1]

Skin creams, medicines, laser treatments, massage, and surgery all are attempts at decreasing cellulite. Despite the hopes and desires of millions, none of these techniques really work. Many treatments, even massage, can temporarily improve cellulite. When the tethering connections are cut, the cellulite improves. Unfortunately, the body doesn't understand that the connections were cut to improve the cosmetic appearance. It acts as though an injury has occurred. In trying to heal the connections, it re-creates them. The real proof that any cellulite treatment works would be an independent evaluation at least six months after the last treatment. But no treatment can claim results after this length of time.

Although Retinol, the vitamin A drug, can increase elasticity in skin, it does not change the appearance of cellulite. Nothing works—short of cutting out the skin, stretching it, and resuspending it.[2]

No cream has ever been shown to reduce cellulite. The most common "active ingredient" is caffeine. While they are not effective, at least the creams are safe. One of the few documented studies showed that aminophylline did absolutely nothing for cellulite.[3]

When analyzing before and after photographs of cellulite treatments, it is important to note that lighting alone can remove cellulite! And the key question is how long a particular treatment lasts. Vigorous massage camouflages cellulite for hours by causing temporary swelling.

Endermologie

The same study also showed that Endermologie did nothing for cellulite. Endermologie has been around for more than a decade. It is a technique that lifts, stretches, and squeezes skin between two rollers—basically a glorified massage. Certain studies have shown a benefit to Endermologie, but analysis of the studies showed various design flaws. Endermologie defies common sense: how and why would massage decrease cellulite and fat? The massage may temporarily increase the blood supply to the fat and skin and stretch the skin. It most likely causes swelling, and that may account for the improvement in the cellulite in some studies and in some photos of patients. In fact, the FDA's approval of Endermologie is for the *"temporary improvement in the appearance of cellulite and circumferential body measurements."* Note the words carefully. If it does temporarily work, shouldn't standard deep massage do the same thing?

Other "Treatments"

Syneron's Velasmooth machine combines bipolar radio frequency (a type of electrical current) and infrared light (heat) with tissue mobilization (basically massage) to attempt to decrease cellulite. The Dermosonic machine combines ultrasound and tissue mobilization. It purportedly decreases cellulite, stretch marks, and scars; rejuvenates the face; and enhances liposuction! All this without any corroborating controlled studies published in peer-reviewed medical journals.

Triactive laserdermology is the latest entry in this field. The new device claims to smooth and tighten cellulite. It does so by cooling and stimulating the skin with light energy and massage. Unfortunately, I was again unable to find any peer-reviewed scientific papers that support these claims.

PhotoActif also claims to treat cellulite. The company calls this treatment photomesotherapy. Red and infrared LEDs presumably force an "anticellulite gel" into the skin. Cellulite appeared improved in the company's small unpublished study of just eleven people. Many more studies with many hundreds of patients need to be performed before we can decide whether this technique actually works.

The Bottom Line Is That There Really Is No Cure for Cellulite

And at this point, there is nothing promising. No machine has been shown to decrease cellulite long term. The short-term results have not passed the muster of the scientific community. No valid studies compare a treated thigh with an untreated thigh. If the machines really worked, these types of studies would be available.

37

Mesotherapy

There are two holy grails in plastic surgery. The first is a noninvasive wrinkle removal method. The second is a noninvasive fat removal method. Neither is possible at this time. That doesn't stop people from claiming they have the wrinkle cure or the fat cure.

Mesotherapy claims to be a holy grail procedure. Phone books in metropolitan areas across the United States are filled with ads for mesotherapy. Several thousand physicians, most of them non–plastic surgeons, in the United States now perform this procedure.

Mesotherapy has been performed in France since 1952. It was originally a technique for delivering small amounts of drugs to the dermal layer of the skin. The skin is supposed to act as a reservoir for the drug, allowing slow release into the rest of the body. Mesotherapy was originally proposed as a method of pain control. It is now used for cellulite reduction, body weight loss, wrinkle and fat reduction, hair loss, and scar correction.

Does Mesotherapy Destroy Fat?

In Europe, the treatment schedule typically is ten to twenty sessions using small amounts of dilute drugs. In the United States, physicians employ three to five sessions using higher amounts of more concentrated drugs. An anesthetic cream is used and small needles are mounted on an injector machine. The medicine is injected in small increments over a large area of fat. Bruising and swelling are expected. The typical charge for a treatment is $1,000 to $1,500.

Although There Is Scant Evidence in *Real* Medical Journals Proving It Works, Thousands of Mesotherapists Are Setting Up Shop

One study in pigs showed that injection of two chemicals under the skin could destroy muscle and fat. Then again, so will sulfuric acid. The chemicals probably do kill fat, but why inject known toxins into healthy tissue? Another small study —not particularly scientific—showed that four monthly injections of chemicals into lower eyelid fat did improve its appearance. The real question is whether the procedure is safe, effective, and generates uniformly satisfactory results. One small study presented at the 2004 meeting of the American Society of Aesthetic Plastic Surgeons suggested that there might be a benefit. And a small study showed that a technique called Lipodissolve may work, but the end point was the patient's opinion. Not too scientific.[1]

I've seen a number of patients who underwent mesotherapy and felt that the procedure had no benefit. If it were harmless, I'd say they were just wasting their money. But a few dozen papers discussing the complications of the technique noted that infections and other adverse reactions abound. Allergic reactions have already been reported with mesotherapy; it is only a matter of time until a patient succumbs to an anaphylactic reaction.

A huge industry is developing that manufactures the injector equipment and the drugs, even without any proof the treatment works. Courses promise full training in just one or two days. There is even a mesotherapy "society." To me it sounds like a giant scam.

The Most Common Drug Injected for Fat Removal Mesotherapy Is Phosphatidylcholine

This drug is supposed to dissolve fat. According to the American Society of Aesthetic Plastic Surgery (ASAPS), the safety and effectiveness of this drug is unproven. Because it is not approved in the United States for any indication, physicians who inject it are stepping far outside the umbrella of safe medicine. Technically, it may be illegal to inject a drug that is not approved by the FDA. In actuality *no* drugs have been approved by the FDA for mesotherapy. Other substances that are injected include aminophylline (an asthma drug), isoproterenol (a heart drug), vitamins, minerals, anesthetics, amino acids, and herbs. Since mesotherapy has never been studied scientifically, many different recipes for drug mixtures have sprung up. Different doctors use different "cocktails." The industry appears to be pushed by the companies selling the drugs, and by physicians of differing specialties looking for a way to earn income beyond fee-limited insurance-covered medicine. It is also pushed by aggressive advertising for expensive one- to two-day courses. Very scary.

The Lure of Money Is So Great That Physicians Are Putting Their Licenses and Personal Fortunes on the Line When They Perform Mesotherapy

If anything goes wrong, they are on thin ice. It is a non-FDA approved procedure, with non-FDA approved drugs, and no scientific proof that it works! In an editorial in the journal *Plastic and Reconstructive Surgery,* editor Rod Rohrich, M.D., stated, "It is mind-boggling to think that a physician would inject patients—or that patients would allow physicians to inject them—with unknown, unproved substances based on hearsay and unsubstantiated clinical findings."[2]

The ASAPS points out that the injection of fat-destroying chemicals cannot be controlled; too much or too little fat may be killed. The toxicity of the drugs and their long-term effects need to be determined.

Until further studies are done on mesotherapy, my advice is this: if it sounds too good to be true, it probably is!

38

Cool Lasers and Other Less Invasive Treatments

The *N-Lite laser* is a pulsed dye laser that was billed to treat wrinkles without destroying the epidermis. This is called nonablative wrinkle removal. The laser was cleared by the FDA in 2002 and has been used by scores of surgeons, despite a lack of proof that it works. In 2004, a study looked at eighty-nine patients after two N-lite treatments. After four months, patients and doctors evaluated photographs. Twenty-nine percent of the patients thought their wrinkles improved, but only 10 percent of the doctors saw improvement. The study noted that, in the best cases, the changes were "subtle and not impressive." It concluded that this laser was *not* an effective way to treat facial wrinkles. Despite these hard scientific data, the Internet is flooded with doctor's websites touting the benefits of N-lite. Its claims include improvement of wrinkles, acne, and stretch marks.[1]

If Plastic Surgeons Don't Do Their Homework, They Will Experiment on Their Patients

We have noted that an expensive machine that does not function up to expectations will rarely be placed in a back closet; it will be used until it is paid for. Understandably, patients believe that if a machine is FDA approved and their doctor uses it, it must work. Once they put down thousands of dollars for their treatments, many patients will claim a benefit, not wanting to upset their doctors or themselves. They may even *believe* they have seen a benefit. The study quoted above is a perfect example: more patients than doctors thought the laser worked.

Now there's a procedure that punches holes in your skin. "Percutaneous collagen induction" is touted as an alternative to laser resurfacing. It is supposed to decrease wrinkles and facial scars. A medieval-looking drum-shaped device with protruding needles is rolled over the skin, kind of like a lawn thatcher. Ouch! It places thousands of fine pricks side by side in the skin, penetrating 1.5 millimeters deep. Allegedly, this injury starts a cascade of healing events that eventually leads to a thicker collagen layer. Honest surgeons call this *scar.* No controlled studies exist that show this technique actually works, although surgeons are out there performing the procedure.[2]

Myontology is a noninvasive system that stimulates the facial muscles with electrical current. A series of eight to twenty-five sessions is needed, according to the manufacturer. The treatment is supposed to increase the muscle tone and skin circulation. While this may indeed happen, I believe that it would *increase* the wrinkles. Logically, if the face is paralyzed, wrinkles improve. Think Botox. No controlled studies prove this technology works; common sense says it doesn't.

39

Microdermabrasion

Microdermabrasion has been touted as a noninvasive wrinkle reducer. Proponents say the technique can decrease scarring, equalize skin tone and pigmentation, and decrease acne. Microdermabrasion is performed both in physicians' offices and by aestheticians.

Sanding the Face

Let's examine exactly what microdermabrasion is and what it can do. Back in the 1960s, the loofah pad (the dried, fibrous part of a loofah fruit) became popular to exfoliate the skin. The human upper epidermis, the *stratum corneum,* is composed of dead skin that keeps out water and germs. The loofah pad strips off this layer, making the skin look better.

Microdermabrasion is really just an expensive loofah pad. It is called "micro" because traditional "dermabrasion" removes not only the dead skin, but also varying layers of the epidermis (upper layer) and even the dermis (lower layer). For a full discussion, see Chapter 10.

As originally described in 1992, microdermabrasion "sandblasts" the skin with aluminum oxide crystals that are immediately suctioned up by the microdermabrasion machine. The crystals exfoliate by knocking off the upper layers of skin. The suction removes the crystals and unclogs oils and dirt from the pores. Clean, smooth-feeling skin results. The crystals are so effective that they are able to completely strip off the upper layers of the skin. But errant crystals can escape the suction system and cause a corneal injury to the eye. They enter the room air and can be breathed in by the patient and the person performing the procedure.

Newer machines have been designed that use common table-salt crystals. Although safer, they can clump and clog the machine.

Many thought the sandblasting concept was not really necessary. Abrasive diamonds were embedded in the handpiece and rubbed across the skin, simultaneously vacuuming debris from pores. Since the purpose of the sandblasting was exfoliation, this safer and simpler method eliminated the crystal spray.

Other new machines use liquids to lubricate the skin while abrading and suctioning. Various medicines, such as the pigment reducer hydroquinone and the acne drug salicylic acid, are sprayed on the skin. Whether or not this makes a difference is anyone's guess. In the world of microdermabrasion, as in so much of this field, scientific studies are few and far between.

The latest version of microdermabrasion does not use crystals or suctioning at all. The Vibraderm machine simply vibrates on the surface of the skin, causing exfoliation. We've come full circle from that one-dollar loofah pad.

Over a million microdermabrasion procedures were performed in plastic surgeons' offices alone in 2005. No one knows how many additional procedures were performed in spas and salons.

With all the hype about microdermabrasion, very few studies show that it does anything at all. In the best study, after eight consecutive weekly treatments, there appeared to be mild improvements in skin quality. But this tiny seventeen-patient study did not compare microdermabrasion to other exfoliant techniques such as the loofah pad. The evaluation was performed immediately after the final treatment, when some temporary swelling would be expected. I would like to have seen if there was a difference a month after the treatments. I would guess not, since there is absolutely no science involved.[1]

Microdermabrasion Is the "Poster Child" for Procedures Performed with Claims of Medical Benefit but Without Scientific Proof

The procedure really can't do anything other than exfoliate, clean the pores, and cause mild, temporary swelling. Although microdermabrasion claims to improve scars, the claim is controversial. Beyond transient swelling around the scar, most doctors feel there is no benefit. Any pigment that can be removed is at the very surface of the skin, pigment that would have rubbed off with any exfoliation technique, including a washcloth.

Zein Obagi, M.D., comments, "To correct wrinkles and scars with any rejuvenation procedure, the dermal layer must be reached; exfoliation procedures that penetrate only to the epidermis will not work." That says it all. Microdermabrasion, by definition, cannot treat wrinkles or scars.

Other studies that have shown improvements and even biochemical changes in the skin are all either short term, poorly controlled, or include very few patients. A worthwhile study would compare the effects of microdermabrasion with vigorous face washing. I would bet there would be no measurable differences.

Having said that, let me add that I am a big believer in microdermabrasion. But let's be honest: the procedure is simply a very good facial. It exfoliates. Exfoliation makes the skin look better, but has no real medical benefit. In fact, exfoliation defies the function of the skin—to be a barrier to the outside world. The thicker the outside layer, the better the protection. But a thick outside layer is unattractive. Perhaps the best benefit of microdermabrasion is the deep cleaning of pores. After the dead skin is removed from the skin, oils and dirt are literally sucked out. Pores that have been held open by this debris stay large for a while, then shrink down.

Microdermabrasion is basically a high-tech facial. It cleans the skin and decreases acne in adults and adolescents. It allows creams and peels to penetrate more effectively. When the skin is pretreated with microdermabrasion, a 35-percent glycolic acid peel penetrates like a 70-percent peel. The result is dramatic.

40

"Anti-Aging" Medicine

A shot in the dark. That's how I view the use of human growth hormone as a so-called anti-aging treatment. Scientists haven't a clue to how many different hormones and chemicals are involved in the aging process. And they certainly don't have any idea how the known and unknown hormones interact. One hormone affects dozens of others, in a scheme that requires a team of supercomputers to unravel. Hormone manipulation is not new. To stall aging, physicians in the nineteenth century tried irradiating women's ovaries and injecting dog testicle extract into men. Monkey testicles were also injected into men to attempt to reverse aging. According to Leonard Hayflick, Ph.D., legendary researcher from the University of California at San Francisco: "There have been 3,500 years of anti-aging fraud. People who claim that humans will live to 200 have always said that this breakthrough will be in 20 years."[1]

Quack, Quack

I went to a plastic surgery anti-aging symposium a few years ago thinking that I'd get in on the ground floor of a new and exciting field. I listened, learned, and came away with disdain for a group of people who reminded me of a poster I had on my wall in college: *If you can't dazzle 'em with brilliance, baffle 'em with bullsh—!*

There were anti-aging institutes that measured hormone levels in men and in women. They administered various hormones, including human growth hormone, thyroid hormone, thymic protein from the thymus gland, testosterone or estrogen and progesterone, DHEA (dehydroepiandrosterone) or androstenedione, and pregnenolone. Increased libido, energy, fat loss, muscle gain, and better mental concentration, as well as lower cholesterol, glucose, fat, blood pressure, and abdominal fat, were the promised outcomes. Even today, some "centers" perform a variety of diagnostic tests and administer nutritional supplements with antioxidants. Some measure DNA damage and free-radical levels, and develop individualized DNA repair programs, stressing a healthy diet and exercise! Interesting, but where's the proof that the tests are meaningful and the plan works?

In 2006, a Senate Special Committee on Aging called this type of DNA testing "modern-day snake oil." The Government Accountability Office said that these tests are not clinically valid, and that they are misleading, exploitative, ambiguous, and meaningless.

In 2004 three noted researchers on aging gave their Silver Fleece Award for exaggerated claims to the founders of the American Academy of Anti-Aging Medicine: Ronald Klatz, M.D., and Ronald Goldman, M.D. Their journal, *International Journal of Anti-Aging Medicine,* also received an award. One of the presenters of the biannual award was Dr. Hayflick. The goal is to warn consumers of anti-aging misinformation. Hayflick, arguably the foremost aging researcher in the world, stated in a 2002 paper in *Scientific American* that "no currently marketed intervention—none—has yet been proved to slow, stop, or reverse human aging, and some can be downright dangerous." He went on to say that anyone who offers an anti-aging product today is either mistaken or lying. "Human longevity has increased by interfering with disease, not the fundamental aging process," the 80-year-old Dr. Hayflick told me. "The problem with growth hormone is that there is no evidence that it interferes with the fundamental process of aging. We have no information now as to who should take the hormone and who should not."[2]

Robert Bernard, M.D., former president of the ASAPS, echoed these statements. He believes that the selling of vitamins and the administration of hormones under the banner of an "anti-aging program" without scientific evidence borders on charlatanism. He prefers to use the terms "wellness, life enhancement, and lifestyle change" as opposed to the imprecise "anti-aging."

Beware of "Homemade" Credentials

A physician can be certified in "age management medicine" after completing a five-day, $9,000 course sponsored by a private organization. As Dr. Hayflick says, "These people are laughing as they run to the bank." I can't help but believe that many of the people in the anti-aging field are motivated by profit. Patients are desperate to not age, to not grow old. Many will do anything to stay young forever. The fountain of youth has been chased for centuries; today it simply looks like a doctor's office. One of the reasons the public doesn't hear too many exposés about the fraudulent practices is the fear of litigation. Dr. Hayflick's two coauthors were subjected to a $200 million lawsuit after their statements. Although the suit has gone nowhere, its effect is purposely chilling.

My faith in anti-aging medicine was restored after I attended another symposium at the University of Pennsylvania in 2005. There, thoughtful researchers in aging, plastic surgery, and related fields presented unbiased, balanced research in aging. More questions than answers arose in this meeting, for we are just beginning to learn about this complex field.

The Star of the Show Is . . . Human Growth Hormone

The focus of many anti-aging gurus is human growth hormone, which is produced by the pituitary gland in the brain. Among other functions, it is responsible for stimulating growth in children. As we age, the amount of this hormone declines. It becomes more difficult for adults to put on muscle mass and easier to accumulate fat, even with the same degree of exercise and the same caloric intake. As the skin thins, sexual and heart function decline and bone density decreases. Energy wanes and depression becomes more common. A third of people older than age 60 have very low levels of growth hormone. The billion-dollar question is whether the administration of human growth hormone to these people, or all people, makes a difference. Even the elderly who have normal growth hormone levels still get old.

In 1990 a landmark study was published in the *New England Journal of Medicine.* It showed remarkable improvement in a dozen elderly men who were injected with growth hormone three times a week for six months. Their muscle mass and bone density increased and their fat decreased. The six months of hormones reduced aging by ten to twenty years.[3]

In another study, published in the *Journal of the American Medical Association* in 2002, human growth hormone with either testosterone in men or estrogen and progesterone in women was given for six months. The good news is that muscle increased and fat decreased, as would be expected with three exercise sessions a week. The bad news is that diabetes increased and participants had leg swelling, acne, weight gain, arthritis, muscle pain, and carpal tunnel syndrome from the hormone. Ten percent of men's breasts grew. The study concluded that growth hormone was not "ready for prime time." It is considered experimental until the complex interactions with other hormones and body systems are better understood. Still other studies with growth hormone have shown that breast and prostate cancers are more likely in people taking growth hormone.[4]

W. Glenn Lyle, M.D., a Raleigh, North Carolina, plastic surgeon, has reviewed the growth hormone literature for the Plastic Surgery Educational Foundation. He says that information supporting human growth hormone usually comes from non–peer-reviewed studies and is often found on websites and in lay books. He concluded that growth hormone replacement doesn't increase lifespan in animals or humans.[5]

The monthly cost of growth hormone injections is $500 to $2,000. Worldwide sales are near $2 billion. Add to this a host of diagnostic medical evaluations of questionable worth, and laboratory tests as unproven as DNA analysis, and we are talking real money. A common monthly fee in some "institutes" is $4,000. The promotional material distributed by one company that sells "genetic health" states that the doctor will "enjoy ongoing revenue streams from testing and product sales long after the visit." That pretty much spells out the motivation.

And Just Maybe Your Doctor Is Committing a Felony

Interestingly, the use of human growth hormone may be illegal unless the treatment is for a recognized medical disease. According to lawyer and plastic surgeon Neal Reisman, M.D., J.D., federal law 21 USC§333(e) restricts the use of this hormone. The FDA allows it only in adult AIDS patients and in people with true growth hormone deficiency. Unlike most other drugs, off-label use for non-approved indications is illegal. The FDA has already warned websites distributing growth hormone that prison penalties and fines may be forthcoming. Physicians who prescribe human growth hormone may be committing a fraudulent and perhaps criminal act by prescribing this drug for cosmetic or anti-aging reasons.[6]

The Other Anti-Aging Hormones

Other hormones that have been prescribed in the so-called anti-aging programs include dehydroepiandrosterone (DHEA), thyroid hormone, testosterone, melatonin, estrogens, and progestins. Growth hormone needs to be given by injection. However, growth hormone releasing factor can be taken by mouth. The amino acids arginine and glutamine, also eaten, can stimulate the release of growth hormone in the brain.

Testosterone declines with aging in both men and women. When men are administered testosterone, their strength and muscle mass, bone density, energy, mood, and sexual and immune functions increase, according to Anne Cappola, M.D., endocrinologist at the University of Pennsylvania. However, there is an increased risk of benign prostate disease (BPH) and prostate cancer, plus an increase in red blood cells (polycythemia), male breast growth (gynecomastia), and heart disease.

At this point in time, we are at the early stages of testosterone research. The National Institutes of Health is still trying to determine whom to study and how. We don't know how much drug to give, how to administer it, when to start and stop the drug, what side effects to watch for, and what other hormones will be affected. For these reasons, we are a decade away from an answer as to whether testosterone should be used.

DHEA

Dehydroepiandrosterone is present in huge quantities in both men and women. To fully understand this hormone, one must know that many chemically similar types of hormones exist in the body. Some are understood and some aren't. The hormones are made in the ovaries and testes, but can be converted to other hormonally active compounds in the brain, thyroid, heart, liver, adrenal glands, gastrointestinal tract, fat, bone, and even skin.

DHEA is an example of a prohormone, a substance that is converted by the body into other more active hormones. Because DHEA levels decline beginning in a person's early 20s, some people think it is a "youth hormone." On a simple level, this makes sense. If a hormone is present when we are young and declines as we age, some think that its replacement may slow aging. Once again, this very simple approach is the one the less-than-honest doctors take. In actuality, the interactions of this hormone with other hormones and tissues in the body are so complex that they are mind-boggling. Only endocrinologists and physiologists truly understand this field. And they state that we don't know enough to be administering hormones to patients. The more than eight thousand DHEA studies in the medical literature report various findings. Many of the studies are poorly done and some are frankly bogus. Side effects such as acne, hair growth in women, loss of scalp hair, mood changes, irregular heart rhythms, and an increase in breast and prostate cancer are possible. Unfortunately, the complex nature of this drug and its possible side effects render its administration dangerous. Also, DHEA has been shown to *not* improve muscle mass or strength in 60- to 80-year-old men and women. But that doesn't stop the erroneous advertisements.[7]

One of the problems in dealing with over-the-counter medications such as DHEA is the lack of uniformity in their production. DHEA is classified as a food supplement, not a drug. As such, it escapes close FDA scrutiny. A study published in the *Journal of the American Medical Association* showed huge variation among manufacturers in the actual doses of DHEA. Some did not include enough DHEA to be effective, some had too much.[8]

Wellness Doesn't Have to Be Expensive

There is solid evidence that exercise, a sensible diet, vitamin supplementation, stress reduction, and even a little wine can improve health and quality of life. Overall, this concept is called wellness, and it can be achieved inexpensively. With some knowledge, common sense, and a mix of old and new world values, wellness can be achieved by anyone.[9]

Perhaps Our Understanding of What Foods to Eat Will Come from Dogs

Dr. Steven Zicker, veterinarian and nutritionist for the Hill's Pet Nutrition Company, has been studying foods that increase the lifespan of dogs. His research is eye opening and centers on the free-radical theory of aging. Free radicals are produced by tiny mitochondria within cells. They increase over time and have harmful effects that eventually kill the cells. In theory, substances called antioxidants decrease the damage caused by free radicals. When a diet high in antioxidants such as vitamins C and E, beta-carotene, dl-alpha lipoic acid, and l-carnitine is administered to dogs, the normal age-related decline in cognitive function slows down. In humans, such a diet translates to eating five to seven portions a day of high-antioxidant fruits and vegetables. Antioxidants and mitochondrial helpers may reduce ongoing damage to the brain caused by free radicals. This process, in turn, may improve brain function as we age. Stimulating activities such as reading this book may improve learning ability and may even help to grow nerve cells in the brain. Keep reading! The real question is, of course, how Dr. Zicker measured cognitive abilities in dogs. For those of you who are dog lovers, the answer is obvious.[10] While it is always difficult and even dangerous to extrapolate animal studies to humans, mounting evidence tells us that diets rich in antioxidants will result in better health.

But Will We Ever Really Find the Fountain of Youth?

I'm sure that in twenty or forty years, we'll have a much better comprehension of the aging process. We'll understand the complex interactions of hormones and their effects. Whether hormones will be the "fountain of youth" is anyone's guess. More likely, as the human genome becomes better understood, the genetic program that defines our ultimate demise will become clearer. DNA manipulation through drugs may someday turn off the messages that cause us to age, and ultimately allow us to live longer.

Epilogue: The Future of Plastic Surgery

Cosmetic surgery is a field in flux. Certainly, as time progresses, procedures and medical treatments to improve one's appearance will continue to evolve. New operations will be described. New medicines and creams will be developed. Aging will be slowed or even stopped. Let me conclude this book by describing what I consider to be the biggest issues and most likely advances in the cosmetic surgery of the future.

Face Transplants

Even the *New York Times* gave front-page coverage to the first face transplant in France in 2005. The concept of a face transplant conjures up images of John Travolta and Nicolas Cage trading faces in the 1997 movie *Face-Off.* In 1998, a complete face and scalp were replanted on a person after trauma. The Cleveland Clinic and the University of Louisville are both ready to perform the first full face transplant in the United States.

Ethicists love to talk about the transplantation of one person's face to another person. From a medical perspective, facial transplantation can preserve quality of life in severe burns, trauma, or cancer. But it requires a lifetime of immunosuppressive drugs. Disaster looms if rejection occurs. Medical ethicists and surgeons question whether the benefit from this surgery is worth the risk of these potent drugs. Face transplants are about to become commonplace; they will spark a flurry of news articles—and perhaps legislation to prevent abuse.

Noninvasive Wrinkle Removal

Sooner or later, a bright plastic surgeon or biomedical engineer will discover a drug or a machine that reverses wrinkles without destroying the skin. The difficulty arises because the dermis is seriously damaged in aging, and current methods of altering the dermis require destruction of the overlying epidermis. It takes weeks for the epidermis and dermis to regrow, and months or years (or never) before the color of the skin returns to normal.

When a technology—whether it be a laser, electrical current, magnetic field, ultrasound, or other not-yet-described technique—is invented that can harmlessly reduce wrinkles with minimal downtime, the technique will be popular beyond belief.

Noninvasive Fat Removal

Current fat removal techniques involve surgery—liposuction or direct removal of the fat. Mesotherapy, injecting drugs into the skin or fat to destroy the fat, is minimally invasive, but is unproven and possibly dangerous. Externally beamed ultrasound is compelling, but still unproven. When a technique truly destroys fat evenly without creating ripples, waviness, or dimples, and doesn't involve surgery or significant risk, that technique will be as popular as the noninvasive wrinkle removal technique that I want to develop.

Female Genital Cosmetic Surgery

A nascent area of cosmetic surgery is surgery of the genitals. Some women are self-conscious about bulky or unattractive labia that create a bulge in their clothing, or inner labia that protrude beyond the outer lips. Laser hair removal has allowed women to permanently remove more and more pubic hair, exposing their previously unseen labia both inside and outside their clothing.

Plastic surgeons and gynecologists, perhaps working together, will be performing more of the procedures called *labiaplasties,* as fashions become progressively more revealing. One Beverly Hills gynecologist charges $5,000 for a labiaplasty, although in other areas of the country, the fee would be less.

Florida plastic surgeon Pam Loftus, M.D., performs all sorts of female genital cosmetic surgery. The list of her procedures gives us a look at the future. She performs labiaplasty, labia reduction, labia majora (outer labial lips) augmentation, vaginal rejuvenation, hymen repair to restore virginity, clitoral hood and adhesion removal, vagina reduction, and mons pubis liposuction. Labial reduction can be performed simply, under local anesthesia with sedation.

Male Genital Cosmetic Surgery

In men, penile augmentation procedures would be popular if they were low risk and really worked. Urologists have actually studied who should be a candidate for a penile augmentation: they state that the man's penis must be shorter than 4 centimeters, or shorter than 7.5 centimeters when erect, to be considered appropriate for penile lengthening.[1]

Penile augmentation procedures inject fat into the penis and elongate it by releasing its supporting structures. Liposuction above the penis can enhance the appearance of the penis in overweight people and skin flaps can be advanced into the penis to increase its flaccid length.

Fat grafting is the most common, and the most notorious, of these procedures. It can result in disasters such as loss of the penis if fat is injected into blood vessels or if infection occurs. When the augmentation does work, the result is temporary. Complications such as nodules in the penis, skin deformity, and scarring and loss of normal contour are common. The injected fat is extremely fragile and needs to remain fairly motionless in order for blood vessels to grow into the tissue. If they don't grow in three days, the fat will die and be absorbed by the body. If the fat is disturbed during the first three weeks, it will lose its new blood supply and be resorbed. The penis cannot stay motionless when urinating and when erections develop. Virtually by definition, fat grafting into the penis is doomed to fail.[2]

Newer techniques for penile augmentation include the injection of hyaluronic acid into the glans penis. Dermal-fat grafting has also been tried. Now that injectable Radiesse is available, I predict that it will be injected into the glans or shaft of the penis in pellets or lines to provide for female sexual stimulation, sort of like a semipermanent ribbed condom.

Unless the penis is extremely small or the labia are extremely large, psychological factors may be the motivation factor behind the desire to alter the genitals. Psychological consultations are appropriate for many people, male or female, who desire altered genitals.

The Perfect Breast Implant

We saw in Chapter 29 that all breast implants have problems. The medical products industry needs to recognize the demand for a better implant—ideally one that is immobile but sways with the breast. The implant should have a low leakage rate and feel like a normal breast. If it leaks, the body should be able to digest or remove the material with no medical consequences. An implant could contain a Jello-like material that, once released from the implant, becomes absorbed. The implant should not be prefilled. If a valve is used it may leak, but it allows insertion of a much smaller implant, and in turn a smaller incision and a smaller scar.

Of course, in order for the implant companies to have proper financial incentive to develop such a product, they must be shielded from product liability. After the implant crisis of 1992, it's no wonder that there have been remarkably few advances in breast implants. Once the FDA or CE approves a device or drug, the manufacturer must be protected from product liability claims unless it has been truly negligent. Liability concerns could quash development of drugs or devices in the future, to the detriment of society.

Hair Regrowth

The answer to hair regrowth will ultimately be medical, not surgical. With DNA manipulation, it should be possible to restimulate the hair follicles to grow anew. When the secrets of stem cells are unlocked and their DNA is successfully manipulated, it is only a matter of time until scientists figure out how to regrow hair.

Dr. George Cotsarelis, dermatologist at the University of Pennsylvania, has identified hair follicle stem cells that can be injected into the skin. He has done this already in animals. Humans can't be too far behind.[3]

Current surgical techniques are primitive, even by today's standards. Single-hair grafting is time consuming and rarely has an excellent result. Even if it does, the genetic program will ultimately override the new hairs. They will follow their destiny and eventually be lost in many people. Both the donor and recipient scars will then be exposed.

Permanent Fillers

The ideal filling material is painless to inject, is permanent, and is fully compatible with the body. Such material would take on the characteristics of the area into which it is injected. Filling the dermis would result in more dermis; filling the fat, more fat; and filling the bone, more bone. We are on the right track with the current generation of fillers, but the problem is lack of permanence. The ideal filler should permanently maintain its effect and become indistinguishable from the native tissue. If it is not, it must be retrievable surgically.

Test Tube Tissues and Organs

In the future, after a tissue is destroyed by disease or trauma, a new tissue or organ will be grown in the laboratory. The new organ will then be surgically attached. Noses, ears, lips, and breasts will be created, and whole faces and scalps with hair. Imagine the day when a new heart, liver, kidney, or other organ is created this way! By taking stem cells from the patient, incubating them with certain growth hormones, and providing them a framework to grow on, these tissues are in our near future. The research is already under way: fat stem cells have been used to grow bone; cartilage cells, to grow ears and pieces of the nose, and amniotic stem cells have grown heart valves.[4]

Pigment Control

One of the difficult challenges of plastic surgery has been the control of pigment. The skin produces pigment to protect against ultraviolet radiation. But this pigment production is unpredictable—most often too much is made, but if the pigment-producing cells are destroyed, too little may be made. Pigment is not a major issue in light-skinned people, but it becomes very important in anyone with darker skin. It is most difficult to control in those who have intermediate amounts of pigment, such as Latin Americans, Mediterranean people, or Asians.

Drugs currently available to inhibit pigment production include the hydroquinones and kojic acid. However, they do not know when to stop; there is no way to perfectly match the existing pigment of the skin. The development of drugs that control pigment will aid postoperative care in people of all skin colors.

Scar Control

Most people heal with good-quality scars—that is, scars that are not particularly wide, red, or noticeable. About 5 percent of people heal with red, raised scars or keloids.

Drugs or devices need to be developed that will control scarring. Someday it will be possible to alter healing so that scarring no longer occurs. We may learn from reptiles, animals that regenerate new limbs instead of creating scar. When scarring becomes less of an issue, the decision to undergo surgery will be far easier. Surgery that leaves large scars (tummy tucks, breast reductions, and face-lifts) will become more common as a major impediment disappears.

Anti-Aging Treatments

Endocrinologists, along with plastic surgeons, will map the complex interrelationships among the hormones that affect aging. This task, nearly as complex as the human genome project, will forge ahead over time. As each new hormone is discovered, the interactions with the hundreds of other hormones will be determined. When this project is complete, anti-aging treatments will truly be safe at last.

While future skin care and cosmetic surgery will maintain a youthful appearance for progressively longer spans, nutritional, genetic, and hormone therapy will allow the body to age less rapidly and with less disease and disability.

The future of plastic surgery is bright. Better, less invasive procedures will improve the quality of life for many of us.

I'D LIKE TO SPEAK TO THE DOCTOR ABOUT GETTING MY EYELIDS DONE.

Appendix A

Fees for Typical Procedures

In 2005, consumers spent $12.4 billion on cosmetic surgery and procedures, according to the American Society for Aesthetic Plastic Surgery (data reproduced here with permission). Fees vary around the world and depend on the local economy and the skills of the surgeon. Experienced surgeons usually charge more than those just out of training: you are paying for their years of experience in the operating room and generally superior results.

The figures below represent physician fees only and do not include fees for the surgical facility, anesthesia, medical tests, prescriptions, surgical garments, or other miscellaneous costs related to the surgery. The amounts shown, averaged to reflect nationwide statistics for each procedure, are based on a survey of doctors who have been certified by boards recognized by the American Board of Medical Specialties (including but not limited to the American Board of Plastic Surgery).

Procedure	Fee
Abdominoplasty	$5,232
Blepharoplasty	2,813
Breast augmentation—saline implants	3,583
Breast augmentation—silicone gel implants	4,005
Breast lift	4,258
Breast nipple enlargement (cosmetic only)	1,538
Breast nipple reduction	1,469
Breast reduction (women)	5,550
Buttock augmentation	4,258
Buttock lift	4,878
Calf augmentation	4,009
Cheek implant	2,720
Chin augmentation	2,095
Face-lift	6,298
Forehead lift	3,148
Gynecomastia, treatment of	3,305
Hair transplantation	5,033
Lip augmentation (other than injectable materials)	1,819
Lipoplasty—suction assisted	2,697
Lipoplasty—ultrasound assisted	2,979
Lower body lift	7,810
Otoplasty	2,951
Pectoral augmentation (male)	3,603
Rhinoplasty	4,188
Thigh lift	4,653

Umbilicoplasty (separate from abdominoplasty)	1,610
Upper arm lift	3,610
Total Expenditure for All Such Surgical Procedures in the United States	**$8,190,485,895**

Botox injection	$382
Calcium hydroxylaptite (Radiesse)	911
Cellulite treatment (mechanical roller massage therapy)	223
Chemical peel	848
Collagen, bovine (includes Zyderm/Zyplast)	398
Collagen, human	488
Dermabrasion	1,376
Hyaluronic acid (Hyalaform, Restylane)	527
Laser hair removal	347
Laser skin resurfacing—ablative	2,484
Laser skin resurfacing—nonablative	888
Laser treatment of leg veins	407
Mesotherapy	475
Microdermabrasion	149
Poly-L-lactic acid (Sculptra)	1,022
Sclerotherapy	326
Soft tissue fillers—autologous fat	1,395
Total Expenditure for All Such Nonsurgical Procedures in the United States	**$4,186,021,182**
TOTAL EXPENDITURE FOR ALL PROCEDURES IN THE UNITED STATES	**$12,376,507,077**

Appendix B

National Rankings of Cosmetic Procedures

The following list ranks the popularity of both surgical and nonsurgical cosmetic procedures in 2005. The rankings of surgical and nonsurgical categories combined are indicated in the column "Overall rank," while the rankings of procedures by category (surgical versus nonsurgical) are indicated in the final columns. Data reproduced with permission of the American Society for Aesthetic Plastic Surgery.

Procedure	No. of procedures	Percentage of total	Overall rank	Rank within category (surgical)
Abdominoplasty	169,314	1.5	13	5
Blepharoplasty	231,467	2.0	10	3
Breast augmentation	364,610	3.2	9	2
Breast lift	120,980	1.1	17	8
Breast nipple enlargement	86	0.0	40	25
Breast nipple reduction	4,113	0.0	34	19
Breast reduction	160,531	1.4	14	6
Buttock augmentation	2,361	0.0	36	21
Buttock lift	3,742	0.0	35	20
Calf augmentation	263	0.0	38	23
Cheek implant	11,820	0.1	32	18

Procedure	No. of procedures	Percentage of total	Overall rank	Rank within category (surgical)
Chin augmentation	31,818	0.3	25	11
Face-lift	150,401	1.3	15	7
Forehead lift	71,751	0.6	19	9
Gynecomastia, treatment of	17,730	0.2	27	13
Hair transplantation	13,519	0.1	29	15
Lip augmentation (other than injectable materials)	50,237	0.4	21	10
Liposuction	455,489	4.0	8	1
Lower body lift	11,871	0.1	31	17
Otoplasty	27,298	0.2	26	12
Pectoral augmentation (male)	172	0.0	39	24
Rhinoplasty	200,924	1.8	12	4
Thigh lift	12,489	0.1	30	16
Umbilicoplasty	2,115	0.0	37	22
Upper arm lift	15,917	0.1	28	14
Total for All Such Surgical Procedures	**2,131,019**	**18.6**		

Procedure	No. of procedures	Percentage of total	Overall rank	Rank within category (non-surgical)
Botox injection	3,294,782	28.8	1	1
Calcium hydroxylapatite (Radiesse)	40,495	0.4	23	13
Cellulite treatment (mechanical roller massage therapy)	53,949	0.5	20	11
Chemical peel	556,172	4.9	5	5
Collagen	220,890	1.9	11	8
Dermabrasion	42,347	0.4	22	12
Hyaluronic acid (Hylaform, Restylane)	1,194,222	10.4	3	3
Laser hair removal	1,566,909	13.7	2	2
Laser skin resurfacing	475,690	4.2	7	7
Laser treatment of leg veins	143,785	1.3	16	9
Mesotherapy	4,773	0.0	33	15
Microderma-brasion	1,023,931	9.0	4	4
Poly-L-lactic acid (Sculptra)	34,887	0.3	24	14
Sclerotherapy	554,251	4.8	6	6

Procedure	No. of procedures	Percentage of total	Overall rank	Rank within category (non-surgical)
Soft tissue fillers—Autologous fat	90,647	0.8	18	10
Total for All Such Nonsurgical Procedures	**9,297,731**	**81.4**		
TOTAL FOR ALL PROCEDURES	11,428,750	100.0		

Appendix C

Age Distribution of Cosmetic Procedures

The following table shows the popularity of different procedures in different age groups for the year 2005. Data reproduced with permission of the American Society for Aesthetic Plastic Surgery.

	Age				
Procedure	18 and Under	19–34	35–50	51–64	65+
Abdominoplasty	299	33,783	98,887	32,807	3,538
Blepharoplasty	98	11,170	90,890	95,682	33,627
Breast augmentation	3,446	184,899	144,106	28,889	3,269
Breast lift	434	27,067	68,609	22,195	2,674
Breast nipple enlargement	0	30	50	5	0
Breast nipple reduction	30	1,370	2,414	280	19
Breast reduction (women)	4,316	59,778	66,874	25,043	4,518
Buttock augmentation	0	1,300	1,031	31	0
Buttock lift	0	762	2,366	600	13
Calf augmentation	6	83	163	10	0
Cheek implant	42	3,822	6,129	1,576	251
Chin augmentation	601	13,382	11,491	5,661	681
Face-lift	0	499	47,144	82,067	20,689
Forehead lift	61	2,882	25,077	35,118	8,612
Gynecomastia treatment	2,151	10,255	4,424	804	96

Procedure	Age				
	18 and Under	19–34	35–50	51–64	65+
Hair transplantation	24	1,713	8,387	3,322	75
Lip augmentation (other than with injectables)	238	13,304	24,272	11,159	1,261
Liposuction	3,793	168,112	207,660	68,353	7,577
Lower body lift	17	3,502	6,505	1,824	23
Otoplasty	13,908	8,431	3,638	1,122	198
Pectoral augmentation (male)	7	114	45	6	0
Rhinoplasty	14,013	100,338	70,159	14,263	2,128
Thigh lift	7	2,825	7,153	2,429	74
Umbilicoplasty	0	516	1,264	335	0
Upper arm lift	21	3,040	7,800	4,643	414
Total for All Such Surgical Procedures	**43,511** 2.0% of total	**652,977** 30.6%	**906,538** 42.5%	**438,224** 20.6%	**89,737** 4.2%
Botox injection	4,504	588,797	1,711,797	849,872	139,808
Cellulite treatment (mechanical roller massage therapy)	994	16,920	27,914	6,877	1,244
Chemical peel	7,580	97,970	245,640	166,163	38,775
Dermabrasion (not including microdermabrasion)	851	7,539	10,954	15,255	7,751
Laser hair removal	67,049	647,138	630,619	185,040	37,064
Laser skin resurfacing	2,474	98,263	222,643	128,467	23,825
Laser treatment of leg veins	209	31,933	74,144	31,277	6,223

Procedure	Age				
	18 and Under	19–34	35–50	51–64	65+
Mesotherapy	0	262	3,220	925	367
Microdermabrasion	44,043	231,812	465,084	230,506	52,484
Sclerotherapy	522	115,191	283,380	131,295	23,824
Soft tissue fillers:					
1. Autologous fat	414	17,455	35,874	32,000	4,900
2. Calcium hydroxylapatite	89	4,295	20,744	12,804	2,563
3. Collagen	774	33,000	110,038	61,434	15,646
4. Hyaluronic acid	1,837	186,624	548,743	373,720	83,298
5. Poly-L-lactic acid (Sculptra)	0	3,039	15,787	13,432	2,629
Total for All Such Nonsurgical Procedures	**131,339** 1.4% of total	**2,080,240** 22.4%	**4,406,580** 47.4%	**2,239,065** 24.1%	**440,401** 4.7%
TOTAL FOR ALL PROCEDURES	**174,850**	**2,733,217**	**5,313,118**	**2,677,289**	**530,138**

Notes

Chapter 1. An Overview

1. B. O. Rogers, "History of the Development of Aesthetic Surgery," in P. Regnault and R. K. Daniel, *Aesthetic Plastic Surgery* (Little, Brown, 1983), 3; S. Romm, *The Changing Face of Beauty,* Mosby Year Book, 1992.
2. M. Rankin, G. Borah, A. W. Perry, P. D. Wey, "Quality-of-Life-Outcomes after Cosmetic Surgery," *Plastic Reconstructive Surgery* 104 (1999): 1209.
3. D. Sarwer, "Physical Appearance in Daily Life: Youthfulness Equals Beauty; Aging Equals What?" paper presented at conference entitled "The Art and Science of Anti-Aging Therapies: Convergence of Theory and Practice," University of Pennsylvania, March 2005.

Chapter 2. The Cosmetic Surgery Consultation

1. S. Romm, *The Changing Face of Beauty,* Mosby Year Book (1992), 177.
2. D. B. Sarwer et al., "Mental Health Histories and Psychiatric Medication Usage among Persons Who Sought Cosmetic Surgery," *Plastic Reconstructive Surgery* 114 (2004): 1927.
3. R. Goldwyn, "Aesthetic Surgery: Basic Principles," in P. Regnault and R. K. Daniel, *Aesthetic Plastic Surgery* (Little, Brown, 1983), 31; R. J. Honigman et al., "A Review of Psychosocial Outcomes for Patients Seeking Cosmetic Surgery," *Plastic Reconstructive Surgery* 113 (2004): 1229.
4. H. J. Brody et al., "Beauty versus Medicine: The Nonphysician Practice of Dermatologic Surgery," *Dermatologic Surgery* 29 (2003): 319.
5. H. S. Byrd et al., "Safety and Efficacy in an Accredited Outpatient Plastic Surgery Facility: A Review of 5316 Consecutive Cases," *Plastic Reconstructive Surgery* 112 (2003): 636.

Chapter 3. The Risks

1. N. Tanna et al., "Correction of Lipodystrophy in HIV-Positive Patients: Surgeon Beware," *Plastic Reconstructive Surgery* 116 (2005): 136.
2. C. E. Hughes, "Reduction of Lipoplasty Risks and Mortality: An ASAPS Survey," *Aesthetic Surgery Journal* 21 (2001): 120.
3. R. A. Cahill et al., "Duration of Increased Bleeding Tendency after Cessation of Aspirin Therapy," *Journal of the American College of Surgeons* 200 (2005): 564.
4. J. F. Reinisch et al., "Deep Venous Thrombosis and Pulmonary Embolus after Face Lift: A Study of Incidence and Prophylaxis," *Plastic Reconstructive Surgery* 107 (2001): 1570.
5. Y. Har-Shai et al., "Intralesional Cryotherapy for Enhancing the Involution of Hypertrophic Scars and Keloids," *Plastic Reconstructive Surgery* 116 (2005): 162.
6. J. M. Atkinson et al., "A Randomized, Controlled Trial to Determine the Efficacy of

Paper Tape in Preventing Hypertrophic Scar Formation in Surgical Incisions That Traverse Langer's Skin Tension Lines," *Plastic Reconstructive Surgery* 116 (2005): 1648.

7. T. D. Rees, D. M. Liverett, and C. L. Guy, "The Effect of Cigarette Smoking on Skin Flap Survival in the Face Lift Patient," *Plastic Reconstructive Surgery* 73 (1984): 911; H. H. A. Schumacher, "Breast Reduction and Smoking," *Annals of Plastic Surgery* 54 (2005): 117; J. K. Krueger and R. J. Rohrich, "Clearing the Smoke: The Scientific Rationale for Tobacco Abstention with Plastic Surgery," *Plastic Reconstructive Surgery* 108 (2001): 1063.
8. R. E. Iverson and the ASPS Task Force on Patient Safety in Office-Based Surgery Facilities, "Patient Safety in Office-Based Surgery Facilities: I. Procedures in the Office-Based Surgery Setting," *Plastic Reconstructive Surgery* 110 (2002): 1337; "II. Patient Selection," *Plastic Reconstructive Surgery* 110 (2002): 1785.
9. J. K. McLaughlin et al., "Increased Risk of Suicide among Patients with Breast Implants: Do the Epidemiologic Data Support Psychiatric Consultation?" *Psychosomatics* 45 (2004): 277.

Chapter 5. Aging

1. E. S. Epel et al., "Accelerated Telomere Shortening in Response to Life Stress," *Proceedings of the National Academy of Sciences of the United States of America* 101 (2004): 17312; R. Glaser and J. K. Kiecolt-Glaser, "Stress-Induced Immune Dysfunction: Implications for Health," *Nature Reviews Immunology* 5 (2005): 243.
2. D. E. Antell and E. M. Taczanowski, "How Environment and Lifestyle Choices Influence the Aging Process," *Annals of Plastic Surgery* 43 (1999): 585.
3. M. F. Roizen, *The RealAge Makeover* (HarperCollins, 2004).
4. S. P. Bartlett, R. Grossman, and L. A. Whitaker, "Age-Related Changes of the Craniofacial Skeleton: An Anthropometric and Histologic Analysis," *Plastic Reconstructive Surgery* 90 (1992): 592.
5. P. Morganti et al., "Protective Effects of Oral Antioxidants on Skin and Eye Function," *Skinmed* 3 (2004): 310; M. Placzek et al., "Ultraviolet B-Induced DNA Damage in Human Epidermis Is Modified by the Antioxidants Ascorbic Acid and D-Alpha-Tocopherol," *Journal of Investigative Dermatology* 124 (2005): 304.

Chapter 6. Skin Care

1. P. Begoun, *The Beauty Bible* (Beginning Press, 2002).
2. A. Moore, "The Biochemistry of Beauty," *EMBO Reports* 3 (2002): 714.
3. J. D. Bos and M. M. H. M. Meinardi, "The 500 Dalton Rule for the Skin Penetration of Chemical Compounds and Drugs," *Experimental Dermatology* 9 (2000): 165.
4. T. E. Moon et al., "Retinoids in the Prevention of Skin Cancer," *Cancer Letters* 114 (1997): 203.
5. M. P. Vienne et al., "Retinaldehyde Alleviates Rosacea," *Dermatology* 199 (Suppl., 1999): 53.
6. E. J. Van Scott and R. J. Yu, "Control of Keratinization with Alpha Hydroxy Acids and Related Compounds," *Archives of Dermatology* 110 (1974): 586.
7. G. E. Pirard et al., "Comparative Effects of Retinoic Acid, Glycolic Acid and a Lipophilic Derivative of Salicylic Acid on Photodamaged Epidermis," *Dermatology* 199 (1999): 50.

8. FDA Guidance for Industry, "Labeling for Topically Applied Cosmetic Products Containing Alpha Hydroxy Acids as Ingredients," January 10, 2005.
9. P. E. Grimes et al., "The Use of Polyhydroxy Acids (PHAs) in Photoaged Skin," *Cutis* 73 (2004): 3.
10. S. S. Traikovich, "Use of Topical Ascorbic Acid and Its Effects on Photodamaged Skin Topography," *Archives of Otolaryngology—Head and Neck Surgery* 125 (1999): 1091; R. B. Carlin and C. A. Carlin, "Topical Vitamin C Preparation Reduces Erythema of Rosacea," *Cosmetic Dermatology,* February 2001, 35.
11. C. Rubino et al., "A Prospective Study of Anti-Aging Topical Therapies Using a Quantitative Method of Assessment," *Plastic Reconstructive Surgery* 115 (2005): 1156.
12. H. Seznec et al., "Idebenone Delays the Onset of Cardiac Functional Alteration without Correction of Fe-S Enzymes Deficit in a Mouse Model for Friedreich Ataxia," *Human Molecular Genetics* 13 (2004): 1017.
13. S. P. Schmidt et al., "The Combined Effects of glycyl-1-histidyl-1-lysine: Copper (II) and Cell Take on the Healing of Linear Incision Wounds," *Wounds: A Compendium of Clinical Research and Practice* 6 (1994): 62; A. A. Abulghani et al., "Studies of the Effects of Topical Vitamin C, a Copper Binding Cream, and Melatonin Cream as Compared with Tretinoin on the Ultrastructure of Normal Skin," *Journal of Investigative Dermatology* 110 (1998): 686; J. D. Pollard et al., "Effects of Copper Tripeptide on the Growth and Expression of Growth Factors by Normal and Irradiated Fibroblasts," *Archives of Facial Plastic Surgery* 7 (2005): 27.
14. H. A. Epstein, "Peptides: The Science behind the Technology," *Skinmed* 3 (2004): 340.
15. A. Perrin et al., "Stimulating Effect of Collagen-like Peptide on the Extracellular Matrix of Human Skin: Histological Studies," *International Journal of Tissue Reactions* 26 (2004): 97; E. Bauza et al., "Collagen-like Peptide Exhibits a Remarkable Anti-wrinkle Effect on the Skin when Topically Applied: In vivo Study," *International Journal of Tissue Reactions* 26 (2004): 105.
16. G. C. Jagetia et al., "Triphala, an Ayurvedic Rasayana Drug, Protects Mice against Radiation-Induced Lethality by Free-Radical Scavenging," *Journal of Alternative Complementary Medicine* 10 (2004): 971.
17. E. R. Rostan et al., "Evidence Supporting Zinc as an Important Antioxidant for Skin," *International Journal of Dermatol* 41 (2002): 606.
18. N. B. Hampson, N. W. Pollock, and C. A. Piantadosi, "Oxygenated Water and Athletic Performance," *Journal of the American Medical Association* 290 (2003): 2408.

Chapter 7. Growths on the Face

1. L. Isikson et al., "Prevalence of Melanoma Clinically Resembling Seborrheic Keratosis: Analysis of 9204 Cases," *Archives of Dermatology* 138 (2002): 1562.

Chapter 8. Tackling Wrinkles

1. M. Edward, "Proteoglycans and Glycosoaminoglycans," in G. C. Priestley, ed., *Molecular Aspects of Dermatology* (John Wiley, 1993), 89.
2. Zein E. Obagi, *Obagi Skin Health: Restoration and Rejuvenation* (Springer, 2000).

Chapter 9. Filling Wrinkles

1. S. Romm, *The Changing Face of Beauty* (Mosby Year Book, 1992), 180.
2. S. Miller et al. (eds.), *Year Book of Plastic Surgery* (2005), 137.
3. L. S. Cooperman et al., "Injectable Collagen: A Six-Year Clinical Investigation," *Aesthetic Plastic Surgery* 9 (1985): 145; R. A. Beauchamp, J. Cukier, and D. E. Trentham, "Autoimmune Disease and Collagen Dermal Implants," *Annals of Internal Medecine* 120 (1994): 524.
4. Centers for Disease Control and Prevention, "Hepatitis C Virus Transmission from an Antibody-Negative Organ and Tissue Donor—United States, 2000–2002," *Journal of the American Medical Association* 289 (2003): 3235; R. Patel and A. Trampuz, "Infections Transmitted through Musculoskeletal-Tissue Allografts," *New England Journal of Medicine* 350 (2004): 2544.
5. S. Burres, "Fascian," *Facial Plastic Surgery* 20 (2004): 149.
6. C. Lindqvist et al., "A Randomized, Evaluator-Blind, Multicenter Comparison of the Efficacy and Tolerability of Perlane versus Zyplast in the Correction of Nasolabial Folds," *Plastic Reconstructive Surgery* 115 (2005): 282.
7. B. Bellman, "Immediate and Delayed Hypersensitivity Reactions to Restylane," *Aesthetic Surgery Journal* 25 (2005): 489.
8. R. J. Havlik and the PSEF Data Committee, "Hydroxylapatite," *Plastic Reconstructive Surgery* 110 (2002): 1176.
9. G. Lemperle et al., "Human Histology and Persistence of Various Injectable Filler Substances for Soft Tissue Augmentation," *Aesthetic Plastic Surgery* 27 (2003): 354; R. C. Beljaards, K. D. Roos, and F. G. Bruins, "NewFill for Skin Augmentation: A New Filler or Failure?" *Dermatologic Surgery* 31 (2005): 772.
10. D. H. Jones et al., "Highly Purified 1000-cSt Silicone Oil for Treatment of Human Immunodeficiency Virus–Associated Facial Lipoatrophy: An Open Pilot Trial," *Dermatologic Surgery* 30 (2004): 1279; G. E. Gurvits, "Silicone Pneumonitis after a Cosmetic Augmentation Procedure," *New England Journal of Medicine* 354 (2006): 211.
11. S. R. Cohen and R. E. Holmes, "Artecoll: A Long-Lasting Injectable Wrinkle Filler Material: Report of a Controlled, Randomized, Multicenter Clinical Trial of 251 Subjects," *Plastic Reconstructive Surgery* 114 (2004): 964.
12. S. R. Coleman, "Avoidance of Arterial Occlusion from Injection of Soft Tissue Fillers," *Aesthetic Surgery Journal* 22 (2002): 555.

Chapter 10. Peeling and Lasering Wrinkles

1. M. Spira et al., "Comparison of Chemical Peeling, Dermabrasion, and 5-Fluourouracil in Cancer Prophylaxis," *Journal of Surgical Oncology* 3 (1971): 367.
2. G. P. Hetter, "An Examination of the Phenol-Croton Oil Peel: Parts I–IV," *Plastic Reconstructive Surgery* 105 (2000): 227.
3. S. J. Dijkema and B. van der Lei, "Long-Term Results of Upper Lips Treated for Rhytides with Carbon Dioxide Laser," *Plastic Reconstructive Surgery* 115 (2005): 1731; J. M. Stuzin et al., "Histologic Effects of the High-Energy Pulsed CO_2 Laser on Photo-aged Facial Skin," *Plastic Reconstructive Surgery* 99 (1997): 2036.
4. M. Alam et al., "Glycolic Acid Peels Compared to Microdermabrasion: A Right-Left Controlled Trial of Efficacy and Patient Satisfaction," *Dermatologic Surgery* 28 (2002):

475; W. J. Kitzmiller et al., "Comparison of a Series of Superficial Chemical Peels with a Single Midlevel Chemical Peel for the Correction of Facial Actinic Damage," *Aesthetic Surgery Journal* 23 (2003): 339.
5. D. Kligman and A. M. Kligman, "Salicylic Acid Peels for the Treatment of Photoaging," *Dermatologic Surgery* 24 (1998): 325.

Chapter 11. Botox

1. M. Li, B. A. Goldberger, and C. Hopkins, "Fatal Case of BOTOX-Related Anaphylaxis?" *Journal of Forensic Science* 50 (2005): 169.
2. L. Nelson, P. Bachoo, and J. Holmes, "Botulinum Toxin Type B: A New Therapy for Axillary Hyperhidrosis," *British Journal of Plastic Surgery* 58 (2005): 228.
3. N. H. Kim et al., "The Use of Botulinum Toxin Type A in Aesthetic Mandibular Contouring," *Plastic Reconstructive Surgery* 115 (2005): 919.

Chapter 12. Eyelid Lifts

1. J. W. May, Jr., J. Fearon, and P. Zingarelli, "Retro-Orbicularis Oculus Fat (ROOF) Resection in Aesthetic Blepharoplasty: A 6-year Study in 63 Patients," *Plastic Reconstructive Surgery* 86 (1990): 682.
2. D. Saadat and S. C. Dresner, "Safety of Blepharoplasty in Patients with Preoperative Dry Eyes," *Archives of Facial Plastic Surgery* 6 (2004): 101.
3. L. K. Rosenfield, "The Pinch Blepharoplasty Revisited," *Plastic Reconstructive Surgery* 115 (2005): 1405.
4. N. T. Iliff and L. Snyder, "LASIK, Blepharoplasty and Dry Eyes," *Aesthetic Surgery Journal* 22 (2002): 382.

Chapter 13. Brow-Lifts

1. E. S. Chiu and D. C. Baker, "Endoscopic Browlift: A Retrospective Review of 628 Consecutive Cases over 5 Years," *Plastic Reconstructive Surgery* 112 (2003): 628.

Chapter 14. Face-Lifts and Lifting with Stitches

1. E. J. Ivy, Z. P. Lorenc, and S. J. Aston, "Is There a Difference? A Prospective Study Comparing Lateral and Standard SMAS Face Lifts with Extended SMAS and Composite Rhytidectomies," *Plastic Reconstructive Surgery* 98 (1996): 1135.
2. O. Barker, "'Lunchtime Beauty Fix' Makes a Splash," *USA Today,* April 14, 2005; S. Isse and N. Isse, "Barbed Polypropylene Sutures for Midface Elevation: Early Results," *Archives of Facial Plastic Surgery* 7 (2005): 55.
3. T. A. Miller, "Excision of Redundant Neck Tissue in Men with Platysma Plication and Z-Plasty Closure," *Plastic Reconstructive Surgery* 115 (2005): 304.
4. D. Marchac and A. L. Greensmith, "Early Postoperative Efficacy of Fibrin Glue in Face Lifts," *Plastic Reconstructive Surgery* 115 (2005): 911.
5. O. Barker, "Acupuncture Gets a Face Lift and Much More," *USA Today,* December 7, 2004.

Chapter 15. Fat Grafting

1. S. K. Kanchwala and L. P. Bucky, "Facial Fat Grafting: The Search for Predictable Results," *Facial Plastic Surgery* 19 (2003): 137; B. Moelleken, "Disarticulated Fascial-Fat Grafts Offer Superior Viability and Constancy Compared to Fat Injection for Facial

Filling: Clinical, Histologic, and 3-D CT Correlation," *Plastic Reconstructive Surgery* 116 (2005): 30.
2. S. R. Coleman, "Hand Rejuvenation with Structural Fat Grafting," *Plastic Reconstructive Surgery* 110 (2002): 1731.
3. A. R. Shamma and R. J. Guy, "Laser Ablation of Unwanted Hand Veins (LAUV)," *Plastic Reconstructive Surgery* 116 (2005): 27.
4. FDA Regulations, 21 CFR 1270, Regulations for Tissue Banking, 1997.

Chapter 17. Rhinoplasty

1. J. P. Gunter, "The Merits of the Open Approach in Rhinoplasty," *Plastic Reconstructive Surgery* 99 (1997): 863.

Chapter 20. Ear Reshaping

1. C. J. W. Porter and S. T. Tan, "Congenital Auricular Anomalies: Topographic Anatomy, Embryology Classification, and Treatment Strategies," *Plastic Reconstructive Surgery* 115 (2005): 1701.

Chapter 21. Hair Restoration

1. O. T. Norwood and T. Shiell, eds., *Hair Transplantation Surgery,* 2nd ed. (Charles C. Thomas, 1984).
2. M. Yamazaki et al., "Linear Polarized Infrared Irradiation Using Super Lizer Is an Effective Treatment for Multiple-Type Alopecia Areata," *International Journal of Dermatology* 42 (2003): 738.

Chapter 23. Laser Hair Removal

1. W. W. Lou et al., "Prospective Study of Hair Reduction by Diode Laser (800 cm) with Long-term Follow-up," *Dermatologic Surgery* 26 (2000): 428.

Chapter 26. New Lasers and Intense Pulsed Light

1. R. A. Weiss, M. A. Weiss, and K. L. Beasley, "Rejuvenation of Photoaged Skin: 5 Years Results with Intensive Pulsed Light of the Face, Neck, and Chest," *Dermatologic Surgery* 28 (2002): 1115.
2. D. Kligman and Y. Zhen, "Intense Pulsed Light Treatment of Photoaged Facial Skin," *Dermatologic Surgery* 30 (2004): 1085; T. Kono et al., "Comparison Study of Smooth Pulsed Light and Long-Pulsed Dye Laser in the Treatment of Facial Skin Rejuvenation," *Plastic Reconstructive Surgery* 116 (2005): 134.
3. L. Fodor et al., "Using Intense Pulsed Light for Cosmetic Purposes: Our Experience," *Plastic Reconstructive Surgery* 113 (2004): 1789.
4. S. N. Doshi and T. S. Alster, "Combination Radiofrequency and Diode Laser for Treatment of Facial Rhytides and Skin Laxity," *Journal of Cosmetic Laser Therapy* 7 (2005): 11.
5. R. Fitzpatrick et al., "Multicenter Study of Noninvasive Radiofrequency for Periorbital Tissue Tightening," *Lasers in Surgery and Medicine* 33 (2003): 232; W. K. Nahm et al., "Objective Changes in Brow Position, Superior Palpebral Crease, Peak Angle of the Eyebrow, and Jowl Surface Area after Volumetric Radiofrequency Treatments to Half of the Face," *Dermatologic Surgery* 30 (2004): 922.
6. B. A. Bassichis, S. Dayan, and J. R. Thomas, "Use of a Nonablative Radiofrequency

Device to Rejuvenate the Upper One-third of the Face," *Otolaryngology—Head and Neck Surgery* 130 (2004): 397.

7. R. A. Weiss et al., "Clinical Trial of a Novel Non-thermal LED Array for Reversal of Photoaging: Clinical, Histologic, and Surface Profilometric Results," *Lasers in Surgery and Medicine* 36 (2005): 85.
8. D. Manstein et al., "Fractional Photothermolysis: A New Concept for Cutaneous Remodeling Using Microscopic Patterns of Thermal Injury," *Lasers in Surgery and Medicine* 34 (2004): 426.
9. J. F. Tremblay et al., "Repeated Non-ablative Plasma Skin Resurfacing (PSR) for Full Face Photorejuvenation: A Double-Center Open-Label Study," *Lasers in Surgery and Medicine* 36 (2005): 17.

Chapter 27. Liposuction

1. A. W. Perry, C. Petti, and M. Rankin, "Lidocaine Is Not Necessary in Liposuction," *Plastic Reconstructive Surgery* 104 (1999): 1900.
2. G. Giugliano et al., "Effect of Liposuction on Insulin Resistance and Vascular Inflammatory Markers in Obese Women," *British Journal of Plastic Surgery* 57 (2004): 190; S. Y. Giese et al., "Improvements in Cardiovascular Risk Profile with Large-Volume Liposuction: A Pilot Study," *Plastic Reconstructive Surgery* 108 (2001): 510; S. Klein et al., "Absence of an Effect of Liposuction on Insulin Action and Risk Factors for Coronary Heart Disease," *New England Journal of Medicine* 350 (2004): 2549.
3. G. P. Maxwell and M. K. Gingrass, "Ultrasound-Assisted Lipoplasty: A Clinical Study of 250 Consecutive Patients," *Plastic Reconstructive Surgery* 101 (1998): 189; S. A. Trott et al., "Sensory Changes after Traditional and Ultrasound-Assisted Liposuction Using Computer-Assisted Analysis," *Plastic Reconstructive Surgery* 103 (1999): 2016.
4. M. Topaz, "Long-Term Possible Hazardous Effects of Ultrasonically Assisted Lipoplasty," *Plastic Reconstructive Surgery* 102 (1998): 280; M. Topaz et al., "The Possible Protective Effects of Antioxidants in Ultrasound-Assisted Lipoplasty," *Plastic Reconstructive Surgery* 113 (2004): 788.
5. M. L. Jewell et al., "Clinical Application of VASER-Assisted Lipoplasty: A Pilot Clinical Study," *Aesthetic Surgery Journal* 22 (2002): 131.
6. E. G. Murray et al., "Evaluation of the Acute and Chronic Systemic and Metabolic Effects from the Use of High Intensity Focused Ultrasound for Adipose Tissue Removal and Non-invasive Body Sculpting," *Plastic Reconstructive Surgery* 116 (2005): 151; A. Glicksman, "Non-invasive Body Contouring by Focused Ultrasound," *Plastic Reconstructive Surgery* 116 (2005): 180.
7. S. A. Brown et al., "Effect of Low-Level Laser Therapy on Abdominal Adipocytes before Lipoplasty Procedures," *Plastic Reconstructive Surgery* 113 (2004): 1796.

Chapter 28. Tummy Tucks and Body Contouring

1. A. S. Aly et al., "Belt Lipectomy for Circumferential Truncal Excess: The University of Iowa Experience," *Plastic Reconstructive Surgery* 111 (2003): 398.
2. D. J. Hurwitz, "Medial Thighplasty," *Aesthetic Surgery Journal* 25 (2005): 180.

Chapter 29. Breast Enlargement

1. L. R. Hölmich et al., "Incidence of Silicone Breast Implant Rupture," *Archives of Surgery* 138 (2003): 801.

2. FDA Report, "Breast Implant Adverse Events During Mammography," 2004.
3. S. E. Gabriel et al., "Risk of Connective-Tissue Diseases and Other Disorders after Breast Implantation," *New England Journal of Medicine* 330 (1994): 1697.
4. M. Angell, *Science on Trial: The Clash of Medical Evidence and the Law in the Breast Implant Case* (Norton, 1996).
5. C. A. Brunner and R. W. Groner, "Carboxy-Methyl-Cellulose Hydrogel-Filled Breast Implants—An Ideal Alternative? A Report of Five Years' Experience with This Device," *Canadian Journal of Plastic Surgery* 14 (2006): 151.
6. S. G. Wallach, "Maximizing the Use of the Abdominoplasty Incision," *Plastic Reconstructive Surgery* 113 (2004): 411.
7. J. B. Tebbetts, "Dual Plane Breast Augmentation: Optimizing Implant-Soft-Tissue Relationships in a Wide Range of Breast Types," *Plastic Reconstructive Surgery* 107 (2001): 1255.
8. L. A. Brinton et al., "Risk of Connective Tissue Disorders among Breast Implant Patients," *American Journal of Epidemiology* 160 (2004): 619.
9. D. L. Miglioretti et al., "Effect of Breast Augmentation on the Accuracy of Mammography and Cancer Characteristics," *Journal of the American Medical Association* 291 (2004): 442.
10. M. Kriege et al., "Efficacy of MRI and Mammography for Breast-Cancer Screening in Women with a Familial or Genetic Predisposition," *New England Journal of Medicine* 351 (2004): 427.
11. R. J. Greco, "Nonsurgical Breast Enhancement—Fact or Fiction?" *Plastic Reconstructive Surgery* 110 (2002): 337.
12. "ConsumerLab.com Finds 'Breast Enhancement' Pills Lack Evidence of Efficacy" and "Review Article: Breast Enhancement Supplements," April 16, 2002. www.consumerlab.com.

Chapter 30. Breast Reductions and Lifts

1. R. Sood et al., "Effects of Reduction Mammaplasty on Pulmonary Function and Symptoms of Macromastia," *Plastic Reconstructive Surgery* 11 (2003): 688.
2. R. J. Wise, "A Preliminary Report of a Method of Planning the Mammaplasty," *Plastic Reconstructive Surgery* 17 (1956): 367.
3. D. Hidalgo, "Vertical Mammaplasty," *Plastic Reconstructive Surgery* 115 (2005): 1179.
4. R. J. Rohrich et al., "Current Preferences for Breast Reduction Techniques: A Survey of Board Certified Plastic Surgeons," *Plastic Reconstructive Surgery* 114 (2004): 1724.
5. J. Noe and M. Keller, "Can Stitches Get Wet?" *Plastic Reconstructive Surgery* 81 (1988): 82.
6. S. L. Spear, C. V. Pelletiere, and N. Menon, "One-Stage Augmentation Combined with Mastopexy: Aesthetic Results and Patient Satisfaction," *Aesthetic Plastic Surgery* 28 (2004): 259.
7. D. D. McGeorge, "The 'Niplette': An Instrument for the Non-surgical Correction of Inverted Nipples," *British Journal of Plastic Surgery* 47 (1994): 46.

Chapter 31. Male Breast Reduction

1. G. J. Wise, A. K. Roorda, and R. Kalter, "Male Breast Disease," *Journal of the American College of Surgeons* 200 (2005): 255.

2. I. M. Wiesman et al., "Gynecomastia: An Outcome Analysis," *Annals of Plastic Surgery* 53 (2004): 97.

Chapter 32. Calf, Buttock, and Pectoral Implants

1. T. L. Roberts et al., "Augmentation of the Buttocks by Micro Fat Grafting," *Aesthetic Surgery Journal* 21 (2001): 311; W. L. Murillo, "Buttock Augmentation," *Plastic Reconstructive Surgery* 114 (2004): 1606.

Chapter 33. Weight Loss and Cosmetic Surgery

1. M. L. Dansinger et al., "Comparison of the Atkins, Ornish, Weight Watchers, and Zone Diets for Weight Loss and Heart Disease Risk Reduction: A Randomized Trial," *Journal of the American Medical Association* 293 (2005): 43.
2. S. Z. Yanovski and J. A. Yanovski, "Obesity," *New England Journal of Medicine* 346 (2003): 591.
3. M. F. Roizen and J. LaPuma, *The RealAge Diet* (Cliff Street Books, 2001).
4. H. Buchwald et al., "Bariatric Surgery: A Systemic Review and Meta-Analysis," *Journal of the American Medical Association* 292 (2004): 1724; L. Sjostrom et al., "Lifestyle, Diabetes, and Cardiovascular Risk Factors Ten Years after Bariatric Surgery," *New England Journal of Medicine* 351 (2004): 2683.

Chapter 34. Spider Veins

1. R. A. Weiss et al., "Post-sclerotherapy Compression: Controlled Comparative Study of Duration of Compression and Its Effects on Clinical Outcome," *Dermatologic Surgery* 25 (1999): 105.
2. F. Lurie et al., "Prospective Randomised Study of Endovenous Radiofrequency Obliteration (Closure) versus Ligation and Vein Stripping (EVOLVeS): Two-Year Follow-up," *European Journal of Vascular and Endovascular Surgery* 29 (2005): 67; J. E. Sybrandy and C. H. Witten, "Initial Experiences in Endovenous Treatment of Saphenous Vein Reflux," *Journal of Vascular Surgery* 36 (2002): 1207.

Chapter 36. Endermologie and the Quest to Cure Cellulite

1. M. Rosenbaum et al., "An Exploratory Investigation of the Morphology and Biochemistry of Cellulite," *Plastic Reconstructive Surgery* 101 (1998): 1934.
2. C. Pierard-Franchimont et al., "A Randomized, Placebo-Controlled Trial of Topical Retinol in the Treatment of Cellulite," *American Journal of Clinical Dermatology* 1 (2000): 369.
3. E. L. Sainio et al., "Ingredients and Safety of Cellulite Creams," *European Journal of Dermatology* 10 (2000): 596; N. Collis et al., "Cellulite Treatment: A Myth or Reality: A Prospective Randomized, Controlled Trial of Two Therapies, Endermologie and Aminophylline Cream," *Plastic Reconstructive Surgery* 104 (1999): 1110.

Chapter 37. Mesotherapy

1. D. I. Duncan and F. Hasengschwandtner, "Lipodissove for Subcutaneous Fat Reduction and Skin Retraction," *Aesthetic Surgery Journal* 25 (2005): 530; A. P. Salas and M. Asaadi, "Aesthetic Application of Mesotherapy: A Preliminary Report," paper presented at ASAPS meeting, Vancouver, British Columbia, April 17, 2004.

2. R. Rohrich, "Mesotherapy: What Is It? Does It Work?" *Plastic Reconstructive Surgery* 115 (2005): 1425.

Chapter 38. Cool Lasers and Other Less Invasive Treatments

1. D. I. Shapiro and J. A. Friedland, "A Prospective Study of the Efficacy of the N-Lite Laser for the Treatment of Facial Wrinkles," *Aesthetic Surgery Journal* 24 (2004): 431.
2. D. Fernandes, "Percutaneous Collagen Induction: An Alternative to Laser Resurfacing," *Aesthetic Surgery Journal* 22 (2002): 307.

Chapter 39. Microdermabrasion

1. M. Coimbra et al., "A Prospective Controlled Assessment of Microdermabrasion for Damaged Skin and Fine Rhytides," *Plastic Reconstructive Surgery* 113 (2004): 1438.

Chapter 40. "Anti-Aging" Medicine

1. S. Romm, *The Changing Face of Beauty* (Mosby Year Book, 1992), 184.
2. S. J. Olshansky, L. Hayflick, and B. A. Carnes, "No Truth to the Fountain of Youth," *Scientific American* 286 (2002): 92.
3. D. Rudman et al., "Effects of Human Growth Hormone in Men over 60 Years Old," *New England Journal of Medicine* 323 (1990): 1.
4. M. R. Blackman et al., "Growth Hormone and Sex Steroid Administration in Healthy Aged Women and Men," *Journal of the American Medical Association* 288 (2002): 2282.
5. W. G. Lyle and the Plastic Surgery Educational Foundation DATA Committee, "Human Growth Hormone and Anti-Aging," *Plastic Reconstructive Surgery* 110 (2002): 1585.
6. T. T. Perls, N. R. Reisman, and S. J. Olshansky, "Provision of Distribution of Growth Hormone for 'Antiaging,'" *Journal of the American Medical Association* 294 (2005): 2086.
7. G. Percheron et al., "Effect of 1-Year Oral Administration of Dehydroepiandrosterone to 60- to 80-Year-Old Individuals on Muscle Function and Cross-Sectional Area: A Double-Blind Placebo-Controlled Trial," *Archives of Internal Medicine* 163 (2003): 720.
8. J. Parasrampuria, K. Schwartz, and R. Petesch, "Quality Control of Dehydroepiandrosterone Dietary Supplement Products," *Journal of the American Medical Association* 280 (1998): 1565.
9. A. Weil, *Healthy Aging* (Knopf, 2005), p. 19.
10. N. W. Milgram et al., "Learning Ability in Aged Beagle Dogs Is Preserved by Behavioral Enrichment and Dietary Fortification: A Two-Year Longitudinal Study," *Neurobiology of Aging* 26 (2005): 77.

Epilogue

1. H. Wessells, T. F. Lue, and J. W. McAninch, "Penile Length in the Flaccid and Erect States: Guidelines for Penile Augmentation," *Journal of Urology* 156 (1996): 995.
2. H. Wessells et al., "Complications of Penile Lengthening and Augmentation Seen at One Referral Center," *Journal of Urology* 155 (1996): 117.
3. M. Ito et al., "Hair Follicle Stem Cells in the Lower Bulge Form the Secondary Germ,

a Biochemically Distinct but Functionally Equivalent Progenitor Cell Population, at the Termination of Catagen," *Differentiation* 72 (2004): 548.

4. J. L. Dragoo et al., "Tissue-Engineered Bone from BMP-2-Transduced Stem Cells Derived from Human Fat," *Plastic Reconstructive Surgery* 115 (2005): 1665; J. W. Xu et al., "Tissue-Engineered Flexible Ear-Shaped Cartilage," *Plastic Reconstructive Surgery* 115 (2005): 1633.

Resources for Further Learning

The Internet has opened up the world to easier study. Unfortunately, there is no way for the average consumer to know which websites are authoritative and unbiased. The following are a sampling of credible sources whose websites will provide more detailed information on cosmetic surgery to the reader.

Alliance for Aging Research, www.agingresearch.org
This website has a calculator that estimates how long you will live based on your current habits and genes. It is particularly effective in analyzing your answers and suggesting lifestyle changes.

American Board of Medical Specialties, www.abms.org
This not-for-profit organization, with its twenty-four medical specialty boards, is the preeminent entity overseeing physician certification in the United States.

American Board of Plastic Surgery, Inc., www.abplsurg.org
This board, which certifies plastic surgeons, is one of the twenty-four members of the American Board of Medical Specialties.

American College of Surgeons, www.facs.org
This is the site of the umbrella organization for all true surgeons—plastic, cardiac, and so on.

American Society for Aesthetic Plastic Surgery, www.surgery.org
This site of the definitive organization for aesthetic (cosmetic) surgery provides information about procedures and plastic surgeons.

American Society of Plastic Surgeons, www.plasticsurgery.org
This recommended site has information about all of plastic surgery.

Arthur Perry, M.D., StraightTalkAboutCosmeticSurgery.com
The cosmetic surgery site of the author features four basset hounds and links to PerryPlasticSurgery.com and PlasticSurgeryInTheAir.com.

Awfulplasticsurgery, www.awfulplasticsurgery.com
This fun site analyzes the cosmetic surgery of celebrities.

New Vitality, www.newvitality.com
This is the site of the manufacturer of Dr. Perry's NightSkin Cream.

RealAge, www.realage.com
This site of RealAge, created by Michael Roizen, M.D., has a wealth of information that can help extend your life. A "real age" calculator can be used to predict your real, as opposed to your chronological, age.

State licensing board To contact your state licensing board, search on the Internet or look at the first entry in the yellow pages under the heading "Physicians and Surgeons." State licensing boards may be fully Internet capable, or they may still read credentials from 3 × 5 cards.

Glossary

Abdominoplasty: a tummy tuck.

Alpha hydroxy acid (AHA): also called fruit acid; a natural chemical exfoliant and antioxidant.

Anaphylactic: a severe and sometimes deadly allergic reaction.

Antioxidant: a chemical that protects against free-radical damage.

Areola: the pigmented area surrounding the nipple.

Augmentation mammaplasty: a breast implant procedure.

Band-lift: surgical procedure that brings the two bands of the neck together.

Barbed suture: a special type of stitch that can lift facial tissue.

Bariatric: relating to the science of weight loss.

Belt lipectomy: a surgical procedure that removes excess skin from around the body at the waist.

Blepharochalasia: overgrowth of eyelid skin due to chronic allergic reactions.

Blepharoplasty: an eyelid lift.

Botulinum toxin (Botox): a chemical derived from the bacteria that cause botulism. It paralyzes overactive muscles and lessens wrinkles.

Brachioplasty: a surgical procedure that removes excess skin and fat from the arm.

Browpexy: a type of brow-lift that sews the brow to a higher level through eyelid incisions.

Bunny line: the oblique wrinkles of the upper nose.

Cannula: a hollow tube that is used to suction fat.

Canthopexy: a procedure that tightens and repositions the lower eyelid.

Cartilage: a plastic-like tissue that is present in the nose and ears.

Cellulite: the "cottage cheese" appearance of the skin that accumulates with age.

Chemical peel: a procedure that removes and rejuvenates the top layer of skin.

Chin augmentation: a chin implant or advancement of the chin bone.

Chromophore: the colored substance that lasers target for destruction.

Collagen: the major structural part of the skin.

Columella: the skin between the nostrils.

Compression boot: a device that keeps the blood flowing in the legs during surgery.

Corrugator muscle: the muscle that produces vertical frown lines between the eyebrows.

Cross-track: one of the marks left by stitches if they are tied too tightly.

Crow's-feet: the wrinkles at the outer corner of the eyelid.

Dalton: the scientific unit of molecular weight. Hydrogen weighs one Dalton and carbon weighs 12.

Deep venous thrombosis (DVT): a blood clot in the legs or thighs.

Dermabrasion: a sanding of the top layers of the skin to reduce wrinkling and scars.

Dermis: the deep, structural layer of skin.

Ectropion: a pull-down of the lower eyelid.

Elastin: the fibers in the skin that allow it to stretch.

Embolism: a clot that travels through the bloodstream.

Endermologie: a type of machine-assisted massage.

Endoscope: a surgical instrument with a camera attached to the end of a tube.

Endoscopy: surgery through small incisions, using an endoscope.

Epidermis: the outer, waterproofing layer of skin.

Exfoliation: removal of the outermost layer of dead skin cells.

Face-lift: a surgical procedure that tightens the skin of the face and neck.

Fibrin: the "biological glue" that holds tissues together.

Fibroblast: the cell that makes collagen.

Flap: tissue moved to an adjacent area, without removal from the body.

Free radical: a chemical that injures the DNA and contributes to aging and cancer.

General anesthesia: a drug-induced state in which breathing is controlled by a respirator and the patient feels no pain and has no memory of surgery.

Genioplasty: a surgical procedure that changes the shape of the chin.

Glabella: the anatomic region between the eyebrows.

Glycolic acid: the most popular of the alpha hydroxy acids; an exfoliant and peeling agent.

Graft: the transfer of tissue—skin or fat, for instance—from one part of the body to another. The tissue is completely detached from the body at some point.

Granuloma: inflammatory tissue that often creates a nodule.

Gynecomastia: breast growth in men.

Hematoma: bleeding in the operated area.

Hexapeptide: a chemical composed of six amino acids alleged to rejuvenate the skin.

Hyaluronic acid: a building block of skin that has become a popular wrinkle filler.

Hydroquinone: a chemical that stops brown pigment production in the skin.

Hydroxylapatite: the building block of bone; also used to fill folds.

Hyperpigmentation: brown discoloration of the skin.

Keloid: a tumor of the scar tissue.

Keratosis: a scaly condition of the skin.

Labiaplasty: a surgical procedure that alters the shape of the female genitals.

Lap-band: a surgical procedure that restricts the amount of food that can enter the stomach.

Laser peel: removal of the outer skin layers with a laser to reduce wrinkles.

Lidocaine: a local anesthetic.

Light chemical peel: rejuvenation of skin with trichloroacetic acid.

Liposuction: fat removal through suction.

Lithotripsy: a surgical procedure that destroys kidney stones.

Local anesthesia: the blocking of pain with lidocaine or other drugs.

Marionette line: a crease between the corner of the mouth and the chin.

Mastopexy: a breast-lift.

Melanin: brown skin pigment.

Melasma: the brown pigment in the face that commonly occurs after pregnancy.

Mesotherapy: a procedure that injects chemicals into the fat.

Microdermabrasion: a procedure that exfoliates the skin and suctions out pores.

Nasolabial fold: a fold between the nose and the corner of the mouth.

Otoplasty: a procedure that sets back protruding ears.

Panniculectomy: a surgical procedure that removes excess skin and fat from the belly.

Periodontia: the field of dentistry that deals with diseases of the gums.

Periosteum: tissue that covers bone.

Phenol peel: a deep chemical skin peel.

Photoaging: sun-induced skin aging.

Platysmaplasty: a "band-lift" procedure that repairs the paired muscles of the neck.

Point mark: one of the tiny marks caused by stitches that are left in too long.

Ptosis: the drooping of tissue.

Reduction mammaplasty: an operation that decreases the size of the breasts.

Retro-orbicularis oculi fat (ROOF): present on the bone above the eyeball.

Rhinoplasty: a procedure that surgically reshapes the nose.

Rhytidectomy: a face-lift.

Saline: saltwater.

Sedation surgery: a surgery performed that uses relaxing intravenous or oral medications.

Seroma: the collection of clear fluid under a wound.

Silicone: the most commonly used implant material.

Skin resurfacing: any procedure that removes old skin to allow new, younger-looking skin to grow back in its place.

Sliding genioplasty: chin augmentation by cutting and moving the chin bone.

Stem cell: cells in the body that have the potential to develop into other types of cells.

Stratum corneum: the outermost layer of the skin.

Telangiectasia: visible capillaries in the skin.

Thrombosis: a clot.

Topical: said of a substance placed *on* the skin, as opposed to *under* the skin.

Tretinoin: a vitamin A–type drug that rejuvenates the skin, known commonly as Retin-A.

Trichloroacetic acid (TCA): a chemical used for face peels.
Umbilicus: the belly button.
Undermining: a lifting of the skin for ease of skin repair.
Valtrex: an antiviral drug.
Zovirax: another antiviral drug.

Index

Page numbers in italic type indicate illustrations.

www.jefftismanphotography.com

Arthur William Perry, M.D., F.A.C.S., is a plastic surgeon who specializes in cosmetic surgery. With offices in the Princeton and Bridgewater areas of New Jersey, Dr. Perry has a conservative approach to plastic surgery. He received his bachelor's degree from Rutgers College in three years, magna cum laude with high distinction in zoology. Graduating in 1981 from the Albany Medical College of Union University with distinction in research, Dr. Perry did his surgical internship and residency at Harvard Medical School's Beth Israel Hospital, then spent a year as a fellow in burn surgery at Cornell University Medical College–New York Hospital. He then completed his residency in plastic surgery at the University of Chicago and, after being chief resident in 1987, took an additional aesthetic surgery fellowship. He joined the faculty of the Robert Wood Johnson Medical School in New Jersey, where he holds the rank of clinical associate professor of plastic surgery, and the University of Pennsylvania, where he is a clinical associate. Now in private practice, actively teaching, and on the staff of four hospitals, he has served as chief of plastic surgery at the Somerset Medical Center in New Jersey.

Dr. Perry is the author of many scientific papers and of several chapters in plastic surgery textbooks. His previous book, *Are You Considering Cosmetic Surgery?* was published by Avon Press in 1997. Appointed by Governor Christine Whitman to the New Jersey State Board of Medical Examiners, Dr. Perry has been instrumental in developing the standards for office sterilization procedures, office operating rooms, electrology, and tattooing and piercing. In 2006, he received the Distinguished Service Award of the New Jersey Society for Plastic Surgery. Dr. Perry hosts a weekly plastic surgery radio show and created the NightSkin Cream for New Vitality Corporation. He lives in the Princeton, New Jersey, area with his wife, three children, and four basset hounds.